BIG DATA
OF COMPLEX
NETWORKS

Chapman & Hall/CRC
Big Data Series

SERIES EDITOR
Sanjay Ranka

AIMS AND SCOPE

This series aims to present new research and applications in Big Data, along with the computational tools and techniques currently in development. The inclusion of concrete examples and applications is highly encouraged. The scope of the series includes, but is not limited to, titles in the areas of social networks, sensor networks, data-centric computing, astronomy, genomics, medical data analytics, large-scale e-commerce, and other relevant topics that may be proposed by potential contributors.

PUBLISHED TITLES

BIG DATA COMPUTING: A GUIDE FOR BUSINESS AND TECHNOLOGY MANAGERS
Vivek Kale

BIG DATA OF COMPLEX NETWORKS
Matthias Dehmer, Frank Emmert-Streib, Stefan Pickl, and Andreas Holzinger

BIG DATA : ALGORITHMS, ANALYTICS, AND APPLICATIONS
Kuan-Ching Li, Hai Jiang, Laurence T. Yang, and Alfredo Cuzzocrea

NETWORKING FOR BIG DATA
Shui Yu, Xiaodong Lin, Jelena Mišić, and Xuemin (Sherman) Shen

Chapman & Hall/CRC
Big Data Series

BIG DATA
OF COMPLEX
NETWORKS

Edited by

Matthias Dehmer

UMIT - The Health & Life Sciences
University, Hall in Tyrol, Austria
and
Nankai University, Tianjin, China

Frank Emmert-Streib

Queen's University Belfast, UK

Stefan Pickl

Universitaet der Bundeswehr
Muenchen, Neubiberg, Germany

Andreas Holzinger

Medical University Graz, Austria

 CRC Press
Taylor & Francis Group
Boca Raton London New York

CRC Press is an imprint of the
Taylor & Francis Group, an **informa** business

A CHAPMAN & HALL BOOK

CRC Press
Taylor & Francis Group
6000 Broken Sound Parkway NW, Suite 300
Boca Raton, FL 33487-2742

© 2017 by Taylor & Francis Group, LLC
CRC Press is an imprint of Taylor & Francis Group, an Informa business

No claim to original U.S. Government works

Printed on acid-free paper
Version Date: 20160609

International Standard Book Number-13: 978-1-4987-2361-9 (Hardback)

Library of Congress Cataloging-in-Publication Data

Names: Dehmer, Matthias, 1968- editor. | Emmert-Streib, Frank, editor. |
Pickl, Stefan, 1967- editor. | Holzinger, Andreas, editor.
Title: Big data of complex networks / editors, Matthias Dehmer, Frank
Emmert-Streib, Stefan Pickl, and Andreas Holzinger.
Description: Boca Raton : CRC Press, 2016. | Series: Chapman & Hall/CRC big
data series | Includes bibliographical references and index.
Identifiers: LCCN 2016003196 | ISBN 9781498723619 (hardback : alk. paper)
Subjects: LCSH: Big data. | Large scale systems. | System analysis.
Classification: LCC QA402 .B485 2016 | DDC 005.7--dc23
LC record available at http://lccn.loc.gov/2016003196

Visit the Taylor & Francis Web site at
http://www.taylorandfrancis.com

and the CRC Press Web site at
http://www.crcpress.com

Contents

Preface

Today, *Big Data* affects practically every scientist in any domain, from astronomy to zoology. Data science is meanwhile seen as key in the investigation of our nature, from the microcosm to the macrocosm. Big Data actually reverses the classical scientific hypothetic-deductive approach, hence data science itself produces unprecedented amounts of Big Data.

However, in certain domains, for example, in the biomedical domain, we are confronted not only with enormously Big Data, but with *complex* data. The increasing trend toward personalized medicine has resulted in an explosion in the amount of *complex data*, for example, from genomics, proteomics, metabolomics, transcriptomics, lipidomics, fluxomics, phenomics, microbiomics, epigenetics, and so on.

Here the science of *networks* can be of great help because much of this Big Data is available in the form of point clouds in arbitrarily high dimensions, which consequently lets us make use of the great benefits of graph theory—a prime object of discrete mathematics with sheer endless application possibilities and many open future research avenues.

The main goal of the book *Big Data of Complex Networks* is to present and demonstrate existing and novel approaches for handling methods from Big Data for analyzing networks. The underlying mathematical methods have been developed with the aid of graph theory, Big Data, general computer science, data analysis, machine learning, and statistical techniques. This book is intended for researchers and graduate and advanced undergraduate students in fields including mathematics, computer science, physics, bioinformatics, and systems biology. Of course, as the potential of Big Data methods has been huge, this list of scientific fields cannot be complete and will hopefully be extended in the future.

The topics addressed in this book cover a broad range of Big Data concepts and methods applied to complex networks, including

- Big Data analysis for biological networks

- Big Data analytics for storage and processing of servers by means of complex networks

- Big Data text analysis by using networks

- Network visualization for Big Data

- Big Data querying in large networks

- Matrix analysis and Big Data analytics

- Legal aspects of Big Data

- Software for Big Data visual analytics

By covering this broad range of topics, the book aims to fill a gap in the contemporary literature of discrete and applied mathematics, computer science, Big Data, data analysis, and related disciplines.

This book is dedicated with all my heart to the sister of Matthias Dehmer, Ms. Marion Dehmer, who unfortunately passed away in 2012. Marion was always helpful, creative, and inspiring when helping with earlier projects. Her positive spirit would have been highly beneficial for this and future projects. This book shall be a memento to Marion.

Many colleagues, whether consciously or unconsciously, have provided us with input, help, and support before and during the preparation of the present book. In particular, we would like to thank Zengqiang Chen, Marion Dehmer, Andrey A. Dobrynin, Boris Furtula, Ivan Gutman, Xueliang Li, D. D. Lozovanu, Alexei Levitchi, Abbe Mowshowitz, Miriana Moosbrugger, Andrei Perjan, Stefan Shetschew, Yongtang Shi, Fred Sobik, Shailesh Tripathi, Kurt Varmuza, Chengyi Xia, and Dongxiao Zhu, and apologize to all whose names have been inadvertently omitted. Also, we would like to thank our editors Sunil Nair and Alex Edwards from CRC Press who have always been available and helpful. Last but not least, Matthias Dehmer thanks the Austrian Science Funds (project P26142) and the Universität der Bundewehr München for supporting this work. Stefan Pickl wants to thank the University of Nottingham Malaysia Campus for their cooperation in the area of complex analytics.

To the best of our knowledge, this is the first book dedicated exclusively to Big Data network analysis. Existing books dealing with graph theory and Big Data have limited scope by mainly explaining the two concepts in isolation. Therefore, we hope that this book will help to establish *Big Data of Complex Networks* as a modern and independent branch of network sciences and graph theory. Also, it should broaden the scope of pure and applied researchers who deal with graph-based techniques and Big Data methods. Finally, we hope this book conveys the enthusiasm and joy we have for this field and inspires fellow researchers in their own practical or theoretical work.

Where available, color versions of the Figures contained in this book can be found at https://www.crcpress.com/Big-Data-of-Complex-Networks/Dehmer-EmmertStreib-Pickl-Holzinger/9781498723619.

Editors

Dr. Matthias Dehmer studied mathematics at the University of Siegen, Germany, and received his PhD in computer science from the Darmstadt University of Technology, Germany. He was a research fellow at the Vienna Biocenter, Austria, Vienna University of Technology, Austria, and the University of Coimbra, Portugal. Currently, he is a Professor at UMIT—The Health and Life Sciences University, Hall in Tyrol, Austria, and also holds a position at Nankai University, Tianjin, China. His research interests include graph theory, data science, complex networks, complexity, statistics and information theory. He has published extensively in data science and structural network analysis.

Prof. Frank Emmert-Streib's research interests include computational biology, biostatistics, and machine learning in the development and application of statistical analysis methods to investigate data from high-dimensional data. He has worked on problems from neurobiology, molecular biology, web mining, chemistry, and physics. He is currently working at the University of Tampere, Finland, as an Associate Professor focusing on the analysis of high-throughput data from complex diseases.

Stefan Pickl studied mathematics and received his PhD in mathematics from the Darmstadt University of Technology, Germany. He was a research fellow at Santa Fe Institute, New Mexico, and held several other research positions. Currently, he is a Professor at Bundeswehr Universität München, Neubiberg, Germany. His research interests include operations research, systems biology, graph theory, and discrete optimization.

Andreas Holzinger studied information engineering and physics at the Graz University of Technology, Austria, and received his PhD in cognitive science from the University of Graz, Austria. He received his habilitation in computer science from the Graz University of Technology, where he teaches biomedical informatics, and supervises engineering students at the Faculty of Computer Science and Biomedical Engineering. Currently, he is the head of research unit HCI-KDD at the Medical University of Graz, Austria. Andreas has been a Visiting Professor in Berlin, Innsbruck, Vienna, London, and Aachen. His research interests include machine learning and he has published extensively in computer and information science.

Contributors

James Abello
Computer Science Department and
  DIMACS
Rutgers University
Piscataway, NJ

Klaus Arto
Universität der Bundeswehr München
Fakultät für Betriebswirtschaft
Neubiberg, Germany

Yiqi Bai
Department of Computer Science
University of Massachusetts
Lowell, Massachusetts

Noureddine Boudriga
Communications Networks and Security
  Research Lab (CN&S)
University of Carthage
Tunis, Tunisia

Zengqiang Chen
Department of Automation, College of
  Computer and Control Engineering
Nankai University
Tianjin, China

David DeSimone
Rutgers University
Piscataway, New Jersey

Shuai Ding
School of Management
Hefei University of Technology
and
Key Laboratory of Process Optimization
  and Intelligent Decision-Making
  (Ministry of Education)
Hefei University of Technology
Anhui Hefei, China

Yacine Djemaiel
Communications Networks and Security
  Research Lab (CN&S)
University of Carthage
Tunis, Tunisia

Ranga Chandra Gudivada
University of Pittsburgh Medical Center,
  Medical Informatics Technology
Pittsburgh, Pennsylvania

Steffen Hadlak
Fraunhofer IGD
Rostock, Germany

Bo Hu
Universität der Bundeswehr München
Fakultät für Betriebswirtschaft
Neubiberg, Germany

Yifan Hu
Yahoo Labs
New York, New York

Anil G. Jegga
Division of Biomedical Informatics
Cincinnati Children's Hospital Medical
  Center
and
Department of Pediatrics
University of Cincinnati College of Medicine
Cincinnati, Ohio

Ming Jia
Department of Computer Science
University of Massachusetts
Lowell, Massachusetts

Qi Liao
Central Michigan University
Mt. Pleasant, Michigan

Robert Caiming Qiu
Center for Manufacturing Research
Tennessee Technological University
Cookeville, Tennessee

Mayur Sarangdhar
Division of Biomedical Informatics
Cincinnati Children's Hospital Medical
    Center
Cincinnati, Ohio

Hans-Jörg Schulz
Fraunhofer IGD
Rostock, Germany

Lei Shi
State Key Laboratory of Computer Science
Chinese Academy of Sciences
Beijing, China

Rasu B. Shrestha
Medical Informatics Technology
University of Pittsburgh Medical Center
Pittsburgh, Pennsylvania

Mika Sumida
Yale University
New Haven, Connecticut

Shiwen Sun
Tianjin Key Laboratory of Intelligence
    Computing and Novel Software
    Technology
and
Key Laboratory of Computer Vision and
    System (Ministry of Education)
Tianjin, China

Toyotaro Suzumura
Thomas J. Watson Research Center
Yorktown Heights, New York

Florent Thouvenin
University of Zurich
Zürich, Switzerland

Joaquin J. Torres
Institute "Carlos I" for Theoretical and
    Computational Physics, and Department
    of Electromagnetism and Matter Physics
University of Granada
Granada, Spain

Jie Wang
Department of Computer Science
University of Massachusetts
Lowell, Massachusetts

Jingwen Wang
Department of Computer Science
University of Massachusetts
Lowell, Massachusetts

Yunguan Wang
Department of Pathobiology
University of Cincinnati College of Medicine
Cincinnati, Ohio

Chengyi Xia
Tianjin Key Laboratory of Intelligence
    Computing and Novel Software
    Technology
and
Key Laboratory of Computer Vision and
    System (Ministry of Education)
Tianjin, China

Wenjing Yang
Department of Computer Science
University of Massachusetts
Lowell, Massachusetts

Hao Zhang
Department of Computer Science
University of Massachusetts
Lowell, Massachusetts

Liang Zhao
GSAIS (Shishu-kan) Kyoto University
Kyoto, Japan

Chapter 1

Network Analyses of Biomedical and Genomic Big Data

Mayur Sarangdhar, Ranga Chandra Gudivada, Rasu B. Shrestha, Yunguan Wang, and Anil G. Jegga

1.1 Introduction

In the past two decades, technological advances in the biomedical and genomic fields have generated enormous volumes of -omics data. These data continue to be scattered in various databases accessible through the World Wide Web (Table 1.1); there have been some significant efforts to integrate these to answer specific questions (e.g., candidate disease gene prioritization [1], tissue-specific interaction networks [2], etc.). Concomitant with the development of specialized, powerful gene and protein and disease variant annotation databases, there has been an explosion in the number of publications in the scientific literature. For example, in October 2015, there were approximately 43,000 journals in PubMed (ftp://ftp.ncbi.nih.gov/pubmed/J_Entrez.txt) and more than 21.5 million PubMed abstracts. Collectively, these resources represent different facets and also our current knowledge of pathophysiological mechanisms. These heterogeneous and humongous data resources have spurred the demand for development of robust and efficient computational approaches to store, process, integrate, analyze, interpret, and generate novel hypotheses.

Biological networks share several features (e.g., scale-freeness and small-world properties) with social and communication networks [3]. The algorithms used for analyzing social networks or Web networks are therefore equally relevant and useful for processing or interpreting biomedical and -omics networks. The various types of -omics data, such as protein interactions, gene expression, gene annotations, drug targets, disease phenotypes, and so on (see Table 1.1 for some examples of biomedical and genomic Big Data repositories and resources) can therefore be viewed from a network perspective, wherein a node represents any of the biomedical or genomic entities while an edge represents the relationship between these entities. Several studies have been published that report the successful application of such algorithms in biomedical and genomic domains [4–13].

Note: All the URLs listed were working at the time of writing this chapter. This list is extensive, but not exhaustive,and we apologize for any oversights.

In summary, the World Wide Web is an indispensable tool for biomedical researchers striving to understand the molecular basis of disease and drug response. However, it presents challenges too in the form of proliferation of heterogeneous data resources, ranging from their content, to functionality, access, and interfaces. Knowledge extraction and hypotheses generation from biomedical concept-based networks, arising from such large-scale, cross-domain, and heterogeneous data aggregation is challenging. In this review, we present some specific examples which used network analysis-based approaches to mine integrated heterogeneous Big Data sets to address different problem areas in the biomedical and genomic domains. In the first half of this review, we present an overview of genomic and biomedical Big Data and some basic concepts related to network analysis. In the second half, we present several use cases representing different published studies that have used genomic and biomedical Big Data to understand biological systems and for translational research. Specifically, we present three case studies focusing on genomic Big Data followed by case studies related to disease gene prioritization and drug repositioning.

1.2   Genomic and Biomedical Big Data

With the recent advent of the next-generation sequencing technologies, there has been an exponential increase in the genomic and clinical data. Two of the largest and "deep" biomedical and genomic Big Data sets produced by large consortia are ENCODE (Encyclopedia of DNA Elements) and TCGA (The Cancer Genome Atlas). The ENCODE project [14,15] was launched to identify all functional elements in the human genome—regions of

TABLE 1.1: Examples of some of the open-source biomedical and genomic Big Data resources

Category	Database	URL	Description
Genes and gene annotations	NCBI Entrez Gene	http://www.ncbi.nlm.nih.gov/gene	Gene-specific connections in the nexus of map, sequence, expression, structure, function, citation, and homology data.
	Gene Ontology	http://geneontology.org/	Gene and gene product annotations across multiple species with supported tools for analytics
	KEGG	http://www.genome.jp/kegg/	Database resource that integrates genomic, chemical, and systemic functional information
	BioSystems	http://www.ncbi.nlm.nih.gov/biosystems/	Integrated access to biological systems and their component genes, proteins, and small molecules, as well as literature describing those biosystems and other related data
	Pathway Commons	http://www.pathwaycommons.org/	Biological pathway information collected from public pathway databases
	Reactome	http://www.reactome.org/	Curated and peer-reviewed pathway database
Gene expression	Gene Expression Omnibus	http://www.ncbi.nlm.nih.gov/geo/	Public functional array- and sequence-based genomics data repository; ~62,000 assays and a million samples
	Array Express	https://www.ebi.ac.uk/gxa	Gene expression patterns under different biological conditions; ~87,000 assays
	The Cancer Genome Atlas (TCGA)	http://cancergenome.nih.gov/	Large-scale genome sequencing platform for multiple cancers led by the NCI and the NHGRI
	BioGPS	http://biogps.org	Extensible and customizable gene annotation portal, a complete resource for learning about gene and protein function
Proteomics	dbDEPC	http://lifecenter.sgst.cn/dbdepc	Database of differentially expressed proteins in human cancer
	PRIDE	http://www.ebi.ac.uk/pride	Public data repository for proteomics data, including protein and peptide identifications, posttranslational modifications and supporting spectral evidence; ~54,000 assays

(Continued)

TABLE 1.1: Examples of some of the open-source biomedical and genomic Big Data resources (*continued*)

Category	Database	URL	Description
	The Human Protein Atlas	http://www.proteinatlas.org/	Millions of high-resolution images showing the spatial distribution of proteins in 44 different normal human tissues and 20 different cancer types, as well as 46 different human cell lines
	UniProt	http://www.uniprot.org/	Comprehensive, high-quality and freely accessible resource of protein sequence and functional information
Protein interactions and gene regulatory networks	BioGRID	http://thebiogrid.org/	~834,000 protein and genetic interactions, ~27,000 chemical associations and ~38,000 posttranslational modifications from major model organism species; searches 55,218 publications
	HPRD	http://www.hprd.org/	~41,000 protein interactions manually extracted from the literature by expert biologists
	STRING	http://string.embl.de/	Known and predicted protein–protein interactions from experimental repositories and computational methods; ~9.6 million proteins from >2000 organisms
	ENCODE	https://www.encodeproject.org/	Comprehensive parts list of functional elements in the human genome
Disease	OMIM	http://omim.org	Online catalog of human genes and genetic disorders
	OrphaNet	http://www.orpha.net	Portal for rare diseases and orphan drugs
	Diseases	http://diseases.jensenlab.org	Disease-gene associations mined from literature
	DisGeNET	http://www.disgenet.org/web/ DisGeNET	Human gene–disease associations (GDAs) from various expert-curated databases and text-mining–derived associations
Drug	BindingDB	https://www.bindingdb.org/bind	Database of binding data, for protein targets and small molecules.
	DGiDb	http://dgidb.genome.wustl.edu	Drug–gene interaction database curated from multiple well-established databases
	ChEBI	https://www.ebi.ac.uk/chebi/	Dictionary of molecular entities focused on "small" chemical compounds.
	ChemBank	http://chembank.broadinstitute. org/	Data derived from small molecules and small-molecule screens

(*Continued*)

TABLE 1.1: Examples of some of the open-source biomedical and genomic Big Data resources (*continued*)

Category	Database	URL	Description
	ChEMBL	https://www.ebi.ac.uk/chembl/	Manually curated bioactive molecules with drug-like properties maintained by the EBI
	DrugBank	http://www.drugbank.ca/	FDA-approved and experimental drugs with drug target, bio- and chemo-informatic data; ~7800 drug entries including ~1630 FDA-approved small molecule drugs, ~170 FDA-approved biotech (protein/peptide) drugs, and over 6000 experimental drugs
	LINCS	http://www.lincscloud.org/	Perturbational profiles across multiple cell and perturbation types, as well as read-outs, at a massive scale
	PharmGkb	https://www.pharmgkb.org/	Pharmacogenomics knowledge resource that encompasses clinical information including dosing guidelines and drug labels, potentially clinically actionable gene–drug associations and genotype–phenotype relationships
	SMPDB	http://smpdb.ca/	Small molecule pathway database with > 600 unique human pathways (~70% of these are not found in other databases)
	Therapeutic Target Database	http://bidd.nus.edu.sg/group/cjttd/	~21 million compounds that are commercially available and prepared for virtual screening
Metabolomics	BiGG	http://bigg.ucsd.edu/	Metabolic reconstruction of human metabolism designed for systems biology simulation and metabolic flux balance modeling. It is a comprehensive literature-based genome-scale metabolic reconstruction that accounts for the functions of 1496 ORFs, 2004 proteins, 2766 metabolites, and 3311 metabolic and transport reactions
	HMDB	http://www.hmdb.ca/	Human small-molecule metabolites with associated chemical, clinical, and molecular biology information; more than 6500 metabolites
	HumanCyc	http://humancyc.org/	Human metabolic pathway/genome bioinformatics database constituting over 28,500 genes

(*Continued*)

TABLE 1.1: Examples of some of the open-source biomedical and genomic Big Data resources *(continued)*

Category	Database	URL	Description
Toxicology (including drug-related side effects)	Comparative Toxicogenomics Database	http://ctdbase.org/	Curation of chemical–gene, chemical–disease, and gene–disease associations and constructs networks
	AERSMine	https://research.cchmc.org/aers/	Web application for mining FDA's adverse event reporting system data
	OpenTox	http://www.opentox.org/	Integration of data from various sources (public and confidential), for the generation and validation of computer models for toxic effects, libraries for the development and seamless integration of new algorithms, and scientifically sound validation routines
	SIDER	http://sideeffects.embl.de/	Information on marketed medicines and their recorded adverse drug reactions
	T3DB	http://www.t3db.ca/	Combines detailed toxin data with comprehensive toxin target information; 3600 toxins (including pollutants, pesticides, drugs, and food toxins) linked to 2000 toxin target records; extracted from over 18,000 sources including scientific literature
	ADReCs	http://bioinf.xmu.edu.cn/ADReCS/	Integrates ADR and drug information collected from public medical repositories like DailyMed, MedDRA, SIDER2, DrugBank, PubChem, UMLS, etc.
	HSDB	http://toxnet.nlm.nih.gov/newtoxnet/hsdb.htm	A toxicology database that focuses on the toxicology of potentially hazardous chemicals
	TOXLINE	http://toxnet.nlm.nih.gov/newtoxnet/toxline.htm	Bibliographic database with an assortment of citations from specialized journals and other sources

transcription, transcription factor association, chromatin structure, and histone modification. The ENCODE project has generated more than 6000 genomic data sets from ChIP-seq, RNA-seq, DNase-seq, and shRNA knockdown followed by RNA-seq (The ENCODE Project Consortium, 2015). The pilot phase of the ENCODE project involved 440 scientists from 32 laboratories worldwide. Following the completion of the pilot phase in 2007 [16], with the advent of new-generation sequencing machines, the raw data generated amounted to about 15 TB! The major contribution of ENCODE Big Data has been high-resolution, highly reproducible maps of DNA segments resulting in about 80% of the human genome now annotated with at least one biochemical function. The biochemical networks or models built by identifying candidate genomic elements and their interconnections based on their shared molecular characteristics can be further used as starting points for testing how these signatures relate to complex organism functions [14]. TCGA is a comprehensive and coordinated effort to decipher the molecular basis and underlying networks of cancer through the application of large-scale genome sequencing and analytical technologies. It contains measurements of mRNA and miRNA expression (array based and sequencing), somatic mutations, copy number variation, protein expression (array based), and histology slides for about 30 types of human cancers (including nine rare tumors) and about 10,000 human tumors (January 2014; http://cancergenome.nih.gov).

We also have the very large data compendia produced in the labs across the world by individual researchers and made available upon publication (Table 1.1). For example, Gene Expression Omnibus (GEO) from the National Center for Biotechnology Information (NCBI) is a data repository of publicly available functional genomics data. At the time of writing this chapter it contains about 62,000 experiments, about 1.6 million samples, and about 15,000 platforms. Likewise, the Array Express, another compendium of publicly available gene expression data contains more than 1.8 million genome-wide assays from more than 60,000 experiments amounting to more than 38 TB of archived data.

The bioinformatics field has been predominantly focused on design, development, and implementation of various robust and novel computational algorithms and approaches to interlink and mine this multitude of data sets. These associations have enabled the emerging studies of network biology, where these relations provide a global perspective of underlying molecular underpinnings between genotype and phenotype data sets. In addition, the applications of network biology may lead to future network medicine, in which a multitude of drugs, drug targets, and biomarkers may replace single drug targets or single molecular biomarkers to improve drug discovery and biomarker development. A rapidly developing trend in network biology is to combine these highly heterogeneous biomedical data sets into coherent biomedical concept networks to facilitate integrated network analysis. Any robust computational technique that supports integration of biomedical or genomic Big Data should be [17,18]

1. Able to capture and model the observed biomedical networks into an interoperable data model

2. Supportive of a data integration framework to map and merge network data across disparate sources

3. A collection of computational services capable of discovering, analyzing, and validating novel associations in an integrated biomedical network

4. Able to make the differences between biomedical concepts and databases explicit

1.3 Biomedical Systems and Network Analysis—Basic Concepts

Several recent studies on biological systems have approached them as complex networks comprising bioentities (genes, biomolecules, etc.) as nodes and their functional interactions as edges [19,20]. The normal and perturbed biomedical systems represented by different types of heterogeneous data are greatly facilitated by constructing them as networks with different biomedical and pharmacological entities (e.g., genes, diseases, drugs, etc.) as nodes and edges being the relationships among them [11,13,21–24]. The concepts from graph theory and computer sciences are widely and successfully adapted to understand the biomedical systems. For example, network analyses–based approaches have been successful in protein function predictions, drugs and target discovery, and novel biomarker identification, to name a few [25–27]. In the following sections, we briefly present some of the basic concepts in network analysis. For additional details and related work, readers are referred to other extensive reviews on biological network analysis [13,28–30].

1.3.1 Overview of basic concepts in network analysis

Network measurements that are typically used to characterize the properties of a network range from node centrality measures to the network modular or community structure. We briefly describe some of the commonly used network features (see Table 1.2 for a summary of commonly used network measurements along with their equations).

1.3.1.1 Node centrality measures

Node centrality refers to the "importance" of a node in a network. Four commonly used node centrality measures in a network are: degree, betweenness centrality, closeness centrality, and eigenvector centrality.

1. *Degree:* Node degree indicates the number of direct connections (edges) that a node has. In a directed network, a node has two measures of degree (in-degree and out-degree). The in-degree represents the number of edges that are directed to a node while the out-degree, as the name indicates, represents the number of connections or edges that are emerging or outgoing directly from a node.

2. *Betweenness:* Also referred to as the shortest path betweenness, it is the ratio of the number of the shortest paths that pass through a node to the number of the shortest path between any pair of nodes in the network [31].

3. *Closeness centrality:* Used to represent how close a node is to all other nodes in a network [32,33]. It is defined as the inverse of the sum of the node's distances to all other nodes in the network. Closeness centrality can be used to estimate how fast the information originating from a node can spread to all other nodes sequentially.

4. *Eigenvector centrality:* Measures the influence of a node in a network. It assigns relative scores to all the nodes in a network based on the concept that connections to high-scoring nodes contribute more to the score of the node than connections to low-scoring nodes. Google's PageRank [34] for instance is a variant of the eigenvector centrality measure.

TABLE 1.2: Summary of network measurements

Network Feature	Equation
Degree (with in-degree and out-degree)	• Undirected networks: D = the number of edges connected to a node • Directed networks: Din = the number of directed-in edges; $Dout$ = the number of directed-out edges of a node • Average degree $<D>= 2E/N$, E is the total number of edges, N is the total number of nodes, and denotes average
Betweenness centrality	$Bci = \sum \frac{\#SP\_through\_node\_i}{\#SP}$ $\#SP$ number shortest paths
Closeness centrality	$Cci = \frac{1}{\sum_{j=1}^{N} d(n_i, n_j)}$ N is the number of nodes, d is the distance from node ni to nj
Eigenvector centrality	$E_i = \frac{1}{\lambda} \sum_{t \in M(i)} E_t = \frac{1}{\lambda} \sum_{t \in G} a_{i,t} E_t$ $AE = \lambda E$ $M(i)$ is a set of the neighbors of node i and λ is a constant, $\underline{A} = (a_{i,t})$ is the adjacent matrix of node i
Shortest path length	Lij = the smallest number of edges between nodes i and j
Clustering coefficient	$Cc = \frac{2n}{E(E-1)}$ n is the number of a node's neighbors in the network, and E is the total number of edges
Density for undirected network	$D = \frac{2E}{N(N-1)}$ N is the number of nodes, and E is the number of edges in the network
Density for directed network	$D = \frac{E}{N(N-1)}$ N is the number of nodes, and E is the number of edges in the network
Edge betweenness centrality	$Bce = \sum \frac{\#SP\_through\_edge\_e}{\#SP}$ $\#SP$ number shortest paths

1.3.1.2 Shortest path

The shortest path length, sometimes also called geodesic distance, is the smallest number of edges or "hops" that are necessary to move from one node to another in a network.

1.3.1.3 Network-clustering coefficient

Clustering coefficient is a measurement of degree of nodes that tend to cluster together in a network [35]. It is an indicator of network density and represents the tendency of nodes in a network to form clusters. It also serves as an indicator of modularity in a network.

1.3.1.4 Network density

Network density measures the extent of the contacts between all pairs of nodes in a network [36]. It is computed as the proportion of contacts that could possibly occur in the network compared with those that are actually observed in the network.

Using statistics of network measurements, influential nodes in the network can be easily identified. For example, nodes with high degree and high global betweenness centrality are defined as hubs and bottlenecks, respectively.

1.3.1.5 Connected components and modules

Connected components, subnetworks, or loosely connected networks are a way to interconnect the components in a system or network so that those components depend on each other to the least extent. Each of the nodes within a connected component is connected to every other node directly or indirectly. A loosely connected network can be easily decomposed into definable modules.

A network community or module, on the other hand, is a group of nodes that are more densely connected to each other than to the nodes outside the group in a network [37]. In the biomedical domain, networks are composed of modules [19] and these modules often represent functional similarity, pathway or a specific biological process or a protein complex [38]. Finding modules or clusters in a network is an active area of research both in computational and bioinformatics fields. The extraction and analysis of clusters in large-scale biomedical and genomic networks have been extensively studied (e.g., protein interaction networks, functional enrichment networks, and gene regulatory networks) [37,39,40].

1.3.1.6 Network clustering

Network-clustering approaches can be classified into four broad categories: (1) connectivity-based clustering (hierarchical clustering), (2) centroid-based clustering, (3) density-based clustering, and (4) centrality-based clustering. In connectivity-based clustering (also known as hierarchical clustering), objects are more related to nearby objects than to objects that are farther away. These algorithms connect "objects" to form "clusters" based on the pair-wise distance [41–43]. In centroid-based clustering (K-means clustering), the number of clusters in the network is fixed to k. K-means clustering tries to find the k cluster centers and allocate the objects to their nearest cluster center, such that the squared distances of each object to its cluster center are minimized [44]. The density based clustering approach tries to find higher densely connected components than the remainder of the network. Sparsely distributed objects in the areas are usually considered to be noise or border points [45]. The most popular density-based clustering method is DBSCAN [46]. Centrality-based clustering methods start by removing an edge that has the highest centrality measure (e.g., the edge betweenness centrality). The edge pruning is repeated iteratively until a network has been partitioned to several clusters [21,47].

Besides these four principal network-clustering methods, several other algorithms have been developed and applied for network analysis, such as Louvain's method for community detection [48]. This method employs a greedy optimization method which attempts to iteratively optimize the "modularity" of a partition from the network until a maximum of modularity is attained with a hierarchy of communities generated. For analyzing biomedical networks, application of a specific network-clustering method is governed by the biological objectives, such as detecting signaling pathways, functional units, complexes, or other defined subnetwork structures.

1.3.2   Genomic Big Data—Understanding multicellular function and disease with human tissue-specific networks

Collecting and integrating data from 987 genome-scale data sets (encompassing approximately 38,000 conditions) from more than 14,000 publications, Greene et al. [2] recently built one of the most comprehensive human tissue- and cell-specific networks (144 human tissues and cell types). Having such large integrated networks enables researchers to use biological information from functionally related tissues and therefore infer gene function in tissues or cell types for which little or no experimental data exist. Using integrative computational analysis, the authors first identified the gene–gene interconnections contained in several heterogeneous and rich data sets for various tissue types. In the next step, tissue-specific interactions were connected and analyzed in the context of relevant genome-wide association studies (GWAS) and identified disease-gene associations that would have been undetectable. This network-guided association study (NetWAS) is a powerful example to demonstrate the utility of integrating quantitative genetics with functional genomics. Apart from increasing the power of GWAS it allows identification of novel genes underlying complex human diseases in an unbiased manner, because the approaches are completely data-driven. On a broader scale, this seminal study has demonstrated that by merging deep biological insight with state-of-the-art computational methods, knowledge-mining and hypothesis generation is possible even from large-scale, noisy, and heterogeneous data sets. Importantly, even though such functional gene interaction networks are well established in model organisms, without "Big Data" from human tissues, this would not have been possible for human tissue-specific interactions. On a more practical note, many human cell types, critical to pathological processes, are not readily amenable to traditional direct experimentation and that is the primary motivation to find a workaround using these rich data sets. To demonstrate the utility of this approach, Greene et al. [2] identified tissue-specific interactions of interleukin-1B (blood-vessel network) and LEF1 transcription factor. Another application of this resource is to investigate candidate gene function in disease-relevant tissues. The authors demonstrate this utility by reanalyzing the genes identified in a GWAS of hypertension. Apart from identifying known disease-causing genes (MTHFR and PPARG), new candidate disease-associated genes for blood pressure regulation are also identified. The results from this study are made available as an online resource and interactive server (GIANT [Genome-scale Integrated Analysis of gene Networks in Tissues]). GIANT can be queried using either individual genes or gene sets to interrogate tissue-specific gene interactions and functions. It also allows researchers to explore the networks, compares how genetic circuits differ in various tissues, and reexamines data from genetic studies to discover and prioritize novel candidate diseases. Studies such as these, and extended versions, that include additional and wider ranges of tissue or cell types and conditions not only help in understanding the plasticity of the interactome in health and disease, but also play a critical role in personalized medicine (e.g., interpretation of patients' genetic and phenotypic information) [49].

1.3.3 ImmuNet—Integrating and mining immune system-related genomic Big Data

Gorenshteyn et al. recently reported a new online tool (http://immunet.princeton.edu) [50] that can be used to predict the molecular basis of immunological disorders from genomic Big Data. Exploiting the advances in computational approaches and relatively inexpensive computing power, the researchers combined information from about 38,000 publicly available data sets. The integrated immune system Big Data were then mined using novel algorithms and models, and tools were developed enabling the discovery of previously unrecognized disease patterns from databases to generate testable hypotheses. Unlike a Big Data project arising from consortia, this resource represents combined biomedical research output from investigators around the world with the goal of accelerating the understanding of immune pathways and genes. This can ultimately help researchers in developing improved interventions for immunological disorders. Since the end users expectedly would be researchers from a noncomputational background, ImmuNet is designed in such a way that it enables biomedical researchers even without computational training to interrogate a huge compendium of valuable albeit heterogeneous public data in a user-friendly way. Additionally, the underlying Bayesian analysis can detect potentially relevant information from vast amounts of often conflicting data obtained from experiments conducted under diverse settings. By supporting the continual incorporation of increasing public data as and when available, the applicability of ImmuNet is expected to impact wide-ranging areas of not only immunological disorders but other diseases including cancers where the immune system is perturbed.

1.3.4 Weighted gene co-expression network analysis—Biological big networks

Since its introduction about a decade ago, weighted gene co-expression network analysis (WGCNA) [51] has been applied to several genomic and biomedical data sets as an exploratory tool, for genetic screening, and to generate testable hypotheses. Briefly, WGCNA uses several node adjacency measures—"soft" thresholding—instead of simple co-expression correlation to represent a weighted gene co-expression. Unlike a traditional gene co-expression network relying on hard thresholding pair-wise gene correlation to assign edges:

$$a_{ij} = \begin{cases} 1 & \text{if } S_{ij} \geq \lambda \\ 0 & \text{otherwise} \end{cases}$$

where S_{ij} is the co-expression similarity and γ is the hard threshold, WGCNA utilizes a soft thresholding method by using S_{ij} directly to a power [51]:

$$a_{ij} = S_{ij}^{\beta}$$

where the determination of soft threshold β is achieved using the scale-free topology criterion. The net result is that WGCNA not only reflects the continuous nature of gene correlation, but also constructs a more robust network [51]. Following network construction, WGCNA calculates dissimilarity between genes using topology overlap measure (TOM) and divides genes into modules using hierarchical clustering and dynamic tree-cut.

WGCNA has been extensively used to find novel gene–gene, gene–disease, and gene–pathway interactions, for example [52–59]. Two of the noteworthy examples of WGCNA are the Allen Brain Project [52] and in understanding the underlying mechanisms in autism spectrum disorder (ASD). In the first study, microarray transcriptome data from 900 microdissection regions of two adult human brains were analyzed and the WGCNA was used to identify gene modules in the integrated brain gene expression map. The module eigengenes (MEs) were calculated and the list of genes with the highest Pearson correlation to ME was compiled, which were then used to relate to known brain cell types and functional enrichment analysis of associated modules. Gene modules discovered by this approach were found to be frequently associated with known primary neural cell types and molecular functions, and the enriched gene ontology (GO) terms are highly relevant to brain biology [52]. These results, backed by differential gene expression analysis, demonstrated a strong correlation between brain-wide gene expression patterns and major brain cell type distribution.

In the second study, Voineagu et al. [57] used transcriptome and gene co-expression network analysis to answer the question of whether ASD is a result of several different pathologies or represents a convergence of a few pathways from different genetic causes. The system-level analysis of the ASD brain transcriptome led to the discovery of the molecular neuropathological basis of ASD, which can be used as a basis for potentially novel therapeutic interventions or to explore the genetic, epigenetic, or environmental etiologies of ASD. For instance, this study found that the gene expression patterns that differentiate the temporal and frontal cortex are less distinct in the ASD brain. Additionally, specific splicing abnormalities and clusters of co-expressed genes associated with ASD were found to be enriched for previously identified genetic association signals. These results suggest that transcriptional and splicing dysregulation could be two of the principal underlying mechanisms of neuronal dysfunction in ASD. The study used brain tissue samples (19 ASD cases and 17 controls) from the Autism Tissue Project (ATP) and the Harvard Brain Bank. Network construction using probes with robust expression was performed using the WGCNA ("blockwiseModules" function) and further details can be obtained from the published study [57].

1.3.5 Biomedical semantic graph mining

To complement the heterogeneous nature of biomedical and genomic networks, the underlying network data model should support a directed labeled graph where nodes can be semantically tagged and relations are explicitly defined. The Semantic Web (SW) [60], proposed by Tim Berners-Lee, supports the required flexible integration, representing the knowledge semantically using formal ontologies. The Resource Description Framework ([RDF]—http://www.w3.org/RDF/), a graph theory-based data model defined as a web technology standard, is fundamental in the SW realm and is ideally suitable for representing big biomedical networks. It provides a generic framework to describe entities, relationships, and constraints and represent the multidimensional data and web resources as directed acyclic graphs. An RDF is made up of "triples" with every triple comprising a subject, predicate (property), and object (property value). Thus, each of the triples in an RDF is a statement and each statement describes a resource, the resource's properties, and the values of those properties. Statements in an RDF can be represented as a network of resources (nodes) connected by properties (edges) to values. For example, the triplet < "TP53" "is a" "Gene" > expresses "TP53" as the subject, "is a" as the property, and "Gene" as the object of the statement. However, to provide semantics (or meaning) that describes specific nodes and properties in the graph, an RDF vocabulary (with classes and properties described) provided by the RDF Schema (RDFS) (http://www.w3.org/TR/rdf-schema/) is required. The (Ontology Web Language ([OWL]—http://www.w3.org/2001/sw/wiki/OWL), with well-defined and more advanced ontology constructs to define classes and properties that support a rich set of semantics derived from description logics, succeeds the RDFS. To access and query data encoded in RDF triples, the W3C recommends SPARQL (http://www.w3.org/TR/rdf-sparql-query/), the standard query language for RD. The two principal advantages of using SW-integrated networks of data from multiple and diverse group of data sets are that it supports the (a) search of a multitude data sets via a single framework, with a collective understanding of terminology (controlled vocabulary or ontology) and (b) search for relationships or paths of relationships that traverse across different data sets [61].

1.3.6   Biomedical network analysis—Semantic web graph mining

Computational ranking or prioritization is commonly used to sort results in genome information retrieval as well as to detect essential nodes in biomedical and genomic networks. For example, identifying essential genes that can guide us to understand their lethality [62] to discovering disease causal genes [61]. Utilizing the RDF graphs for ranking has distinctive characteristics over traditional homogeneous graphs, as they support semantically labeling of nodes and edges, and also allow the addition of contextualization parameters such as weights into the relations. Several approaches have been developed to rank nodes in RDF graphs, such as concept structure analysis based on structural density [63], and Page Rank/HITS–based algorithms [64,65]. Here, we describe a semantic graph mining (SGM) case study as applied to the problem of computational ranking of disease-causal genes. In this study, Gudivada et al. [61] created a network of different biomedical and genomic entities as nodes and the association between genes and different features of the genes as the edges. The gene–feature relationships were based on several gene annotations or features [e.g., gene ontology [GO] annotations, pathways, mouse phenotypes, human clinical features, etc.]. Based on the assumption that disease-causal genes are functionally important and tend to share similar features with other genes causing diseases with similar clinical features, the authors used a Google-like search and ranking algorithm [66] to prioritize candidate disease genes. Compared with a traditional web, this algorithm takes into account specific aspects of SW such as information complexity since it contains different kinds of resources and relations between them apart from the ontological relations and corresponding references. In principle, the algorithm is similar to that of Google's page ranking wherein the search results are determined not just by specific queries, but by the relative importance of the results in the total information space (in this case, the RDF graph). The solution proposed by Gudivada et al. surpasses many limitations of the previous methods such as the requirement of prior knowledge (training data), which is either not available or is difficult to adapt to the specifications of the problem in hand. The algorithm is applied on top of an integrative genomic and phenomic knowledge network that was accomplished by RDF conversion of data sets from multiple, heterogeneous, biomedical, and genomic data sources such as GO (http://www. geneontology.org/), gene-pathway annotations (compiled from KEGG pathways [67]; BioCarta Pathways—http://www.biocarta.com; BioCyc Pathways [68]; and Reactome Pathways [69]), Mouse Genome Informatics (MGI— http://www.informatics.jax.org), Online Mendelian Inheritance in Man (OMIM) [70], and the Multiple Congenital Anomaly/Mental Retardation database (https://www.nlm.nih. gov/archive/20061212/mesh/jablonski/syndrome_ toc/toc_ a.html). In this algorithm, two important metrics were defined to calculate the importance of each resource, namely, subjectivity score (SS) and objectivity score (OS) corresponding to Kleinberg's [71] hub and authority scores, respectively, for the WWW graph. Kleinberg not only takes into account the number of links to and from a node, but also considers the relevance of linked nodes. Accordingly, if a resource in RDF is pointed to by a resource with high SS, its OS increases. On the other hand, if a resource points to a resource with a high OS, its SS is increased. Another important feature of this method is contextualized weights added into the graph properties. For instance, RDF graphs are complex with varying importance of properties which in turn depend on the subject and object it is associated with. The study assigns distinct weights to different types of properties, based on the type of the property and the nodes it's connecting to. Each property is assigned with a subjectivity weight (weight that depends on the subject of the property) and an objectivity weight (weight that depends on the object of the property). These weights play a critical role when calculating the centrality or importance of genes.

The feasibility of this approach in candidate gene prioritization was validated using 60 randomly selected diseases from a total of about 420 cardiovascular disorders with each of the diseases having a clinical synopsis and at least one implicated gene. For validation purposes, no information about any explicit links between the target gene and the disease were part of the input to the algorithm to estimate whether the method detects the true functional relationship between the disease and the gene. The benchmark results were encouraging with the related gene ranked among the top 10 and top 5 74% and 55% of the time, respectively.

There are several advantages to leverage SW, RDF, and OWL standards for knowledge integration and the application of network centrality analysis for mining disease genes. One of the important features of this framework is the ability to include multiple data sources and related features for modeling and ranking. RDF provides a very flexible way to integrate several layers of diverse knowledge bases and also provide the much needed scalability to add new ones. Biological networks are context sensitive and this can be easily incorporated into the mining process via semantic weighting to produce context-sensitive results. Also, as every resource is ranked in the integrated BioRDF graph, query expansions can be applied to retrieve and prioritize other entities (apart from disease genes), such as pathways. For a more complete in-depth analysis, semantic metrics, explanation of the algorithm, and for the experimental results, refer to the original paper [61].

1.3.7 Pharmacology Big Data—Drug target discovery and repositioning

The last two decades have seen a major paradigm shift from the "one drug–one target" and "one drug–one disease" paradigm to a "many-to-many" relationship between drugs and target proteins and drugs and diseases. The underlying complex relationships that form the basis for this polypharmacology and "poly-indications" can be represented as networks with interconnected drugs, their targets, and indications [24,72–74]. Integrating systems biology models with the drug–target network topological information can potentially improve and accelerate the outcomes of computational and virtual drug discovery approaches. For example, identifying a protein target of an approved drug to be topologically important in a pathway or network related or relevant to another disease can provide a basis for a potential novel therapeutic option [75]. There have been several studies which utilize the topological features of a protein–protein interaction network to identify potential drug targets for parasitic [47,76,77] and bacterial diseases such as tuberculosis [78] and other complex diseases (e.g., Alzheimer's disease [79]). For more details on network pharmacology, an emerging paradigm in therapeutic discovery and drug repositioning, readers are referred to [27,80,81].

1.3.7.1 Drug repositioning candidate discovery—Integrative framework across multiple big heterogeneous networks

The time and cost associated with bringing a new drug to market are enormous. To save the *de novo* drug development cost and reduce the risk of failure, drug repositioning (using existing/approved drugs for new therapeutic indications of diseases) provides an efficient and economical route in today's drug development research [82]. The main advantage of drug repositioning is that in *reusing* drugs that have previously passed clinical trials, the risk of failure in future late-stage clinical trials is relatively minimized with potentially faster drug approvals. Current *in silico* drug repositioning strategies are built around integration and mining of large data sets from heterogeneous but related biomedical domains [83]. However, identifying "useful" information from such large and heterogeneous data silos is not trivial. The ability to manage, integrate, and mine drug-related data generated by all -omic technologies—from the discovery phase to real-world use after regulatory body approval—is critical to derive maximum benefit from the technological advances and accumulating data. Thus, there is a paramount need for novel, efficient, and robust computational approaches to effectively manage, integrate, and mine such big heterogeneous data to facilitate knowledge extraction and generate testable hypotheses to enable drug repositioning and drug discovery.

Since a network provides a systems-level view of the corresponding system, network-based approaches are a powerful option for discovering meaningful patterns guiding drug discovery and repositioning [84]. For instance, by integrating drug, gene, disease, and other biological entity relationships and representing them as different types of networks, and applying appropriate network analysis, interesting drug and disease combinations for repositioning candidates may be discovered [74,85–88]. By integrating known disease, drug, and protein interactions and applying a shared neighborhood scoring (SNS) algorithm, Lee et al. [89] identified a potential novel indication for a hypertension drug. Similarly, Li et al. [90] constructed a Chinese herb network to measure the strongly connected herbs and herb pairs through a co-module analysis across herb–biomolecule–disease multilayer networks. The SW standards and technologies [60], such as OWL, RDF, and SPARQL, have also been shown to be a powerful platform to integrate and query various biomedical and genomic entities (e.g., approved drugs, drug targets, genes, pathways, indications, and their interactions). As described previously, SW standards allow complex relations from data to be modeled as semistructured graphs from which inferences can be drawn [61,91–93]. For example, Qu et al. [93] created a large RDF graph by interlinking multiple heterogeneous biomedical and genomic databases. This semantically integrated database was then used to explore the implicit drug-pathway associations and infer new indications for approved drugs.

Another interesting example that employs network-based approaches on a large amount of integrated heterogeneous genomic, biomedical, and high-throughput compound screening data was presented by Iorio et al. [94]. In this study, the researchers developed a bioinformatics tool (mode of action by network analysis) to identify new indications for approved drugs. A new distance measure was computed between the compounds using the data from the connectivity map [95,96]. A drug–drug network was then built using these distance measures and analyzed to identify highly interconnected communities of compounds. In the last step, using known knowledge and literature searches, communities enriched for compounds or drugs with a similar mechanism of action were identified. These communities are used to (a) classify new compounds or drugs when they are co-localized in the network with other drugs whose mode of action is known and (b) generate novel hypotheses on known drugs. The authors used this integrated network and the community structures to discover a novel indication (neurodegenerative disorders) for the rho-kinase inhibitor fasudil.

1.3.7.2 Approved drug-related side-effects Big Data—Post-marketing data analysis

Another large biomedical data source that represents a large and extremely valuable resource to explore the relations of multiple factors with respect to approved drug safety issues is the US Food and Drug Administration's adverse event reporting system (FAERS) data. The FAERS database is growing exponentially each year ($\sim 250,000$ new reports are added each year). The huge size of the database and the data content are two of the principal factors that complicate the data-mining task of FAERS. For instance, FAERS is not a database but a collection of reports reported by anyone who is administering or taking an approved drug. The crowd-reported data has a direct impact on the data content. The drug nomenclature is therefore nonstandard (e.g., brands and generic names of drugs) and replete with typographical errors. In addition, there can be inaccuracy of reporting (e.g., underreporting or biased reporting due to potential media influences). All these factors make it extremely complicated for the biomedical researcher to obtain analytical results from FAERS data. We [97], and others [98–102], have addressed some of these challenges and developed computational approaches that permit mining this noisy and Big Data either to understand the molecular basis of drug-related adverse events or to discover potential drug repositioning candidates. Almost all of these efforts, however, focus predominantly on using traditional pharmacovigilance-based approaches to highlight correlations between drugs and side effects. Further, none of these resources supports the ability for drug or side-effect related ontological aggregations nor permit a more granular cohort-based analyses (e.g., identifying drug-class-specific differential risks or comparing side-effect risk in population subgroups). To address this, Sarangdhar et al. recently designed AERSMine (http://research.cchmc.org/aers) to mine 5.9 million patient reports from the FAERS. It generates drug adverse events, drug indications, and other data matrices that are scored by a variety of approaches (e.g., the WHO signal-reporting algorithms). The matrices can be further filtered and clustered to obtain a more segmented view of the differential reporting patterns of approved drug usage and their side effects. Sarangdhar et al. [97] used AERSMine to understand individual factors that could potentially affect the efficacy and safety of anti–tumor necrosis factor drugs.

Electronic health records (EHRs) represent another biomedical Big Data resource which is complex and incomprehensible, but can be critical for drug repositioning candidate discovery [103]. Combining EHRs with data from FAERS, Tatonetti et al. systematically predicted associations between drugs and side effects. Any of the computationally generated hypotheses need to be validated experimentally or through any other independent verification method or by mining published literature to find reports in support of a hypothesis. Combining computational hypothesis with a preclinical or clinical hypothesis has the potential to accelerate the translational path toward a clinical trial.

1.4 Conclusion

Large heterogeneous biomedical and genomic data repositories are fueling core initiatives of academic, industry, and academic–industry collaborative efforts driving Big Data and translational ventures. To leverage the biomedical and genomic Big Data deluge efficiently for knowledge discovery and testable hypothesis generation, centralization and standardization of biomedical and genomic data collection is essential. Apart from facilitating collaboration, this will accelerate effective mining of the integrated data to enable plausible pharmacological discoveries. To realize the full potential of biomedical and genomic Big Data for personalized and precision medicine, there is an urgent need for biomedical and bioinformatics researchers to approach the Big Data deluge problem in an interdisciplinary fashion. The ready availability of cloud-based parallel high-performance computing can further catalyze and accelerate these efforts.

References

1. Chen, J., E. E. Bardes, B. J. Aronow, and A. G. Jegga. 2009. ToppGene Suite for gene list enrichment analysis and candidate gene prioritization. *Nucleic Acids Res* 37 (Web Server issue):W305-11.

2. Greene, C. S., A. Krishnan, A. K. Wong, et al. 2015. Understanding multicellular function and disease with human tissue-specific networks. *Nat Genet* 47 (6):569–576.

3. Junker, B. H., D. Koschutzki, and F. Schreiber. 2006. Exploration of biological network centralities with CentiBiN. *BMC Bioinformatics* 7:219.

4. Zhou, X., J. Menche, A. L. Barabasi, and A. Sharma. 2014. Human symptoms-disease network. *Nat Commun* 5:4212.

5. Ryan, C. J., P. Cimermancic, Z. A. Szpiech, A. Sali, R. D. Hernandez, and N. J. Krogan. 2013. High-resolution network biology: Connecting sequence with function. *Nat Rev Genet* 14 (12):865–879.

6. Furlong, L. I. 2013. Human diseases through the lens of network biology. *Trends Genet* 29 (3):150–159.

7. Bebek, G., M. Koyuturk, N. D. Price, and M. R. Chance. 2012. Network biology methods integrating biological data for translational science. *Brief Bioinform* 13 (4):446–459.

8. Ideker, T., and N. J. Krogan. 2012. Differential network biology. *Mol Syst Biol* 8:565.

9. Zhang, M., C. Zhu, A. Jacomy, L. J. Lu, and A. G. Jegga. 2011. The orphan disease networks. *Am J Hum Genet* 88 (6):755–766.

10. Lowe, J. A., P. Jones, and D. M. Wilson. 2010. Network biology as a new approach to drug discovery. *Curr Opin Drug Discov Devel* 13 (5):524–526.

11. Goh, K. I., M. E. Cusick, D. Valle, B. Childs, M. Vidal, and A. L. Barabasi. 2007. The human disease network. *Proc Natl Acad Sci U S A* 104 (21):8685–8690.

12. Friedman, A., and N. Perrimon. 2007. Genetic screening for signal transduction in the era of network biology. *Cell* 128 (2):225–231.

13. Barabasi, A. L., and Z. N. Oltvai. 2004. Network biology: Understanding the cell's functional organization. *Nat Rev Genet* 5 (2):101–113.

14. Kellis, M., B. Wold, M. P. Snyder, et al. 2014. Defining functional DNA elements in the human genome. *Proc Natl Acad Sci U S A* 111 (17):6131–6138.

15. Consortium, Encode Project. 2004. The ENCODE (ENCyclopedia Of DNA Elements) Project. *Science* 306 (5696):636–640.

16. Consortium, Encode Project, E. Birney, J. A. Stamatoyannopoulos, et al. 2007. Identification and analysis of functional elements in 1% of the human genome by the ENCODE pilot project. *Nature* 447 (7146):799–816.

17. Chen, H., L. Ding, Z. Wu, T. Yu, L. Dhanapalan, and J. Y. Chen. 2009. Semantic web for integrated network analysis in biomedicine. *Brief Bioinform* 10 (2):177–192.

18. Livingston, K. M., M. Bada, W. A. Baumgartner, and L. E. Hunter. 2015. KaBOB: Ontology-based semantic integration of biomedical databases. *BMC Bioinformatics* 16:126.

19. Hartwell, L. H., J. J. Hopfield, S. Leibler, and A. W. Murray. 1999. From molecular to modular cell biology. *Nature* 402 (6761 Suppl):C47–52.

20. Newman, M. 2003. The structure and function of complex networks. *SIAM Review* 45 (2):167–256.

21. Girvan, M., and M. E. Newman. 2002. Community structure in social and biological networks. *Proc Natl Acad Sci U S A* 99 (12):7821–7826.

22. Tong, A. H., G. Lesage, G. D. Bader, et al. 2004. Global mapping of the yeast genetic interaction network. *Science* 303 (5659):808–813.

23. Balazsi, G., A. L. Barabasi, and Z. N. Oltvai. 2005. Topological units of environmental signal processing in the transcriptional regulatory network of *Escherichia coli*. *Proc Natl Acad Sci U S A* 102 (22):7841–7846.

24. Yildirim, M. A., K. I. Goh, M. E. Cusick, A. L. Barabasi, and M. Vidal. 2007. Drug-target network. *Nat Biotechnol* 25 (10):1119–1126.

25. Chuang, H. Y., E. Lee, Y. T. Liu, D. Lee, and T. Ideker. 2007. Network-based classification of breast cancer metastasis. *Mol Syst Biol* 3:140.

26. Sharan, R., I. Ulitsky, and R. Shamir. 2007. Network-based prediction of protein function. *Mol Syst Biol* 3:88.

27. Hopkins, A. L. 2008. Network pharmacology: The next paradigm in drug discovery. *Nat Chem Biol* 4 (11):682–690.

28. Pavlopoulos, G. A., M. Secrier, C. N. Moschopoulos, et al. 2011. Using graph theory to analyze biological networks. *BioData Min* 4:10.

29. Barabási, A. L., N. Gulbahce, and J. Loscalzo. 2011. Network medicine: A network-based approach to human disease. *Nat Rev Genet* 12 (1):56–68.

30. Vidal, M., M. E. Cusick, and A. L. Barabási. 2011. Interactome networks and human disease. *Cell* 144 (6):986–998.

31. Freeman, L. 1977. A set of measures of centrality based upon betweenness. *Sociometry* 40:35–41.

32. Beauchamp, M. A. 1965. An improved index of centrality. *Behav Sci* (10):161–163.

33. Sabidussi, G. 1966. The centrality index of a graph. *Psychometrika* 31:581–603.

34. Brin, S., and L. Page. 1998. The anatomy of a large-scale hypertextual Web search engine. *Compu Networks ISDN* (30):107–117.

35. Watts, D. J., and S. H. Strogatz. 1998. Collective dynamics of 'small-world' networks. *Nature* (393):440–442.

36. Friedkin, N. E. 1984. Structural cohesion and equivalence explanations of social homogeneity. *Sociol Method Res* 12:235–261.

37. Palla, G., I. Derenyi, I. Farkas, and T. Vicsek. 2005. Uncovering the overlapping community structure of complex networks in nature and society. *Nature* 435 (7043):814–818.

38. Eisenberg, D., E. M. Marcotte, I. Xenarios, and T. O. Yeates. 2000. Protein function in the post-genomic era. *Nature* 405 (6788):823826.

39. Guimera, R., and L. A. Nunes Amaral. 2005. Functional cartography of complex metabolic networks. *Nature* 433 (7028):895–900.

40. Rives, A. W., and T. Galitski. 2003. Modular organization of cellular networks. *Proc Natl Acad Sci U S A* 100 (3):1128–1133.

41. Székely, G. J. and M. L. Rizzo. 2005. Hierarchical clustering via joint between-within distances: Extending Ward's minimum variance method. *J Classif* 22:151–183.

42. Samanta, M. P., and S. Liang. 2003. Predicting protein functions from redundancies in large-scale protein interaction networks. *Proc Natl Acad Sci U S A* 100 (22):12579–12583.

43. Xenarios, I., L. Salwinski, X. J. Duan, P. Higney, S. M. Kim, and D. Eisenberg. 2002. DIP, the database of interacting proteins: A research tool for studying cellular networks of protein interactions. *Nucleic Acids Res* 30 (1):303–305.

44. Kanungo, T., D. M. Mount, N. S. Netanyahu, C. D. Piatko, R. Silverman, and A. Y. Wu. 2002. An efficient k-means clustering algorithm: Analysis and implementation. *EEE Trans. Pattern Anal Mach Intell* 24:881–892.

45. Kriegel, H.-P., P. Kröger, J. Sander, and A. Zimek. 2011. Density-based clustering. *WIREs Data Mining Knowl Discov* 1 (3):231–240.

46. Ester, M., H.-P. Kriegel, J. Sander, and X. Xu. 1996. A density-based algorithm for discovering clusters in large spatial databases with noise. In *Proceedings of the Second International Conference on Knowledge Discovery and Data Mining (KDD-96)*.

47. Yu, H., P. M. Kim, E. Sprecher, V. Trifonov, and M. Gerstein. 2007. The importance of bottlenecks in protein networks: Correlation with gene essentiality and expression dynamics. *PLoS Comput Biol* 3 (4):e59.

48. Blondel, V. D., J.-L. Guillaume, R. Lambiotte, and E. Lefebvre. 2008. Fast unfolding of communities in large networks. *J Stat Mech Theor Exp* 10.

49. Gross, A. M., and T. Ideker. 2015. Molecular networks in context. *Nat Biotechnol* 33 (7):720–721.

50. Gorenshteyn, D., E. Zaslavsky, M. Fribourg, et al. 2015. Interactive Big Data resource to elucidate human immune pathways and diseases. *Immunity* 43 (3):605–614.

51. Zhang, B., and S. Horvath. 2005. A general framework for weighted gene co-expression network analysis. *Stat Appl Genet Mol Biol* 4:Article17.

52. Hawrylycz, M. J., E. S. Lein, A. L. Guillozet-Bongaarts, et al. 2012. An anatomically comprehensive atlas of the adult human brain transcriptome. *Nature* 489 (7416):391–399.

53. Udyavar, A. R., M. D. Hoeksema, J. E. Clark, et al. 2013. Co-expression network analysis identifies spleen tyrosine kinase (SYK) as a candidate oncogenic driver in a subset of small-cell lung cancer. *BMC Syst Biol* 7 Suppl 5:S1.

54. Spiers, H., E. Hannon, L. C. Schalkwyk, et al. 2015. Methylomic trajectories across human fetal brain development. *Genome Res* 25 (3):338352.

55. Kogelman, L. J., and H. N. Kadarmideen. 2014. Weighted interaction SNP Hub (WISH) network method for building genetic networks for complex diseases and traits using whole genome genotype data. *BMC Syst Biol* 8 Suppl 2:S5.

56. Clarke, C., S. F. Madden, P. Doolan, et al. 2013. Correlating transcriptional networks to breast cancer survival: A large-scale coexpression analysis. *Carcinogenesis* 34 (10):2300–2308.

57. Voineagu, I., X. Wang, P. Johnston, et al. 2011. Transcriptomic analysis of autistic brain reveals convergent molecular pathology. *Nature* 474 (7351):380–384.

58. Miller, J. A., S. Horvath, and D. H. Geschwind. 2010. Divergence of human and mouse brain transcriptome highlights Alzheimer disease pathways. *Proc Natl Acad Sci U S A* 107 (28):12698–12703.

59. Maze, I., L. Shen, B. Zhang, et al. 2014. Analytical tools and current challenges in the modern era of neuroepigenomics. *Nat Neurosci* 17 (11):1476–1490.

60. Berners-Lee, T., and J. Hendler. 2001. Publishing on the semantic web. *Nature* 410 (6832):1023–1024.

61. Gudivada, R. C., X. A. Qu, J. Chen, A. G. Jegga, E. K. Neumann, and B. J. Aronow. 2008. Identifying disease-causal genes using semantic web-based representation of integrated genomic and phenomic knowledge. *J Biomed Inform* 41 (5):717–729.

62. Palumbo, M. C., A. Colosimo, A. Giuliani, and L. Farina. 2005. Functional essentiality from topology features in metabolic networks: A case study in yeast. *FEBS Lett* 579 (21):4642–4646.

63. Alani, H., and C. Brewster. 2005. Ontology ranking based on the analysis of concept structures. In *Proceedings of the 3rd International Conference on Knowledge Capture (K-Cap)*. New York, NY: ACM Press.

64. Page, L., S. Brin, R. Motwani, and T. Winograd. 1999. *The PageRank Citation Ranking: Bringing Order to the Web*. Stanford InfoLab.

65. White, S., and P. Smyth. 2003. Algorithms for estimating relative importance in networks. In *KDD '03: Proceedings of the Ninth ACM SIGKDD International Conference on Knowledge Discovery and Data Mining*. New York, NY: ACM Press.

66. Mukherjea, S. 2005. Information retrieval and knowledge discovery utilising a biomedical semantic web. *Brief Bioinform* 6 (3):252–262.

67. Kanehisa, M., S. Goto, M. Hattori, et al. 2006. From genomics to chemical genomics: New developments in KEGG. *Nucleic Acids Res* 34 (Database issue):D354–357.

68. Karp, P. D., C. A. Ouzounis, C. Moore-Kochlacs, et al. 2005. Expansion of the BioCyc collection of pathway/genome databases to 160 genomes. *Nucleic Acids Res* 33 (19):6083–6089.

69. Jupe, S., J. W. Akkerman, N. Soranzo, and W. H. Ouwehand. 2012. Reactome–A curated knowledgebase of biological pathways: Megakaryocytes and platelets. *J Thromb Haemost* 10 (11):2399–2402.

70. Amberger, J. S., C. A. Bocchini, F. Schiettecatte, A. F. Scott, and A. Hamosh. 2015. OMIM.org: Online Mendelian Inheritance in Man (OMIM(R)), an online catalog of human genes and genetic disorders. *Nucleic Acids Res* 43 (Database issue):D789–798.

71. Kleinberg, J. M. 1999. Authoritative sources in a hyperlinked environment. *J ACM* 46 (5):604–632.

72. Sardana, D., C. Zhu, M. Zhang, R. C. Gudivada, L. Yang, and A. G. Jegga. 2011. Drug repositioning for orphan diseases. *Brief Bioinform* 12 (4):346–56.

73. Kaimal, V., D. Sardana, E. E. Bardes, R. C. Gudivada, J. Chen, and A. G. Jegga. 2011. Integrative systems biology approaches to identify and prioritize disease and drug candidate genes. *Methods Mol Biol* 700:241–259.

74. Wu, C., R. C. Gudivada, B. J. Aronow, and A. G. Jegga. 2013. Computational drug repositioning through heterogeneous network clustering. *BMC Syst Biol* 7 Suppl 5:S6.

75. Pujol, A., R. Mosca, J. Farres, and P. Aloy. 2010. Unveiling the role of network and systems biology in drug discovery. *Trends Pharmacol Sci* 31 (3):115–123.

76. Jeong, H., S. P. Mason, A. L. Barabasi, and Z. N. Oltvai. 2001. Lethality and centrality in protein networks. *Nature* 411 (6833):41–42.

77. Florez, A. F., D. Park, J. Bhak, et al. 2010. Protein network prediction and topological analysis in Leishmania major as a tool for drug target selection. *BMC Bioinformatics* 11:484.

78. Raman, K., K. Yeturu, and N. Chandra. 2008. targetTB: A target identification pipeline for *Mycobacterium tuberculosis* through an interactome, reactome and genome-scale structural analysis. *BMC Syst Biol* 2:109.

79. Yang, L., J. Chen, L. Shi, M. Hudock, and L. He. 2010. Identifying unexpected therapeutic targets via chemical-protein interactome. *PLoS One* 5 (3):e9568.

80. Hopkins, A. L. 2007. Network pharmacology. *Nat Biotechnol* 25 (10):1110–1111.

81. Berger, S. I., and R. Iyengar. 2009. Network analyses in systems pharmacology. *Bioinformatics* 25 (19):2466–2472.

82. Ashburn, T. T., and K. B. Thor. 2004. Drug repositioning: Identifying and developing new uses for existing drugs. *Nat Rev Drug Discov* 3 (8):673–683.

83. Wishart, D. S. 2007. Discovering drug targets through the web. *Comp Biochem Physiol Part D Genomics Proteomics* 2 (1):9–17.

84. Arrell, D. K., and A. Terzic. 2010. Network systems biology for drug discovery. *Clin Pharmacol Ther* 88 (1):120–125.

85. Mei, H., T. Xia, G. Feng, J. Zhu, S. M. Lin, and Y. Qiu. 2012. Opportunities in systems biology to discover mechanisms and repurpose drugs for CNS diseases. *Drug Discov Today* 17 (21–22):1208–1216.

86. Hu, G., and P. Agarwal. 2009. Human disease-drug network based on genomic expression profiles. *PLoS One* 4 (8):e6536.

87. Li, Y., and P. Agarwal. 2009. A pathway-based view of human diseases and disease relationships. *PLoS One* 4 (2):e4346.

88. Wu, M., X. Yang, and C. Chan. 2009. A dynamic analysis of IRS-PKR signaling in liver cells: A discrete modeling approach. *PLoS One* 4 (12):e8040.

89. Lee, H. S., T. Bae, J. H. Lee, et al. 2012. Rational drug repositioning guided by an integrated pharmacological network of protein, disease and drug. *BMC Syst Biol* 6:80.

90. Li, S., B. Zhang, D. Jiang, Y. Wei, and N. Zhang. 2010. Herb network construction and co-module analysis for uncovering the combination rule of traditional Chinese herbal formulae. *BMC Bioinformatics* 11 Suppl 11:S6.

91. Antezana, E., M. Kuiper, and V. Mironov. 2009. Biological knowledge management: The emerging role of the semantic web technologies. *Brief Bioinform* 10 (4):392–407.

92. Dumontier, M., and N. Villanueva-Rosales. 2009. Towards pharmacogenomics knowledge discovery with the semantic web. *Brief Bioinform* 10 (2):153–163.

93. Qu, X. A., R. C. Gudivada, A. G. Jegga, E. K. Neumann, and B. J. Aronow. 2009. Inferring novel disease indications for known drugs by semantically linking drug action and disease mechanism relationships. *BMC Bioinformatics* 10 Suppl 5:S4.

94. Iorio, F., R. Bosotti, E. Scacheri, et al. 2010. Discovery of drug mode of action and drug repositioning from transcriptional responses. *Proc Natl Acad Sci U S A* 107 (33):14621–14626.

95. Lamb, J. 2007. The Connectivity Map: A new tool for biomedical research. *Nat Rev Cancer* 7 (1):54–60.

96. Lamb, J., E. D. Crawford, D. Peck, et al. 2006. The Connectivity Map: Using gene-expression signatures to connect small molecules, genes, and disease. *Science* 313 (5795):1929–1935.

97. Sarangdhar, M., A. Kushwaha, J. Dahlquist, A. Jegga, and B. Aronow. 2013. Using systems biology-based analysis approaches to identify mechanistically significant adverse drug reactions: Pulmonary complications from combined use of anti-TNFα agents and corticosteroids. *AMIA Jt Summits Transl Sci Proc* 2013:151–155.

98. Tatonetti, N. P., P. P. Ye, R. Daneshjou, and R. B. Altman. 2012. Data-driven prediction of drug effects and interactions. *Sci Transl Med* 4 (125):125ra31.

99. Böhm, R., J. Höcker, I. Cascorbi, and T. Herdegen. 2012. Openvigil–free eyeballs on AERS pharmacovigilance data. *Nat Biotechnol* 30 (2):137–138.

100. Grigoriev, I., W. zu Castell, P. Tsvetkov, and A. V. Antonov. 2014. AERS spider: An online interactive tool to mine statistical associations in adverse event reporting system. *Pharmacoepidemiol Drug Saf* 23 (8):795–801.

101. Kuhn, M., I. Letunic, L. J. Jensen, and P. Bork. 2015. The SIDER database of drugs and side effects. *Nucleic Acids Res* 44 (D1):D1075-9.

102. Kuhn, M., M. Campillos, I. Letunic, L. J. Jensen, and P. Bork. 2010. A side effect resource to capture phenotypic effects of drugs. *Mol Syst Biol* 6:343.

103. Jensen, P. B., L. J. Jensen, and S. Brunak. 2012. Mining electronic health records: Towards better research applications and clinical care. *Nat Rev Genet* 13 (6):395–405.

Chapter 2

Distributed or Network-Based Big Data?

Bo Hu and Klaus Arto

2.1 Introduction

Big Data is a term closely related with the development of the Internet. Due to the existence of such a global network it is possible to share and collect this huge amount of data [1] as well as to provide Big Data processing and analytics as a service to a broader audience throughout the entire network.

Following [2], Big Data "means data that's too big, too fast, or too hard for existing tools to process." The enormous amount of data could no longer be handled by conventional scale-up database systems. With the rapidly growing demands such system designs do not only cause exploding costs but also reach quickly the technical limits of performance.

Today's common design of Big Data is to scale-out, or to use thousands of inexpensive commodity servers to achieve unprecedented scalability, performance, and availability [3,4]. The topology of the network connecting those servers with each other and with the clients is expected to have a significant impact on these target characteristics.

The powerfulness of Big Data also raises "privacy concerns which could stir a regulatory backlash dampening the data economy and stifling innovation" [5].

In this chapter we propose a concept of decentralized network-based Big Data architecture to provide both high scalability and protection of privacy. A network of loosely connected Big Data analytic server nodes allows not only decentralized storage and processing, but also individual and privacy-preserving response policies. In Section 2.2 we will first give some background information from a literature research review. Our network-based Big Data architecture will be described in Section 2.3.1, followed by a computational estimation of the performance of this architecture in comparison with two other concepts in Section 2.3.2. Section 2.3.3 shows the advantages of the network-based Big Data architecture. Section 2.3.4 addresses further applications before we conclude this chapter in Section 2.4.

2.2   Background

Data generation and collection has increasingly grown over the last 10 years. At the same time storage has become more and more affordable [1,6]. The appearance of smartphones and tablets enabled people to be connected to the Internet almost anywhere at any time. Along with the expansion of broadband networks, those devices are enabled by integrated sensors to generate additional usable data, for example, motion profiles. Health and fitness tracking is now also possible due to so-called wearables. These developments opened a huge market for new companies in software development but also generated new growth opportunities for established companies (Internet of things, smart metering, etc.). With respect to analytics and enabling new products, Big Data may create significant value to the organizations [8].

Big Data is classified mostly with the three V's of volume, velocity, and variety [8]. Regarding quality and accuracy, there is a fourth V for veracity [9]. These four V's present the—mostly technical—challenges. In addition, the three F's of fast, flexible, and focused and other functions and software components are aspects that need to be considered to find a holistic approach for a Big Data platform [10].

The beginning of Big Data can be dated to 2004 when Google worked on their big table project [3]. In the same year Facebook was founded. Since then there have been a lot of new more or less popular developments which are capable of coping with the vast amount of data [4,11,12].

It is desirable that the performance increases linearly with the number of servers [13]. However, this linear scalability is not achieved by the implemented server clusters. As shown by [3], the performance per server drops by a factor of 2–5 when increasing the number of servers from 1 to 500.

There are also other approaches addressing scalability. Reference [14] presented an information integration system based on a mediator/wrapper or virtual data integration approach [15, p. 62] which enables central information access without requiring central data storage.

No less than seven "big benefits" of Big Data have been presented in [5]: health care, mobile applications, smart grid, traffic management, retail, payments, and online applications. In addition, Big Data may provide significant advantages in law enforcement or terrorism prevention (see, e.g., [16,17]). At the same time increasing concerns about privacy are raised. From the point of view of privacy protection, it is necessary to provide k-anonymity [18,19], that is, "any one individual in the dataset cannot be distinguished from at least $k-1$ other individuals in the same dataset" [20]. Given its comprehensive data collection a centralized Big Data solution can never achieve the k-anonymity for any $k > 1$

(see, e.g., [21]). The fact that the storage of the data can be carried out decentralized is significant, insofar as the fact that private companies and government organizations collect and hold more and more information about individuals is regarded increasingly as critical and dangerous to the society. The German Federal Constitutional Court [22] and the European Court of Justice (ECJ) [23] decided that certain laws and regulations by the German Federal Government were unconstitutional, and the ECJ recently ruled that the safe harbor principles do not meet the requirements for the protection of personal data, especially the transfer of those data to a third country [24]. The German data protection authority requires a balance between the legitimate interests for the processing of data and the legitimate interests of the persons concerned and calls on all parties to get together and discuss the impact of recent developments of Big Data, taking into account data protection aspects [25].

Based on the decentralized concept, meta-search engines have been introduced with a certain success [26,27] while diverse decentralized social networks like Safebook (see, e.g., [28]). Diaspora [29], Friendica [30], and many others have also mushroomed since 2009 [31–33]. It seems that the lack of interoperability (particularly between the social networks) and the incomplete business models (all network-based Big Data services including meta-search engines) prevent the breakthrough of decentralized concepts. From a technical point of view especially in relation to telemetry-based Big Data services in the field of transportation, health, environmental protection, and numerous others will soon be or are already available. Enhancement of both power and privacy protection by networking different Big Data services is the central theme of this chapter.

2.3 Network-Based Big Data

2.3.1 A concept of network-based Big Data

We propose a concept of decentralized network-based Big Data structures. According to this concept the entire network is partitioned into multiple logical Big Data subnetworks. The Big Data service is provided by multiple decentralized Big Data server nodes, each single node in a single subnetwork. It is aimed to provide benefits both for the data creators regarding their demand for protection of privacy and the users of diverse Big Data solutions.

We define the logical structure of data exchange for "network-based Big Data" with the underlying undirected graph given by

$$G_i := (V, \ E) = \left(\bigcup_{i=1}^{n} V_i, \bigcup_{i=0}^{n} E_i \right)$$

As mentioned previously the entire network is partitioned into n subnetworks, denoted as

$$G_i := (V_i, E_i), \ i = 1, 2, \ldots, n$$

where

$$V_i = \{\nu_{i0}, \nu_{i1}, \ldots, \nu_{im_i}\}, E_i = \{\{\nu_{i0}, \ \nu_{ij}\} \, | \, j \in [1, m_i] \subset \mathbb{N}\}$$

and

$$\forall i, j \in \{1, 2, \ldots, n\} \wedge i \neq j \Rightarrow G_i \cap G_j = \varnothing$$

At the same time, all server nodes form a complete graph

$$G_0 := (\{\nu_{10}, \nu_{20}, \ldots, \nu_{n0}\}, E_0) \, , E_0 := \{\{\nu_{i0}, \nu_{j0}\} \, | \, i, \ j \in \{1, 2, \ldots, n\} \wedge i \neq j\}$$

Within any given subnetwork G_i, ν_{i_0} is the only Big Data server node which may collect data there. As a prerequisite to each

$$e_{ij} = \{\nu_{i0}, \nu_{ij}\} \in E_i$$

a possibly individually negotiated Big Data policy is to be applied. This policy defines how Big Data may be collected, processed, retrieved, and deleted in case of termination.

Within G_0 no sharing of Big Data may take place. Instead, the servers cooperate with each other and build a mediator/wrapper architecture(s) (see Section 2.2).

Figure 2.1 shows an example of a network with four logical subnetworks G_1, G_2, G_3, G_4. Each of them consists of one logical Big Data server node and multiple client nodes from which the Big Data may be collected.

The server nodes v_{10}, v_{20}, v_{30}, v_{40} in this example are connected logically to each other, and thus form a complete graph

$$K_4 = (\{v_{10}, v_{20}, v_{30}, v_{40}\}, \{\{v_{i0}, v_{j0}\} : 1 \leq i < j \leq 4\})$$

A Big Data server node, v_{20} for instance, is only allowed to collect Big Data from the client nodes within the logical subnetwork G_2. v_{20}, for example, has no direct access to the data produced by v_{41}.

2.3.1.1 Collecting information from a creator

- Creator v_{41} sends information (tweets, geo information, consumption data, documents, etc.) to server node v_{40}. As a prerequisite, server v_{40} ensures the creator applies its individual Big Data policy, which can be considered as a contract between the creator v_{41} and the provider of the Big Data server node v_{40}.

- Server node v_{40} processes and stores data using standard Big Data techniques.

- Server node v_{40} **does not share data** with the other server nodes, v_{10}, v_{20}, v_{30}.

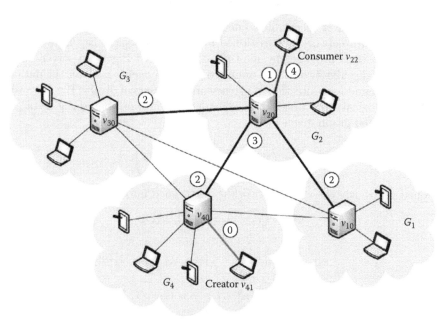

FIGURE 2.1: An example of decentralized network-based Big Data. (0) Collecting information from a creator; (1) Consumer requests data within its network; (2) Server v_{20} broadcasts the request to all other servers it knows; (3) Server v_{40} responds to the request of server v_{20}; (4) Server v_{20} responds to the initial request of Consumer.

2.3.1.2 Consumer requests data within its network

- Consumer v_{22} sends a request to server node v_{20} searching for relevant data (tweets of the originator whom he or she follows, etc.)

2.3.1.3 Server v_{20} broadcasts the request to all other servers it knows

- Server node v_{20} has to have a trusted connection with the other servers.

- Server node v_{20} has to authenticate itself to the server nodes v_{10}, v_{20}, v_{40}.

- Server node v_{20} may provide the identity of the consumer subject to the policy of the requested Big Data service and the individual Big Data policy agreement between the consumer v_{22} and the server node v_{20}.

2.3.1.4 Server v_{40} responds to the request of server v_{20}

- Following the policy defined by the creator v_{41} and based on the requested information provided by v_{20} and the collected Big Data, server node v_{40} responds to the request.

- Server node v_{40} does not provide any Big Data information to v_{20}.

2.3.1.5 Server v_{20} responds to the initial request of consumer

- Server node v_{20} collects all the answers of the other server nodes, if provided.

- Server node v_{20} sends the results to the consumer v_{22}.

2.3.2 Considerations of the network traffic and the computational effort

Based on the architecture discussed in Section 2.3.1, the network traffic and the computational effort for the entire network can be estimated.

2.3.2.1 Network traffic

The network traffic B is given by

$$B = \sum_{i=1}^{n} B_i = \sum_{i=1}^{n} \sum_{j=1}^{m_i} (c_{ij} + r_{ij}) + \sum_{i,k=1}^{n} f_{i,k}$$

where

$$B_i = \sum_{j=1}^{m_i} (c_{ij} + r_{ij}) + \sum_{k=1}^{n} f_{ik}$$

is the network traffic which is related to a single Big Data server node v_{i_0} and consists of two parts, the client-related traffic and the intra-server traffic, c_{ij} the network traffic for collecting data from a single client v_{ij} (⓪ in Figure 2.1), r_{ij} the network traffic to respond to the requests by v_{ij} (① and ④ in Figure 2.1), m_i the number of clients within the subnetwork G_i, and f_{ik} the network traffic for forwarding a request from v_{i_0} to another server node v_{k_0} and collecting answers (② and ③ in Figure 2.1). Notice that

$$\forall i : f_{ii} = 0$$

The intra-server traffic can be estimated for a search engine network and a social network application as explained in the following.

Today Google processes about 50,000 search queries every second [34]. Assuming each query may cause a data traffic of 200 KB we have

$$\sum_{i=1}^{n}\sum_{j=1}^{m_i} r_{ij} \approx 10 \text{ GB/s}$$

If using a decentralized network-based Big Data concept with n server nodes, each server node has to handle all client requests, either directly or forwarded by the other $n-1$ server nodes. The intra-server network traffic can be estimated as follows:

$$\sum_{i,k=1}^{n} f_{ik} = (n-1)\sum_{i=1}^{n}\sum_{j=1}^{m_i} r_{ij} \approx 1 \text{ TB/s}$$

for $n = 101$. This amount of data flow is acceptable for today's Internet, since the estimated total Internet traffic is about 30 TB/s [34].

WhatsApp claims to be able to receive more than 300,000 and send more than 700,000 messages per second for its users [35]. In a decentralized network-based version, each message is (1) first sent to the server node to which the originator is related, (2) then forwarded to possibly another server node which hosts a specific chat, and (3) finally received by all server nodes to which at least one in the chat-registered user is related. The intra-server traffic is approximately 1,000,000 messages per second or about 1 GB/s.

2.3.2.2   Computational effort

The total computational effort C is given by

$$C = \sum_{i=1}^{n} C_i = \sum_{i=1}^{n} P_1\left(\sum_{j=1}^{m_i} c_{ij}, A_i\right) + \sum_{i=1}^{n} P_2\left(\sum_{k=1}^{n} f_{ki} + \sum_{j=1}^{m_i} r_{ij}, A_i\right)$$

where C_i is the computational effort of a single Big Data server node v_{i_0}, A_i the data amount collected by this server node, P_1 a function of the data collection flow and the data amount, and P_2 a function of the data request flow and the data amount.

The sublinear scalability of Big Data server nodes (see Section 2.2, [39]) can be expressed as

$$\lambda \sum_{i=1}^{n} P_l(F_i, A) \geq P_l\left(\sum_{i=1}^{n} F_i, A\right) \geq \sum_{i=1}^{n} P_l(F_i, A), l \in \{1, 2\}$$

and

$$\lambda \sum_{i=1}^{n} P_l(F, A_i) \geq P_l\left(F, \sum_{i=1}^{n} A_i\right) \geq \sum_{i=1}^{n} P_l(F, A_i), l \in \{1, 2\}$$

where F_i is the data flow from and to a Big Data server node v_{i_0}, F and A are, respectively, the sum of the data flow and the data amount of all server nodes or

$$F = \sum_{i=1}^{n} F_i, A = \sum_{i=1}^{n} A_i$$

and

$$\lambda \geq 1$$

is a characteristic scalability factor depending on the technique and the number of the server nodes used. In the case described in [3, Table 6] cited in Section 2.2, λ has approximately the value 5; in an ideal case of full linear scalability its value is 1.

2.3.2.3 Comparison to a centralized architecture

Centralized Big Data architecture is actually a special case in which there is only a single subnetwork with $\hat{m}$ clients. The network traffic and the computational effort are, respectively:

$$\widehat{B} = \sum_{j=1}^{\hat{m}} (\hat{c}_j + \hat{r}_j) = \sum_{i=1}^{n} \sum_{j=1}^{m_i} (c_{ij} + r_{ij}) + \sum_{i,k=1}^{n} f_{ik} - \sum_{i,k=1}^{n} f_{ik} = B - \sum_{i,k=1}^{n} f_{ik}$$

$$\widehat{C} = P_1 \left(\sum_{j=1}^{\hat{m}} \hat{c}_j, A \right) + P_2 \left(\sum_{j=1}^{\hat{m}} \hat{r}_j, A \right) \geq C$$

since

$$P_1 \left(\sum_{j=1}^{\hat{m}} \hat{c}_j, A \right) \geq P_1 \left(\sum_{i=1}^{n} \sum_{j=1}^{m_i} c_{ij}, A_{max} \right) \geq \sum_{i,k=1}^{n} P_1 \left(\sum_{j=1}^{m_i} c_{ij}, A_i \right)$$

and

$$P_2 \left(\sum_{j=1}^{\hat{m}} \hat{r}_j, A \right) \geq P_2 \left(\sum_{i=1}^{n} \sum_{j=1}^{m_i} r_{ij}, A_{max} \right) \geq \sum_{i=1}^{n} P_2 \left(\sum_{j=1}^{m_i} r_{ij}, A_i \right)$$

where

$$A_{max} = \max_{1 \leq i \leq n} (A_i)$$

In other words, compared to the decentralized architecture discussed previously, a centralized architecture will certainly reduce the intra-server network traffic to zero, but increase the total computational effort for data collection because of the sublinear scalability of Big Data server clusters.

2.3.2.4 Comparison with a distributed architecture

An alternative concept to the network-based Big Data architecture is to have distributed server nodes of which the data have to be kept "eventually consistent" [36]. In such an architecture, the network traffic and the computational effort are given by

$$\tilde{B} = \sum_{i=1}^{n} \left(\sum_{j=1}^{m_i} (c_{ij} + r_{ij}) + S_i \right) = B - \sum_{i,k=1}^{n} f_{ik} + \sum_{i=1}^{n} S_i$$

and

$$\tilde{C} = \sum_{i=1}^{n} \left(P_1 \left(\sum_{j=1}^{m_i} c_{ij}, A \right) + P_2 \left(\sum_{j=1}^{m_i} r_{ij}, A \right) + R_i \right) > \frac{C}{\lambda}$$

since

$$P_1 \left(\sum_{j=1}^{m_i} c_{ij}, A \right) \geq P_1 \left(\sum_{j=1}^{m_i} c_{ij}, A_i \right) \geq \frac{1}{\lambda} P_1 \left(\sum_{j=1}^{m_i} c_{ij}, A_i \right)$$

and

$$\sum_{i=1}^{n} P_2 \left(\sum_{j=1}^{m_i} r_{ij}, A \right) \geq \frac{1}{\lambda} P_2 \left(\sum_{k=1}^{n} \sum_{j=1}^{m_k} r_{kj}, A \right) \geq \frac{1}{\lambda} \sum_{k=1}^{n} P_2 \left(\sum_{k=1}^{n} \sum_{j=1}^{m_k} r_{kj}, A_k \right)$$

where

$$S_i > 0$$

is the network traffic and

$$R_i > 0$$

the computational effort to keep the data on the server node v_{i_0} eventually consistent. Notice that both S_i and R_i can be parameterized to a relative low value as a trade-off for more performance.

The computational effort in a distributed architecture is higher than the computational effort in a network-based architecture divided by a certain factor λ. The network traffic in the network-based architecture is again about up to the amount of the intra-server traffic higher than in the distributed architecture.

2.3.3 Privacy, interoperability, and competition

Section 2.3.2 shows that compared with both the centralized and the distributed architecture, the network-based Big Data architecture reduces the computational effort for data collection because of the sublinear scalability, but increases the request-related network traffic by factor n, in the form of intra-server traffic. From a technical point of view, a choice in favor of the network-based architecture and the determination of the number of server nodes may be a result of diverse considerations of specific applications and the costs of server or network deployment.

When considering privacy protection, however, the network-based Big Data architecture becomes a must. Even outside the context of [24], the network-based Big Data architecture provides better privacy protection.

The k-anonymity limits the probability of a random identification of an individual based on any existing quasi-identifier [19] within the data set under consideration:

$$w \leq \frac{1}{k}$$

As mentioned in Section 2.2, the data stored on a Big Data server node cannot be considered k-anonymous. But if a specific Big Data application is built based on n network-based server nodes, a similar effect of k-anonymity can be achieved since the probability that the personal data of an individual is on a specific server node i is given by

$$w_i\left(n\right) \approx \frac{1}{n}$$

Just like the configurable eventual consistency, the privacy protection can also be configured

$$\forall k > 0, \exists n : w_i\left(n\right) \leq \frac{1}{k}$$

The higher the number of network-based Big Data server nodes the higher the privacy protection. In addition, in such an architecture each Big Data server node faces much smaller potential damage from hacker attacks.

Interoperability in mobile telecommunications is a fundamental principle. Nobody wants to register himself in several provider networks only to be enabled to communicate with members of the other networks. Social networks do not support interoperability because they want to bind the users in order to achieve the greatest possible diversification of advertising. If one could choose only to be member of one specific network, the amount of possible members of each network would decrease. This fact means that the mass of the user focus on the network of the market leader to reach most of the peers (in 2015Q2 there were almost 1.5 billion active users [37]). This in turn leads to more market power for the market leader (on July 21, 2015 the market capitalization reached its maximum of almost $280 billion [38]) and more personal information concentrated on one server node.

Interoperability would lead to better competition, since the social networks would have to advertise to their users. The accessibility of other peers would no longer be in the foreground but rather other quality characteristics, for example, privacy and security matters. There are different approaches to decentralized social networks—none of them are really successful. They suffer from the fact that they cannot find their peers or like-minded people so they can learn about a particular subject within the network.

2.3.4 Further applications

As discussed in Section 2.3.2, both **search engines** and **messengers** can be built on a network-based Big Data architecture, which also allows the implementation of the further applications mentioned in Section 2.2.

A **smart grid fault indicator** will be capable of helping consumers to manage and optimize their day-to-day energy use, even at the appliance level. However, "The information collected on a smart grid will form a library of personal information, the mishandling of which could be highly invasive of consumer privacy" [39]. In this context, a multilevel network-based Big Data architecture may contribute decisively to provide both privacy and transparency for better grid-wide load management. In such an architecture appliance-specific data remains in the subnetwork of a private household while aggregated real and forecasted information is forwarded to the Big Data server node, at a higher level, say, the neighborhood server node, which in turn provides aggregated information from several households in the neighborhood to an even higher level server network. In this way a sufficient anonymization of consumer's information takes place.

The network-based Big Data architecture may help to establish **health care** data services which may collect health data of a certain individual only with his or her prior explicit permission. The data collection may even include personal search engine logs [5]. The collected data must not be shared with any other services or persons due to medical confidentiality rules. However, medical analytics can be carried out and the results can be provided to optimize health care management as long as it conforms to the agreement between the data owner and the service provider.

GPS and other physical or even chemical sensors are increasingly becoming relevant data sources. From a technical point of view the data streams from such sources may be used for many different applications, from motion tracking providing evidence of a healthy lifestyle or a defensive driving style to obtain the maximal discount on health care or vehicle insurance, through traffic management [40] to epidemic control and dragnet investigations [16]. A decentralized network-based Big Data architecture obviously provides an exceptional option to ensure the data is only used for the purposes desired by the owner.

2.4 Conclusion

In this chapter, we proposed a concept called "network-based Big Data." A network of loosely connected Big Data analytic server nodes allows not only decentralized storage and processing, but also individual and privacy-preserving response policies.

The sublinear scalability of Big Data systems makes decentralized data storage more favorable with regard to performance. The amount of extra network traffic between the server nodes turns out to be acceptable. The core benefit of a network-based Big Data architecture, however, is in the possibility to implement individual policies of privacy protection on behalf of the owners of the collected data.

A key issue is to establish standards for intra-server communications which shall be supported by different providers. Such interoperability facilitates not only the network-based Big Data architecture on the operational level but also competition, which in the long run will continuously promote improvements in service quality and especially the protection of privacy.

References

1. John Grantz and David Reinsel. The digital universe in 2020: Big Data, bigger digital shadows, and biggest growth in the Far East—United States. *IDC View*, 2013. http://www.emc.com/collateral/analyst-reports/idc-digital-universe-united-states.pdf (August 17, 2015).

2. Sam Madden. From databases to Big Data. *IEEE Internet Computing*, 2012;3:4–6.

3. Fay Chang, Jeffrey Dean, Sanjay Ghemawat, Wilson C. Hsieh, Deborah A. Wallach, Mike Burrows, Tushar Chandra, Andrew Fikes, and Robert E. Gruber. Bigtable: A distributed storage system for structured data. 2006. http://research.google.com/archive/bigtable.html (July 12, 2015).

4. Giuseppe DeCandia, Deniz Hastorun, Madan Jampani, Gunavardhan Kakulapati, Avinash Lakshman, Alex Pilchin, Swaminathan Sivasubramanian, Peter Vosshall, and Werner Vogels. Dynamo: Amazon's highly available key-value store. *www.allthingsdistributed.com*, 2007. http://www.allthingsdistributed.com/files/amazon-dynamo-sosp2007.pdf (July 13, 2015).

5. Omer Tene and Jules Polonetsky. Big Data for all: Privacy and user control in the age of analytics. *Northwestern Journal of Technology and Intellectual Property*, 2013;11(5):239–273.

6. John Grantz and David Reinsel. The digital universe in 2020—Big Data, bigger digital shadows, and biggest growth in the Far East. *IDC View*, 2012. http://www.emc-technology.com/collateral/analyst-reports/idc-the-digital-universe-in-2020.pdf (August 17, 2015).

7. Matthew Komorowski. A history of storage cost. 2014. http://www.mkomo.com/cost-per-gigabyte-update (August 06, 2015).

8. Edd Dumbill. What is Big Data? *beta.oreilly.com*, 2012. https://beta.oreilly.com/ideas/what-is-big-data (July 12, 2015).

9. Michael Schroeck, Rebecca Shockley, Janet Smart, Dolores Romero-Morales, and Peter Tufano. Analytics: The real-world use of Big Data. IBM Institute for Business Value, 2012.

10. Johann-Christoph Freytag. Grundlagen und Visionen großer Forschungsfragen im Bereich Big Data. *Informatik Spektrum,* 2014;37(2):97–104.

11. Avinash Lakshman and Prashant Malik. Cassandra—A decentralized structured storage system. *www.cs.cornell.edu,* 2009. https://www.cs.cornell.edu/projects/ladis2009/papers/lakshman-ladis2009.pdf (July 13, 2015).

12. Brian F. Cooper, Raghu Ramakrishnan, Utkarsh Srivastava, Adam Silberstein, Philip Bohannon, Hans Arno Jacobsen, Nick Puz, Daniel Weaver, and Ramana Yerneni. PNUTS: Yahoo!'s hosted data serving platform. *www.mpi-sws.org,* 2008. http://www.mpi-sws.org/~druschel/courses/ds/papers/cooper-pnuts.pdf (July 13, 2015).

13. Neil Gunther, Paul Puglia, and Kristofer Tomasette. Hadoop superlinear scalability. *Queue,* 2015;13(5):20–42.

14. Jihwan Song, Sanghyun Yoo, Chang-Sup Park, Dong-Hoon Choi, and Yoon-Joon Lee. The design of a grid-enabled information integration system based on mediator/wrapper architectures. *The 2006 International Conference on Grid Computing & Applications, Proceedings,* 2006, pp. 114–120.

15. Erhard Rahm, Gunter Saake, and Kai-Uwe Sattler. *Verteiltes und Paralleles Datenmanagement Von verteilten Datenbanken zu Big Data und Cloud.* eXamen.press, Springer-Verlag, Berlin, Heidelberg Germany 2015. ISBN 978-3-642-45242-0.

16. Elizabeth E. Joh. Policing by numbers: Big Data and the fourth amendment. *Washington Law Review* 2014;89(35):35–68.

17. Wilson Jeberson and Lucky Sharma. Survey on Big Data for counter terrorism. *International Journal of Innovations and Advancement in Computer Science,* 2015;4(Special Issue):197–205.

18. Pierangela Samarati and Latanya Sweeney. Generalizing data to provide anonymity when disclosing information. *PODS,* 1998;98:188.

19. Latanya Sweeney. k-Anonymity: A model for protecting privacy. *International Journal on Uncertainty, Fuzziness and Knowledge-based Systems,* 2002;10(5):55–570.

20. Jon P. Daries, Justin Reich, Jim Waldo, Elise M. Young, Jonathan Whittinghill, Andrew Dean Ho, Daniel Thomas Seaton, and Isaac Chuang. Privacy, anonymity, and Big Data in the social sciences. *Communications of the ACM,* 2014;57(9):56–63.

21. Yves-Alexandre de Montjoye, César Hidalgo, Michel Verleysen, and Vincent D. Blondel. Unique in the crowd: The privacy bounds of human mobility. *Scientific Reports,* 2013;3:1376.

22. Bundesverfassungsgericht. Data retention unconstitutional in its present form. *www.bundesverfassungsgericht.de,* 2010, http://www.bundesverfassungsgericht.de/SharedDocs/Pressemitteilungen/EN/2010/bvg10-011.html (July 13, 2015).

23. curia.europa.eu. Judgment of the court (grand chamber) C-293/12 and C-594/12. *curia.europa.eu,* 2014. http://curia.europa.eu/juris/document/document.jsf?text=&docid=150642&pageIndex=0& doclang=EN&mode=lst&dir=&occ=first&part=1& cid=141836 (July 13, 2015).

24. curia.europa.eu. Judgment of the court (grand chamber) C362/14. *curia.europa.eu,* 2015. http://curia.europa.eu/juris/document/document.jsf?text=&docid=169195& pageIndex=0&doclang=EN&mode=req&dir=&occ=first&part=1&cid=83687 (October 05, 2015).

25. Thilo Weichert. Big Data und datenschutz. *www.datenschutzzentrum.de,* 2013. https://www.datenschutzzentrum.de/bigdata/20130318-bigdata-und-datenschutz. pdf (July 13, 2015).

26. Steve Lawrence and Clyde L. Giles. Inquirus, the NECI meta search engine. *Seventh International World Wide Web Conference,* 1998, pp. 95–105. Elsevier Science, Brisbane, Australia.

27. Bernard J. Jansen, Amanda Spink, and Sherry Koshman. Web searcher interaction with the Dogpile.com Metasearch engine. *Journal of the American Society for Information Science and Technology,* 2007;58(5):744–755.

28. Antonio Cutillo, Revic Molva, and Thorsten Strufe. Safebook: A decentralized answer to current on-line social networks threats leveraging real life trust. *www.eurecom.fr,* 2009. http://www.eurecom.fr/util/publidownload.fr.htm?id=2908 (March 31, 2010).

29. Ames Bielenberg, Lara Helm, Anthony Gentilucci, Dan Stefanescu, and Honggang Zhang. The growth of diaspora—A decentralized online social network in the wild. *Computer Communications Workshops (INFOCOM WKSHPS), 2012 IEEE Conference on,* March 2012, pp. 13–18.

30. Friendica.com. 2015. http://friendica.com/ (August 25, 2015).

31. Shihabur R. Chowdhury, Arup R. Roy, Maheen Shaikh, and Khuzaima Daudjee. A taxonomy of decentralized online social networks. New York, NY: Springer Science+Business Media. 2014. https://cs.uwaterloo.ca/sr2chowd/papers/ ppna_dosn_survey.pdf.

32. Thomas Paul, Sonja Buchegger, and Thorsten Strufe. Decentralizing social networking services. *www.csc.kth.se,* 2010. http://www.csc.kth.se/tcs/publications/ 2010/dosnITWDC.pdf (September 12, 2015).

33. Lorenz Schwittmann, Matthaus Wander, Christopher Boelmann, and Torben Weis. Privacy preservation in decentralized online social networks. *ieee.org,* 2013. http://ieeexplore.ieee.org/xpl/articleDetails.jsp?arnumber=6685803 (December 09, 2013).

34. Internet Live Stats. *internetlivestats.com,* 2015. http://www.internetlivestats.com/ (September 03, 2015).

35. How WhatsApp grew to nearly 500 million users, 11,000 cores, and 70 million messages a second. *highscalability.com,* 2014. http://highscalability.com/blog/2014/3/31/ how-whatsapp-grew-to-nearly-500-million-users-11000-cores-an.html (September 03, 2015).

36. Werner Vogels. Eventually consistent. *Communications of the ACM,* 2009;53(1): 40–44.

37. Number of monthly active Facebook users worldwide as of second quarter 2015. *statistica.com,* 2015. http://www.statista.com/statistics/264810/number-of-monthly-active-facebook-users-worldwide/ (September 12, 2015).

38. Facebook, Inc. (FB). *finance.yahoo.com/*, 2015. http://finance.yahoo.com/echarts?s =FB+Interactive# "allowChartStacking":true (September 12, 2015).

39. Ann Cavoukian, Jules Polonetsky, and Christopher Wolf. SmartPrivacy for the smart grid: Embedding privacy into the design of electricity conservation. *privacybyde-sign.ca*, 2009. https://www.ipc.on.ca/images/Resources/pbd-smartpriv-smartgrid.pdf (September 09, 2015).

40. Carlo Ratti, Ricardo M. Pulselli, Sarah Williams, and Dennis Frenchman. Mobile landscapes: Using location data from cell phones for urban analysis. *Environment and Planning B: Planning and Design*, 2006;33(5):727–748.

Chapter 3

Big Data Text Automation on Small Machines

Ming Jia, Yiqi Bai, Jingwen Wang, Wenjing Yang, Hao Zhang, and Jie Wang

Dealing with Big Data may require powerful machines and vast storage. It may seem that only large companies or organizations with lucrative budgets can afford such undertakings. Focusing on a specific type of data, however, it is possible to use commodity computers in a small lab environment to handle Big Data generated from large complex graphs such as online social networks (OSNs) and other types of Internet applications, and produce useful information.

We are focusing on text data in this chapter. Text data appears everywhere on the Internet, including OSN postings, news articles, commentaries, reports, and reviews, to name just a few. The main contributions of this chapter are to use two text automation systems we recently developed, which are referred to, respectively, as the Microblog Processing Toolkit (MPT) [1,2] and "Kuai-Wen Bao" (KWB), meaning "quick news system" in Chinese [3], to describe

1. How MPT and KWB collect text data from OSNs and media websites using commodity computers.

2. How MPT and KWB generate useful information.

Microblogs are counterparts of tweets in mainland China; they are quick and short texts posted by users. MPT collects microblog posts (MBPs) from application programming interfaces (APIs) provided by popular OSN sites, identifies interesting topics, and carries out statistical analysis on each topic, including gender and location distributions, and discovering popular words and trends. MPT collects on average approximately 4.5 million (sometimes over 10 million) MBPs a day, and stores them in a database running Mongo DB on a commodity computer.

KWB crawls about 200 media websites in mainland China, collects news items, eliminates duplicates, and retrieves a summary of up to 600 characters for each news article using a proprietary summary engine. It then uses a labeled-latent Dirichlet allocation (LLDA)-based classifier to classify the news items into 19 categories, computes popularity ranks called PopuRank of the newly collected news items in each category, and displays the summaries of news items in each category sorted according to PopuRank on a website at http://www.kuaiwenbao.com and on mobile apps. If a picture comes with a news item, it will also be displayed.

This chapter is organized as follows: in Section 3.1 and 3.2 we will describe, respectively, how MPT and KWB acquire and store data. We will describe how MPT and KWB process text data and generate useful information automatically in, respectively, Section 3.3 and 3.4. We conclude the chapter in Section 3.5

3.1 Data Acquisition

Using APIs provided by the content providers and crawling the Internet are common methods for acquiring data from the Internet.

3.1.1 Microblog APIs

OSNs often provide open APIs to developers. MPT collected MBPs continuously for over a year from three popular OSN sites in mainland China. They were Sina, Tencent, and Renren. These OSN sites provided different interfaces for different purposes. We were interested in topic discovering from daily MBPs, rather than in a specific user or a specific topic, and so we used the public-timeline interface [4] to collect MBPs (see Figure 3.1 for an example of such an interface). These MBPs were semistructured JavaScript object notation style records.

3.1.2 Web crawling

Web-crawling technologies are important mechanisms for collecting data from the Internet (see, e.g., [5–11]). The general framework of crawling is given in the following:

1. Provide the crawler a seed URL.

```
[
    {
        "created_at" : "Tue Nov 30 14:34:35 +0800 2010",
        "text" : "吃力不讨好的事情我是坚决不会再做了，RI你个仙人！发飙~~~~我只想说档
次和素质在那里去了，你也就只能在这种地方混！",
        "truncated" : false,
        "in_reply_to_status_id" : "",
        "annotations" : [ ],
        "in_reply_to_screen_name" : "",
        "geo" : null,
        "user" :
        {
            "name" : "习惯寂寞吗",
            "domain" : "",
            "geo_enabled" : true,
            "followers_count" : 5,
            "statuses_count" : 61,
            "favourites_count" : 0,
            "city" : "1",
            "description" : "",
            "verified" : false,
            "id" : 1676792942,
             "gender" : "f",
            "friends_count" : 26,
            "screen_name" : "习惯寂寞吗",
            "allow_all_act_msg" : false,
            "following" : false,
            "url" : "http://1",
            "profile_image_url" :"http://tp3.sinaimg.cn/1676792942/50/1284648784",
            "created_at" : "Wed Dec 30 00:00:00 +0800 2009",
            "province" : "51",
            "location" : "四川 成都"
        },
        "favorited" : false,
        "in_reply_to_user_id" : "",
        "id" : 3978753419,
        "source" : "<a href=\"http://t.sina.com.cn\" rel=\"nofollow\">新浪微博</a>"
    },
    ...
]
```

FIGURE 3.1: JavaScript object notation example of public_timeline API.

2. The crawler grabs and stores the target page's content.

3. Enter the URLs contained in the target page in a waiting queue.

4. Process one URL at a time in the queue.

5. Repeat Steps 2 through 4.

A crawler is responsible for doing the following tasks:

1. **URL fetching.** There are three approaches to grabbing URLs at the target site (initially the target site is the seed URL):

   a. Grab all the URLs in the target site. This approach may waste computing resources of the machines running the crawler on materials that are not useful for the applications at hand.

   b. Grab a portion of the URLs and ignore certain URLs.

   c. Grab only what is needed for the current application.

2. **Content extraction.** Parse the web page and extract the content for the given application. There are two ways to parse a page:

   a. Write specific rules for each website, then use a web parsing tool such as Jsoup to extract content.

   b. Write common rules for all websites (Google's content extractor is a typical example of this approach).

3. **Fetching frequency.** If a crawler visits a target website frequently over a short period of time, then it may trigger the website into blocking the crawler's IP. Thus, it is important for a web crawler not to visit the target website too often over a short period of time to avoid being blocked.

4. **Monitoring.** Monitor whether the target website blocks a crawler's requests and whether the website changes the structures of its web pages.

3.1.3   Crawler framework for news websites

The KWB crawler framework is a vertical crawling mechanism. It can be reused and customized according to the specific layout of a web page. We observe that news websites tend to have the same components: an index page and a number of content pages for news items. When grabbing the index page, we set the crawling depth to one to stop the crawler from grabbing URLs contained in the content page. Meantime, we also want to remove repeating URLs in the URL queue. The KWB crawler framework uses both specific rules and common rules, depending on the individual crawler for a given website.

The KWB crawler framework consists of the following modules (see Figure 3.2):

1. **Visual input module.** This module allows the user to specify the patten of the target web page's layout. The user may specify the following two types of patterns:

   a. The first type is a regular expression representing what content the user wants to extract. For example, the regular expression `<TAG\b[^>]*>(.*?)</TAG>` matches the opening and closing pair of a specific HTML tag `TAG`, and the text wrapped within would be the content the user wants to extract.

   b. The second type is an XPath structure of the content that the user wants to extract. For example, suppose that the user wants to select the content enclosed in all the `<h2>` tags. Then the user can specify an XPath query as `h:h2`.

2. **Web page rule management.** It manages the web page rules entered by users, including the following operations: delete, check, and update.

3. **The core crawler cluster.** This cluster consists of the following components:

   a. Thread pool. This is the set of threads in a multitask system.

   b. URL pool. This is the database with all the pending URL information when a URL was grabbed. We use a Bloom filter to detect duplicate URLs and remove them. The crawler will visit and remove a URL one at a time from the remaining URLs in this pool.

   c. Pattern pool. This is the database of all the web page rules entered by users.

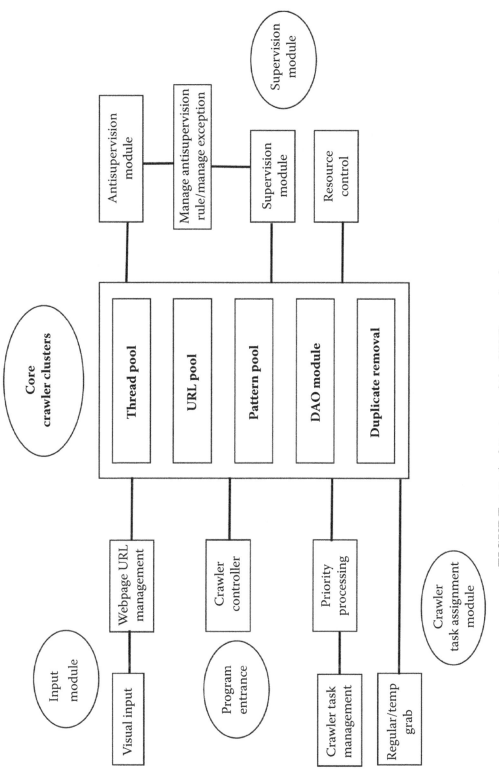

FIGURE 3.2: Architecture of the KWB crawler framework.

   d. DAO module. The data access object (DAO) contains the interface for further operations, including data export and data interface.

   e. Duplicate removal. This removes duplicate URLs in the URL pool and the patterns in the pattern pool.

4. **The crawler task module**. This module consists of the following submodules:

   a. Priority processing. Some websites are updated more frequently than others. This module determines which sites need more frequent visits.

   b. Temp grab. Sometimes the user may just want to fetch a website once without paying a return visit. This component handles this type of crawling.

   c. Regular grab. For most websites, the user sets up a schedule to grab them periodically. This component handles this type of crawling.

5. **The supervision module**. This module consists of the following submodules:

   a. Resource control (proxy/account). This is a pool containing all the proxy information and account information. The proxy is used to avoid IP blocking problems, and the account is used to log on to certain websites that require signing in, such as Twitter and Facebook.

   b. Monitoring. This monitors if the crawler functions normally. For example, it monitors whether the target website has blocked the crawler.

   c. Antiblocking. When the monitoring submodule detects that a crawler is blocked, it decides whether to restart the crawler, change the pattern, or change proxy to avoid being blocked.

   d. Managing antiblocking, exception, and restore rules. This submodule allows the user to manage and change patterns of a website rules. It also determines how often to test if a crawler is still functioning normally.

6. **The program entrance**. This component consists of a crawler controller/entrance submodule, which is responsible for starting the entire system.

The KWB crawler framework is implemented using Java. It uses httpclient to connect to a website and construct the document object model (DOM) tree of the page, and uses CSS and Jsoup to parse and extract content. DAO is implemented using MySQL and JDBC.

3.1.4   Generic article extractions

We explore in this section how to extract the main content from a web page without writing specific rules for the page. Human readers can easily distinguish on a web page the main content from irrelevant content by looking at the page layout displayed on the web browser. Any good UI design must guarantee this effect. Irrelevant content is also referred to as junk content or noise, yet such a simple task is very difficult for a machine to do well, as the machine cannot "visualize" the web page displays as easily as humans do. What a machine can do is process the hypertext markup language (HTML) code of the web page.

The main content may occur anywhere in the source code, which is structured with HTML DOM tags, styles, and scripts (e.g., see Figure 3.3). These tags, styles, and scripts are considered junk content for the purpose of extracting the main content. Other types of junk content are codes for displaying commercial messages, navigation lists, and irrelevant links, among other things, making it more difficult to detect the main content area.

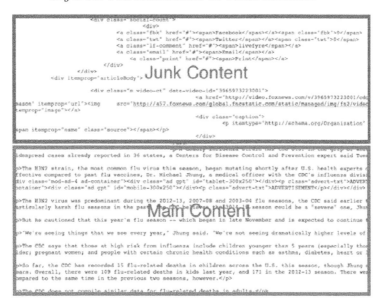

FIGURE 3.3: Junk content (noise) and main content in HTML code.

3.1.4.1 Early methods

Early approaches can be categorized into the following three directions:

1. Devise empirical rules.

2. Construct a DOM tree based on the HTML structures and DOM features.

3. Devise machine learning models, all for the purpose of detecting the main content and eliminating noise.

3.1.4.1.1 Empirical rules

Empirical rules are formed based on the HTML tags contained in the source code of the web page [12] such as `<div>`, `<p>`, and `<span>`; or tag labels such as "header," "nav," "menu," and "footer." In 2009, Joshi et al. [13] pointed out that the article areas are often wrapped in `<div>` or `<p>` tags. On the other hand, not everything wrapped in these tags is part of the main content, and so one cannot simply extract everything wrapped in these tags to form the main content.

Another common HTML tag is the comment tag `<!-->`. Comments are used by programmers to explain the source code, which will not be displayed on the browser; that is, the reader will not see the content wrapped in the comment tags. Thus, it is straightforward to remove all comment tags and everything wrapped in them.

The style tag `<style>` and the script tag `<script>` are used to instruct the browser how to display the web page, which can be removed, together with everything wrapped in them. The following is an example:

```
<style> p {color:green} </style>
<script> var i=0; </script>
```

In 2010, Kohlschutter et al. [14] started to examine shallow text features. In 2013, Uzun et al. [15] developed a model for creating rules to infer informative content from simple HTML pages based on the following two block tags: `<div>` and `<td>`.

While using tag-based empirical rules to extract the main content makes sense in some situations, this method has limitations. Thus, empirical rules are often combined with other methods to achieve a better extraction result.

3.1.4.1.2 DOM trees

Constructing DOM trees from the corresponding HTML DOM structure of a web page is one of the most popular methods to extract the main content [16–20].

In 2003, Yi et al. [21] captured the common presentation styles at that time and the actual contents of the web pages in a given website by sampling the pages of the site and built a site style tree to detect the noise and main content areas.

In 2009, Joshi and Liu [13] devised a method of HTML DOM analysis to find subtrees with large bodies of text by summing up all the text size and combining with natural language processing techniques for automated extractions of the main content with associated images.

In 2010, Guo et al. [22] presented ECON, a system that uses a DOM tree to represent a news page and leverages the substantial features of the DOM tree to extract news content from news pages.

Different HTML tag structures will generate different DOM trees. When a web page has a complex structure, constructing a DOM tree is time consuming. Weninger and Hsu [23] noted that it is not necessary to use the HTML structure or DOM features. They presented a method to extract the main content from diverse web pages using the web page's text-to-tag ratio. This ratio helps distinguish the main content from the noise content.

3.1.4.1.3 Machine-learning methods

Machine-learning methods, classification methods in particular, have been used to train and extract the main content on a web page [24].

In 2007, Gibson et al. [25] explored nonrepeating content, boundary, and other factors, and used the conditional random field sequence labeling model to label every divided unit as "Content" or "NotContent," where "Content" means the main content and "NotContent" means noise. Chakrabarti et al. [26] also developed a novel framework to train the classifier using automatic generated training data by assigning a score to every node of the DOM-tree for the given page.

In 2009, Pasternack and Roth [27] applied the maximum subsequence segmentation (MSS) method to the domain of news websites using a small set of training data to identify the main content. The MSS method is a global optimization method over token-level local classifiers.

In 2010, Kohlschutter et al. [14] derived a stochastic model based on a classification model that combines word length and link density to extract the main content.

In 2013, Uzun et al. [15] presented a hybrid approach to discovering the informative content using the decision tree method and automatic rule creation.

The machine-learning approach, however, often incurs high complexity, requires many assumptions, and tends to make simple things unnecessarily more complicated.

3.1.4.2 qRead: A new method

We present a new algorithm called qRead devised by Wang and Wang [28], which achieves higher efficiency and accuracy in extracting the main content by finding computable measures on the text structure on a web page, without relying on HTML structures or DOM features. In particular, qRead removes the HTML tags and text that are clearly irrelevant to the main content, partitions the text into segments, uses the highest ratio of word counts over the number of lines in each segment, and compares the similarity between each segment with the title of the main content to determine the main content area.

3.1.4.2.1 Filtering

According to the HTML standard, the HTML source code would typically contain two parts, which are wrapped, respectively, inside the `<head> </head>` tags and the `<body> </body>` tags. The information included inside the `<head> </head>` tags contains web page configuration or description details and so everything in it can be removed in the preprocessing step. Moreover, qRead filters out comments, styles, and scripts.

Most of the navigation lists are wrapped in the `<div>` and `</div>` tags with attributes `class="menu"` or `class="nav"`, allowing qRead to filter out most of the navigation lists using these features.

3.1.4.2.2 Removing HTML tags

HTML tags consist of the tag name, id, class name, or other tag properties. The main content text typically appears between two HTML tags, for example,

```
<p> This is an article. </p>
```

qRead removes all `<div>`, `<p>`, `<span>`, `<table>`, and `<ul>` tags, together with their properties such as tag id, class, styles, and width.

After this step, the text of web content that contains only readable text without HTML tags is generated (see Figure 3.4). We can see that Segment 2 is the article area and

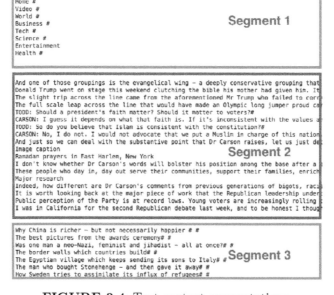

FIGURE 3.4: Text content segmentations.

Segment 1 looks like navigational text. We can see from Figure 3.4 by comparing Segments 1 through 3 that every line in Segment 2 is longer than the lines in Segments 1 and 3.

An article may contain multiple paragraphs. In the web page source code, the web page title or the article title are typically contained in heading tags such as the `<h1>` tag and the `<h2>`. Tables are contained in the `<table>` tag. The `<p>` and `<span>` tags usually contain plain text. A paragraph is typically wrapped in the `<p>` and `</p>` tags, or the `<br>` tags. qRead marks the `</p>` tags and the `<br>` tags to indicate paragraph break points before removing them.

3.1.4.2.3   Text segmentations

Navigational text is junk content. There are two types of navigational text. One type is a menu list with keywords in each line, for example,

Home
Politics
Opinion
Entertainment
Tech
Science
Health
Big Data Text Automation on Small Machines
Travel
Lifestyle
World
Sports
Weather

The other type of navigational text is a list of article titles with a phrase or sentence in each line. The following is an example:

Trending in U.S.
St. Louis police allege hate crime in latest attack on Bosnian resident
Federal autopsy released in Ferguson shooting
Fire in downtown Los Angeles may have been intentionally set
Police shoot, kill knife-wielding man in New York synagogue
See all Trends

Wang and Wang [28] observed that a text line in the noise area is typically much shorter than a text line in the main content where a text line ends with a carriage return in the source code, an end-of-sentence mark placed in the filtering phase, or some other symbols; define line length to be the number of words contained in the line. Figure 3.5a and b depicts the line length distributions of two different news web pages obtained from, respectively, the ABC News website and the BBC News website, where the main content (the news) in the ABC web page appears in lines 492–494 and in the BBC web page appears in lines 178–188.

Thus, the area of main content would typically include more words in each line than those in the noise area. Let L be a line. Denote by $w(L)$ the number of words contained in L. Partition the remaining content into a sequence of text segments, separated by at

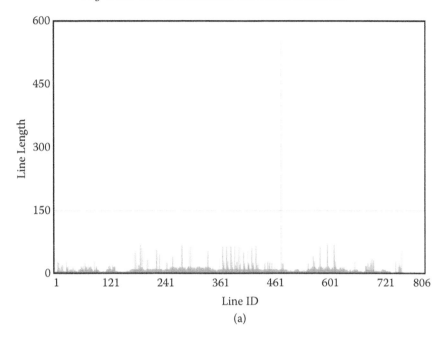

(a)

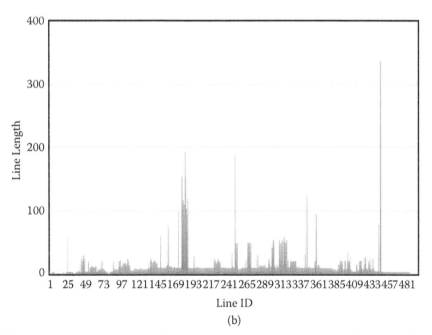

(b)

FIGURE 3.5: Line length distribution in news web pages. (a) Line length distribution for the ABC news web page: http://abcnews.go.com/International/wireStory/ukraine-attempts-newscease-fire-east-27464144, article line #: 492 ∼ 494; (b) Line length distribution for the BBC news web page: http://www.bbc.com/news/world-us-canada-30383922, article line #: 178 ∼ 188.

least λ empty lines, where λ is a threshold value, determined empirically. Paragraphs in the main content are typically separated by HTML tags such as `<p>` and `<br>` (where no empty lines are necessary), or by a single empty line or double empty lines. In practice, setting $\lambda = 3$ is sufficient. In other words, inside each segment S, the first line and the last line are nonempty text lines, and there are at most $\lambda - 1$ consecutive empty lines in between. Let

$$S_1, S_2, \cdots, S_k$$

denote the sequence of segments.

Let L denote a text line. Then the size of S, denoted by $w(S)$, is the number of words contained in S. That is,

$$w(S) = \sum_{L \in S} w(L)$$

Let $l(S)$ denote the number of nonempty text lines contained in S. Let $d(S)$ denote the word desnity of S, which is the ratio of the number of words contained in S over the number of nonempty text lines contained in S. Namely,

$$d(S) = \frac{w(S)}{l(S)}$$

Wang and Wang [28] further observed that the main content is typically the text contained in the segments with the highest density, and they hypothesized that any reasonably well-organized web page with reasonably long text in the main-content area has this property. They carried out extensive experiments and confirmed this hypothesis.

Figure 3.6a and b shows the segment density distribution for each of the web pages depicted in Figure 3.5a and b. In each figure, the segment with the highest word density is exactly the area of the main content (i.e., the news article contained in the web page).

3.1.4.2.4 Similarity process

Using segment density alone, however, does not work well on the following two types of web pages:

1. The main content is a short article with one or two sentences.

2. The main content consists of multiple text segments with the same density.

The text area of the highest density is not the article area in either of these two web page types. To improve extraction accuracy, qRead applies the measure of cosine similarity between a text segment and the article title to determine if the text segment should be part of the main content.

3.1.4.2.5 Accuracy and efficiency

Extensive experiments were carried out [28] on a laptop computer with a 2.5 GHz Intel Core i5 processor and 8 GB 1600 MHz DDR3 memory for 13,327 web pages written in English and 10,382 web pages written in Chinese collected from a large number of websites, where the article contained in each web page is substantially shorter than the page itself. This means that every web page in the data sets contains lots of junk content.

There are three different types of extraction: accurate extraction, extra extraction, and missed extraction.

1. An extraction is *accurate* if it contains exactly the complete article without any junk content.

2. An extraction is *extra* if it contains the whole article but with some junk content.

3. An extraction is *missed* if it contains an incomplete article or nothing related to the article.

The extraction results are shown in Table 3.1, which are superior over previously published results.

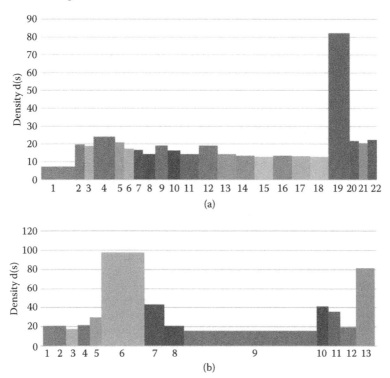

FIGURE 3.6: Segment word density distribution in news web pages. (a) Segment world density distribution for the ABC news web page: http://abcnews.go.com/International/wireStory/ukraineattempts-news-cease-re-east-27464144, Article segment ID: 19; (b) Segment world density distribution for the BBC news web page: http://www.bbc.com/news/world-us-canada-30383922, Article segment ID: 6.

TABLE 3.1: Accuracy and average extraction time

	Accurate	Extra	Missed	Average Extraction Time
qRead (English)	93.6%	5.8%	1.6%	11.37 milliseconds
qRead (Chinese)	96.8%	3.2%	1.0%	13.20 milliseconds

3.2 Data Storage

Text data acquired from OSN APIs and the KWB crawlers are unstructured data. To handle unstructured text data in large quantity, we would need a high-performance, steady, and flexible database system. Mongo DB [29] is a common choice for this purpose.

3.2.1 Database for microblog posts

Different MBPs from different sources use different data structures. With Mongo DB, it is possible to store data in different data structures in the same collection. This makes it convenient to manage the data. Moreover, Mongo DB is a database scheme with high performance on the operations of both read and write, which meets the need of intensive writing and querying of MBPs.

It is straightforward to store in one collection all the MBPs posted on the same date, and store all the collections in the same database. However, Mongo DB would write data in the same database into the same file and load all the files that are accessed frequently into the main memory, and so as more MBPs are stored in the same database, the file will grow larger quickly, causing Mongo DB to consume almost all the RAM and crashing the system. To solve this problem, Jia and Wang [1] (see also [2]) divided the collections of MBPs according to a fixed time interval into different databases. Since writing to the database is the main operation of MBT and MBT only collects real-time data, Mongo DB would load the file of the most recent week into RAM, which consumes much less RAM. This would solve the problem of crashing.

Note that we may also consider using an SQL-based database for reliability.

3.2.2 Database for an automatic news system

The KWB crawler eliminates duplicate URLs, but the same news article may be posted on different URLs, and so a duplicate article cannot be eliminated by removing duplicate URLs. For each news article KWB needs to retrieve a summary of different length (depending on applications) using a proprietary text summary engine. Eliminating duplicate articles and retrieving summaries are both time consuming. To reduce computations, KWB adopts a new database called central DB (see Figure 3.7) to perform these two tasks on raw data collected every 20 minutes.

There are two different types of duplicates in the raw data:

1. News items that are exactly the same due to reposting.

2. Different news items that report the same event.

We will keep the second type of news items, for they report the same event from (possibly) different prospectives, which are useful. To identify the first type of duplicates we may compute cosine similarities for all the raw data collected by the KWB crawlers. This approach, however, is time consuming. Instead, we take a greedy approach to reducing the number of news items that we need to retrieve summaries for by eliminating duplicates posted in

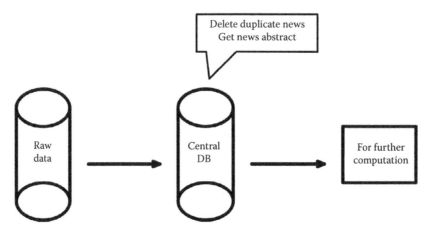

FIGURE 3.7: Central DB for KWB.

a small window of time. We will further remove duplicates later before computing news classifications.

The central DB retrieves article summaries and detects duplicates in a parallel fashion. In particular, it sorts all the unprocessed raw data in increasing order according to their IDs. These are incremental IDs given to the news items based on the time they are fetched by the KWB crawler framework. Starting from the first news article, repeat the following:

1. Send a request to the summary engine to retrieve summaries of required lengths.

2. Compute the cosine similarities of the article with the news items whose IDs fall in a small fixed time window after this article. If a duplicate is found, remove the one whose ID is in the time window (i.e., with a larger ID), for it is likely a reposting and the news article with a smaller ID may have already had the summaries generated from the summary engine running on a different server.

3. Move to the next news article in the sorted list.

3.3   Data Retrieval Tuned for Statistical Analysis

MPT performs the following three types of search:

1. Given a keyword, it retrieves all MBPs that contain the keyword.

2. Given a set of keywords, it retrieves all MBPs that contain at least one of these keywords (this is the logical OR operation). Note that a phrase can be treated as a set of keywords.

3. Given a set of keywords, it retrieves all MBPs that contain all of the keywords (this is the logical AND operation).

These tasks can be carried out using the following three methods.

3.3.1   Mongo DB regular expressions

Mongo DB provides a built-in regular expression search method. Given the regular expression we want the text to match, Mongo DB returns all the records that match the regular expression. Then the system will go over all the MBPs and get the count of MBPs for each statistical term, like for each gender. This search method, however, is inefficient and does not meet our needs. To speed up the search process, we developed an indexing system based on the nextword indexing scheme [30].

3.3.2   Nextword indexing

The nextword index consists of a vocabulary of distinct words and, for each word w, a nextword list and a position list. The nextword list consists of each word s that succeeds w anywhere in the database, interleaved with pointers into the position list. For each word in the nextword list, a string of attributes contained in the corresponding MBP needed for statistical analysis is also stored. Figure 3.8 depicts an example of the nextword structure of MBPs.

Based on the nextword indexing, MPT consists of an index server and a search server. The index server indexes in real time the MBPs collected by MPT's data acquisition

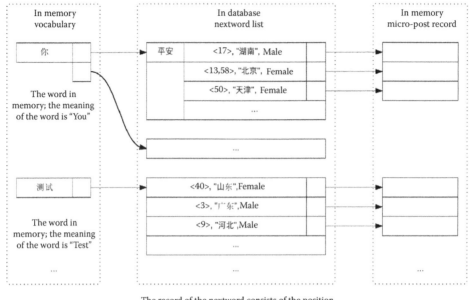

The record of the nextword consists of the position
of the word following the previous word, the
province of the poster, and the gender of the poster

FIGURE 3.8: Nextword index structure.

component. In particular, MPT stores the MBPs for the current time interval in the main memory for fast retrieval and stores the word pair list in the main memory for fast search. To save memory it stores the position of each word pair and the information of corresponding MBPs in the database.

The nextword indexing approach, however, is RAM consuming. Since RAM is a precious and limited resource in a small lab environment, a low-cost RAM solution would be more desirable, which can be accomplished using Apache Lucene.

3.3.3 Lucene-based indexing

Apache Lucene is a high-performance, full-featured text search engine library. It uses a small amount of RAM for indexing and searching, and generates an index file with reasonable size. Jia and Wang [1] customized the Lucene core and built a data retrieval system that solved the problem of memory blowup. The system consisted of two parts: (1) a real-time index server and (2) a search server. The real-time index server refreshes the database frequently and indexes dynamically the newly collected MBPs on the fly. The search server returns the MBPs that contain all the keywords entered by the user. These two parts may work simultaneously without conflicting each other, and so the search server can return real-time posts the system collects. Figure 3.9 shows the structure of the Lucene-based indexing system in MPT. In particular, for each MBP, the system connects its poster's location, gender, and posting time in one string. This string is set as the index field of the MBP.

3.3.4 Statistical analysis

In addition to displaying all the MBPs that contain the search keywords (or phrases) to the user, MPT also returns statistical results of these MBPs, including gender proportions, posters' locations, the sources of the MBPs, the distribution of the posting time, and top

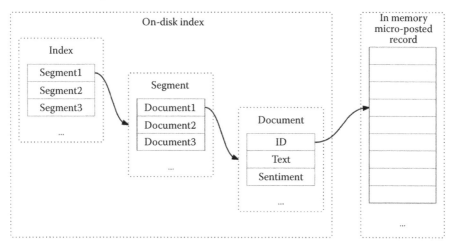

FIGURE 3.9: Lucene-based index structure.

keywords. An example of statistical results on the phrase search "Yibin Bus Explosion" is shown in Figure 3.10 (Yibin is a small city in China).

To carry out statistical analysis on relevant MBPs of a search, Jia and Wang [1] grouped these MBPs according to the following three attributes: (1) posting time, (2) poster's gender, and (3) poster's location.

Using the built-in regular expressions of Mongo DB to carry out the statistical analysis we need to traverse all the MBPs in the database, which is extremely time consuming.

Using the nextword indexing structure shown in Figure 3.8, the statistical analysis can be completed in just a few steps without the need of traversing the MBPs. As shown, the nextword index file contains not only the position of each word pair but also a string of the attributes that need to be counted in a single MBP. For each query, a list of positions and the string of attributes of each MBP are returned. Then the statistical analysis component of MPT just needs to traverse the position list to carry out statistical analysis without actually querying the database for the MBPs.

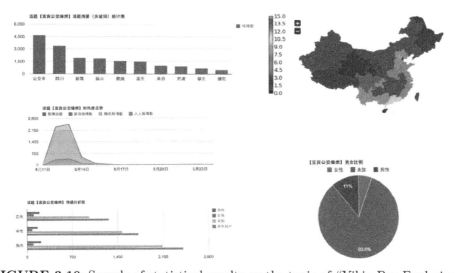

FIGURE 3.10: Sample of statistical results on the topic of "Yibin Bus Explosion."

Using Lucene-based indexing, we need to customize certain functions. Recall that for each MBP, its poster's location, gender, and posting time are included in one string, which is served as the key of the MBP. This means that two MBPs that are posted at the same hour, in the same location, and by posters of the same gender will share the same key. When indexing the MBP, we include this key in the index. When answering a search query, we first carry out the group search by the key. After obtaining the groups with the same key, we traverse each group and perform the statistical analysis in the group.

To ensure that all data are counted, another statistical analysis server, which is not a real-time server, is running synchronously. This server is used for analyzing the keywords generated by a hot-topic server (details are omitted).

3.3.5   Efficiency and accuracy

To test the efficiency and accuracy of the three methods for data retrieval and statistical analysis mentioned in the previous sections, we will refer to these methods of using Mongo DB's built-in regular expressions, nextword indexing, and Lucene-based indexing as Mongo, Nextword, and Lucene, respectively. The experiments were carried out on the MBPs collected in one day (the day was randomly chosen) from Sina and Tencent as the data set for comparing these methods. There were about 4,250,000 MBPs in this one-day collection. All experiments were executed on a commodity computer with a quad-core CPU and 16 GB RAM.

The experiment consists of two parts. In the first part, we examine how fast each method responds to user queries, as well as how fast the method carries out statistical analysis and returns the results. In the second part, we compare the accuracy of each method.

To run the experiment, we precounted all the keyword phrases in the data set and randomly selected a number of keyword phrases as our testing phrases, such that these phrases were made up of two or three keywords and appeared in the test data set more than 100 times.

For each phrase we selected, we executed a query with each method. We recorded the time that each method incurred to respond to the query and carry out the statistical analysis. We tested Mongo, Nextword, and Lucene separately. For each test, we repeated the process twice and calculated the average time. The results are shown in Figure 3.11 and 3.12, where the horizontal axis represents the actual frequency of keywords, and the vertical axis represents the running time of returning the MBPs that contain the keywords.

From Figure 3.11 we can see the following:

1. Lucene offers the fastest response time, and the response time will increase as the phrase frequency increases.

2. Nextword is faster, which is slower but close to Lucene; but its time complexity is not as steady as Lucene.

3. Mongo is the slowest, and does not meet the real-time search requirement.

From Figure 3.12 we can see that, in terms of running time for carrying out statistical analysis, Lucene is still the fastest, Nextword is slower than Lucene, and Mongo is substantially slower than Nextword. The time interval between responding to the search query and finishing up statistical analysis is quite short for Lucene, and is much longer for the other two methods. We note that the mechanism for carrying out the statistical analysis of each method is different. By querying Mongo, we would need to traverse all the MBPs returned on the query, which would incur significant computing time when the returning data set is large. Nextword is a well-structured indexing mechanism, where no traverse is

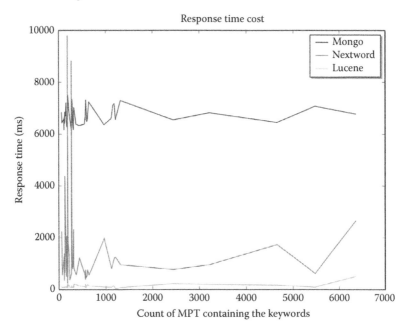

FIGURE 3.11: Response time.

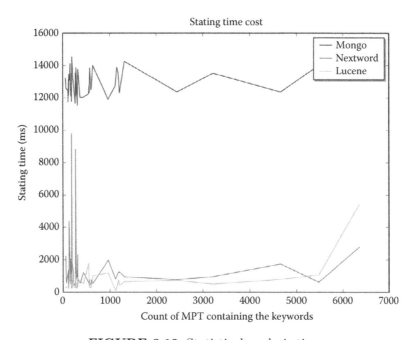

FIGURE 3.12: Statistical-analysis time.

needed to perform statistical analysis, and so it can finish the statistical analysis quickly. Lucene finishes statistical analysis by traversing all the groups instead of all the posts.

We can also see from Figure 3.12 that the time complexity of the Mongo search does not increase much when the frequency of the keywords is increased, while the time complexity of the Nextword search and the Lucene search is clearly correlated to the frequency of the keywords.

Let P denote the precision rate, R the recall rate, TP the number of true positives, FP the number of false positives, and NF the number of false negatives. Then

$$P = \frac{TP}{TP + FP},$$
$$R = \frac{TP}{TP + FN}.$$

The accuracy of a search mechanism is often measured by precision and recall rates. Figure 3.13 shows the accuracy of each method, where the horizontal axis represents the actual frequency of the phrase, and the vertical axis represents the counts of the returning results. We can see that the accuracy of each method differs, with Mongo 100% accurate. In other words, its precision and recall rates are both equal to 1.

Nextword may miss some MBPs. The main reason for missing MBPs is due to the segmentation error of keywords in the Chinese language. The keyword segmentation in the Chinese language is different from that of English, as standard Chinese writing contains no space between characters. Thus, different segmentation tools may return different segmentation results. Even for the same keyword using the same segmentation tool, the results may still be different in different text. For the keywords we queried in the experiments, the segmentation result in the MBPs could differ from that in the search query. The MBPs with different segmentation results would therefore be missed.

Lucene, on the other hand, seems to have the worst precision and recall rates, where the number of returned MBPs is usually larger than the actual number that contain the phrase. This is caused by the Lucene indexing structure, where all the MBPs that contain a subset of the search keywords are returned. For example, if phrase X is made up of words A **and** B, when querying X, Lucene will return all the posts that contain A **or** B. So the returned result usually has a larger-than-actual count.

Table 3.2 shows the precision and recall rates of the search results using the three different methods. We can see from the table that Mongo performs well on accuracy. Both

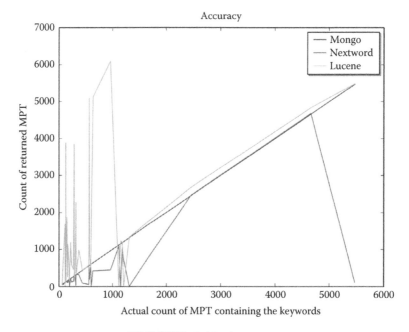

FIGURE 3.13: Accuracy.

TABLE 3.2: Experiment results on precision and recall rates

Method	Precision	Recall
Mongo	1	1
Nextword	1	0.72
Lucene	0.53	0.81

Mongo and Nextword have a value of 1 on the precision rate, which means that the MBPs Mongo returned were exactly those that contain the search phrase. Since Nextword may miss some MBPs, this negatively affected its recall value. Lucene suffers precision loss compared to the other two methods. But it offers better performance in the recall value than Nextword, which means that it may miss fewer MBPs than Nextword.

To conclude this section, we note that each method has its pros and cons. Mongo provides the regular expression search method with the perfect accuracy. But it is too slow to meet the needs of real-time querying and statistical analysis. Nextword has the best performance in statistical analysis, and the response time can meet the need of real-time search. But its memory consumption is intensive. Lucene, on the other hand, has the fastest response time and the fastest statistical-analysis time, but it incurs low accuracy. This method is good for building a fast search engine but may not meet the requirement of high accuracy.

3.4 Automatic Quick News

KWB consists of five components (see Figure 3.14): (1) crawlers, (2) central DB, (3) summary engine, (4) core processing unit, and (5) web display.

Brief descriptions of each of these components are given in the following:

1. The crawler component is responsible for collecting news items around the clock from over 120 news websites in mainland China.

2. The central DB is responsible for processing the raw data collected from the crawlers, including removing duplicated news items and fetching summaries for each news article.

3. The summary engine is responsible for returning summaries for each new article with different lengths required by applications. This is preparatory technology.

4. The core processing unit consists of three parts: (1) Chinese text fragmentation, (2) news article classifications, and (3) ranking each document according to PopuRank.

5. The web display component is responsible for displaying on a website the news items in each category according to their PopuRanks in each day, their summaries, pictures (if there is any), and links to the original news items.

Figure 3.15 describes the data flow in the KWB system in which each module will operate data and save new attributes.

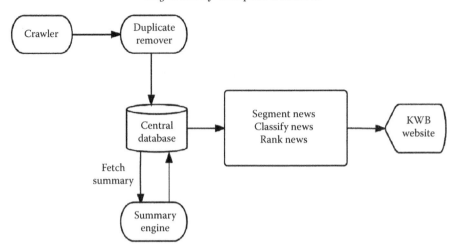

FIGURE 3.14: The architecture of KWB.

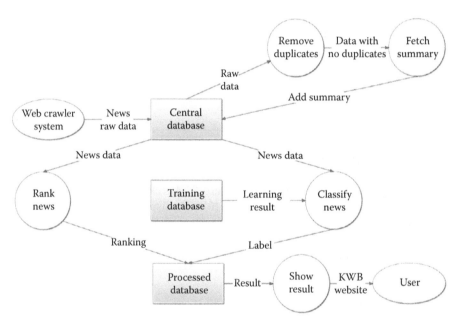

FIGURE 3.15: The data-flow diagram of KWB.

3.4.1 Labeled-LDA-based classifiers

KWB needs to automatically classify news articles. Text classification has been studied intensively and extensively during the last 20 years. A number of supervised [31], semisupervised [32], and unsupervised [33] machine-learning techniques have been investigated (see, e.g., [34] for a survey). In particular, Naive Bayes is a simple, supervised text classifier, but its performance is sensitive to feature selections of the text data in hand [35].

A support vector machine (SVM) is a widely used text classifier that separates data with maximal margins to hyperplanes for reducing misclassification on training data [36]. It performs better than Naive Bayes. We use linear SVM as the baseline classifier and assume that the reader is familiar with SVM. Note that SVM does not provide a word-to-category distribution.

The LDA method computes a word-to-category distribution [37]. LDA models the underlining topics for a corpus of documents, where each topic is a mixture over words and each document is a mixture over topics. It is natural to associate a topic to a class. However, LDA is an unsupervised model and it cannot label classes; that is, it cannot give a meaning to class. Giving meaning to a class would need human interpretation.

LLDA, a natural extension of both LDA and Multinomial Naive Bayes [38], offers a solution which overcomes a number of drawbacks in previous attempts of using LDA to perform classifications, including Supervised LDA [39], DiscLDA [40], and MM-LDA [41].

Unlike SVM, which puts a document in exactly one category, that is, SVM associates each document with exactly one label, LLDA can classify a document with multiple labels, which is useful to KWB. It was shown [38] that LLDA beats SVM on a tagged web page and a corpus from a Yahoo directory.

To verify the accuracy of classification results on news articles we would need a large corpus of classified news articles as the ground truth. The data set we used consists of news articles collected from over 120 national and regional media websites in mainland China, classified by human editors into a number of categories. Bai and Wang [42] constructed LLDA-based classifiers called LLDA-C and simplified LLDA-C (SLLDA-C) to classify these news articles. To compare with SVM, it is necessary to restrict these classifiers to classify a document with exactly one label.

Bai and Wang [42] showed that, through extensive experiments, both LLDA-C and SLLDA-C outperform SVM (our baseline classifier) on precision, particularly when only a small training data set is available. SSLDA-C is also much more efficient than SVM. While LLDA-C is moderately better than SLLDA-C, it incurs higher time complexity than SVM. In terms of recall, we show that LLDA-C is better than SVM, which is better than SLLDA-C. To further explore classifications of news articles we introduce the notion of content complexity, and study how content complexity would affect classifications.

They further showed that, among the news articles correctly classified by LLDA-C, SLLDA-C, and SVM, the number of documents with one significant topic in each category correctly classified by either LLDA-C or SLLDA-C is larger than that by SVM. This may indicate that SVM would do better on documents with multiple significant topics. However, for the news articles incorrectly classified by LLDA-C, SLLDA-C, and SVM, this result does not hold.

In any case, for a document with multiple significant topics, it would be natural to assign it multiple labels using an LLDA classifier, instead of just one label as restricted by SVM.

3.4.1.1 Labeled LDA

LLDA is a probabilistic graphical model based on LDA devised by Ramage et al [38]. It models a document in a corpus as a mixture of topics and topic-generated words, and constructs a one-to-one correspondence between latent topics and labels, from which a word-label (i.e., word-category) distribution could be learned, where a label represents a class. We provide a brief description of LLDA in this section for the convenience of describing LLDA-C and SLLDA-C in the next section. For more details of LLDA the reader is referred to [38].

LLDA uses two Dirichlet distribution priors, one for generating document-topic distribution with hyperparameter α, and one for topic-word distribution with hyperparameter β. LLDA also employs a Bernoulli distribution prior with hyperparameter Φ, which generates topic presence/absence indicators Λ for a document. In other words, θ is a document distribution over topics constrained by Λ for mapping a topic to a label, and φ is the topic distribution over words that affect the generation of words with parameter z_w, sampling from θ.

Let D be a corpus of M documents to be classified, indexed from 1 to M. We view each document d as a bag of words $\boldsymbol{w}^{(d)} = (w_1, \cdots, w_{N_d})$, where N_d is the number of words in document d. Then $D = \{\boldsymbol{w}^{(1)}, \cdots, \boldsymbol{w}^{(M)}\}$. Each word belongs to a fixed vocabulary $\boldsymbol{V} = \{w_1, w_2, \cdots, w_V\}$. Let

$$\boldsymbol{\Lambda}^{(d)} = (\Lambda_1^{(d)}, \cdots, \Lambda_K^{(d)})$$

denote the topic presence/absence indicator for document d, where K is the total number of unique labels in the training data and $\Lambda_k^{(d)} \in \{0, 1\}$ indicates whether document d contains topic k. Thus, $|\boldsymbol{\alpha}| = K$, $|\boldsymbol{\Phi}| = K$, and $|\boldsymbol{\beta}| = V$.

3.4.1.1.1 Mixture model

The number of topics K under the LLDA model is the number of unique labels. In what follows, by "generate $g \sim G$" we mean to draw (sample) g with distribution G, where g may also be a distribution. Let Mult denote a multinomial distribution, Ber a Bernoulli distribution, and Dir a Dirichlet distribution. A labeled document can be generated as follows (Figure 3.16 is a standard graphical representation of the model), where d represents a document, k and $z_i \in \{1, \cdots, K\}$ represent topics, and $w_i \in V$ represent words:

1. (Topic-word generation): For each topic k, generate $\varphi_k \sim \text{Dir}(\cdot \mid \boldsymbol{\beta})$.

2. (Document-topic generation): For each d do the following:

   a. For each topic k, generate $\Lambda_k^{(d)} \sim \text{Ber}(\cdot \mid \phi_k)$.

   b. Compute $\boldsymbol{\alpha}^{(d)} = \boldsymbol{L}^{(d)} \cdot \boldsymbol{\alpha}$, where $\boldsymbol{L}^{(d)}$ is an $L_d \times K$ matrix $\left[l_{ij}^{(d)} \right]$,

$$\begin{aligned} L_d &= |\boldsymbol{\lambda}^{(d)}| \\ \boldsymbol{\lambda}^{(d)} &= \{k \mid \Lambda_k^{(d)} = 1\} \\ l_{ij}^{(d)} &= \begin{cases} 1, & \text{if } \lambda_i^{(d)} = j \\ 0, & \text{otherwise} \end{cases} \end{aligned}$$

   c. Generate $\boldsymbol{\theta}^{(d)} \sim \text{Dir}(\boldsymbol{\alpha}^{(d)})$.

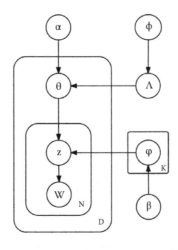

FIGURE 3.16: Graphical model of LLDA.

d. For each w_i in d, generate

$$z_i \in \boldsymbol{\lambda}^{(d)} \quad \sim \quad \text{Mult}(\cdot \mid \boldsymbol{\theta}^{(d)})$$
$$w_i \in \boldsymbol{V} \quad \sim \quad \text{Mult}(\cdot \mid \boldsymbol{\varphi}_{z_i})$$

Let $\boldsymbol{z}^{(d)} = (z_1, \cdots, z_{N_d})$.

In the topic-word generation process, a multinomial topic distribution over the vocabulary for each topic k is generated, denoted by

$$\boldsymbol{\varphi}_k = (\varphi_{k,1}, \cdots, \varphi_{k,V})$$

In the document-topic generation process, a multinomial mixture distribution over the topics for each document d is generated, denoted by

$$\boldsymbol{\theta}^{(d)} = (\theta_1^{(d)}, \cdots, \theta_{L_d}^{(d)})$$

which is restricted on its labels $\boldsymbol{\Lambda}^{(d)}$. The vector $\boldsymbol{\lambda}^d = \{k|\Lambda_k^{(d)} = 1\}$ and the document-specific projection matrix $\boldsymbol{L}^{(d)}$ restrict the parameter of a Dirichlet distribution prior $\boldsymbol{\alpha} = (\alpha_1, \cdots, \alpha_K)$ to a lower dimension $\boldsymbol{\alpha}^{(d)} = \boldsymbol{L}^{(d)} \cdot \boldsymbol{\alpha}$ with length $L_d = \sum_{k=1}^{K} \Lambda^{(d)}$.

With a training data set in hand, where each document is properly labeled, we can obtain $\boldsymbol{\Lambda}$ directly.

3.4.1.1.2 Learning and inference

Suppose that the values of parameters $\boldsymbol{\alpha}$ and $\boldsymbol{\beta}$ are given. For each document $\boldsymbol{w}^{(d)}$, we want to obtain a label-word distribution and determine which category this document would belong to. This means that we would need to infer $\boldsymbol{z}^{(d)}$ from $\boldsymbol{w}^{(d)}$, and we can do so using collapsed Gibbs sampling for the probability $p(\boldsymbol{z}^{(d)} \mid \boldsymbol{w}^{(d)})$.

Let $\boldsymbol{z}_{-i}^{(d)}$ denote $\boldsymbol{z}^{(d)} - \{z_i\}$ and $\boldsymbol{w}_{-i}^{(d)}$ denote $\boldsymbol{w}^{(d)} - \{w_i\}$. Let $n_{-i,j}^{(w_i)}$ denote the total number of word w_i assigned to topic j excluding the current assignment z_i [43]. Following standard computations [44] we have

$$p(z_i = j \mid \boldsymbol{z}_{-i}^{(d)}, \boldsymbol{w}^{(d)}) \quad \propto \quad p(z_i = j, w_i = t \mid \boldsymbol{z}_{-i}^{(d)}, \boldsymbol{w}_{-i}^{(d)})$$
$$= \quad E(\theta_{dj}) \cdot E(\varphi_{jt})$$

where

$$E(\theta_{dj}) \quad = \quad \frac{n_{-i,j}^{(t)} + \beta_t}{\sum_{t=1}^{V} n_{-i,j}^{(t)} + \sum_{l=1}^{V} \beta_l} \tag{3.1}$$

$$E(\varphi_{jt}) \quad = \quad \frac{n_{-i,j}^{(d)} + \alpha_j^{(d)}}{\sum_{k \in \boldsymbol{\lambda}^{(d)}} n_{-i,k}^{(d)} + \sum_{k=1}^{L_d} \alpha_k^{(d)}} \tag{3.2}$$

We can establish $\boldsymbol{\varphi}$ from the training data and then use it to classify test data by calculating the new topic distribution.

3.4.1.2 LLDA-C and SLLDA-C

LLDA-C consists of the following four steps:

1. Each document in the corpus has exactly one label, from which we can learn $\boldsymbol{\Lambda}$ directly (that is, we can bypass $\boldsymbol{\Phi}$). It was noted that when $\boldsymbol{\Lambda}$ is known, $\boldsymbol{\Phi}$ is d-separated from the model [38].

2. Learn $\boldsymbol{\varphi}$ using collapsed Gibbs sampling on the training data with the specified values of $\boldsymbol{\alpha}$ and $\boldsymbol{\beta}$, and the values of $\boldsymbol{\Lambda}$ learned in Step 1.

3. Inference on a new unlabeled document d using Gibbs sampling. We have the following two cases for calculating the sampling probability:

$$p(z_i = j, w_i = t | \boldsymbol{z}_{-i}^{(d)}, \boldsymbol{w}_{-i}^{(d)})$$

where d is the new document and w is a word that appears in d.

Case 1: Word w is in the training data. Let $p(w)$ be the highest probability of word w under $\boldsymbol{\varphi}$. Then the sampling probability is the product of Equation (3.1) and $p(w)$.

Case 2: Word w is not in the training data, that is, word w is not in $\boldsymbol{\varphi}$. Then the sampling probability is the product of Equations (3.1) and (3.2).

Finally, infer $\boldsymbol{\theta}^{(d)}$ from the sampling probability $E(\theta_{dj})$.

4. Assign a label k to document d if k is the topic with the highest probability in $\boldsymbol{\theta}^{(d)}$; that is, the summation of probabilities of words under topic k is the largest.

SLLDA-C consists of the following four steps:

1. Obtain $\boldsymbol{\Lambda}$ from the training data in the same way as Step 1 in LLDA-C.

2. Learn $\boldsymbol{\varphi}$ in the same way as Step 2 in LLLDA-C.

3. After $\boldsymbol{\varphi}$ is learned from the training data, extract the top 20% the highest probability words for each topic from $\boldsymbol{\varphi}$ as label related words.

4. Assign a label k to a document if the document contains the most words related to with topic k.

3.4.1.3 Content complexity

Given a document d, its content complexity is measured by its document-topic distribution $\boldsymbol{\theta}^{(d)}$. It would be interesting to know how content complexity may affect the classification results.

The following definition is first introduced in [42]. A topic t contained in document d is said to be *significant* if the probability of t under the topic distribution $\boldsymbol{\theta}^{(d)}$ is greater than a threshold value v, where v should be at least equal to $1/K$ and K is the fixed number of topics for the corpus.

In this chapter we choose $v = 1/K$.

If d contains only one significant topic, then it has a *straightforward content complexity*, and d is referred to as an SCC document. If d contains two or more significant topics, then it has a *high content complexity*, and d is referred to as an HCC document.

3.4.1.4 Accuracy and efficiency

While we choose the data set of Chinese news articles, we note that the selection of a particular language should not affect the accuracy of the LLDA classifiers, as the accuracy is determined by the topic-word and document-topic distributions learned by LLDA with the training data.

Table 3.3 lists the number of articles we selected for the following 10 categories: politics, technology, military, sports, entertainment, health, history, real estate, automobiles, and games. A total of 5000 news articles were selected in these 10 categories as training data.

3.4.1.5 Parameters

For a given corpus of labeled documents, we view the total number K of different labels as the total number of topics for the corpus. It is conventional to let [44]

$$\alpha_k = 50/K, \ k = 1, \cdots, K$$
$$\beta_i = 0.1, \ i = 1, \cdots, V$$

These seem to be the best empirical values for these two hyperparameters.

For a labeled document d in the training data, we set $\Lambda^{(d)}$ to indicate which labels d belongs to. The collapsed Gibbs sampling method is used to sample each topic to learn φ and $\theta^{(d)}$ [45] by counting the total number of words for each topic in each document and the total number of each word under each topic.

3.4.1.5.1 Experiment framework

We use a linear SVM as a baseline classifier, where words with high TF-IDF scores are used as features. Since SVM can only classify each news article into exactly one category, for each document, LLDA-C classifies it into the category by the label with the highest probability in the document-topic distribution, and SLLDA-C classifies it into the category by the label for which the document has the most top topic words.

Experiments are carried out on a server running QEMU Virtual CPU version 1.2.0 with 2.6 GHz and 16 GB RAM.

For each experiment on a given training data set S, which may be the entire training data set of 5000 news articles or a random subset of it, we select 80% of S uniformly at random as the training set and the remaining 20% as the testing set. We run each experiment on the same data set S for $M = 10$ rounds and take the average result for each of the following measurements: precision, recall, Micro-F_1 score, and Micro-F_2 score.

The experiments consist of three parts. In the first part, we compare the overall precisions, overall recalls, and the running time of LLDA-C, SLLDA-C, and SVM. In the second part, we compare the Macro-F_1 and Micro-F_1 scores for each category under LLDA-C,

TABLE 3.3: Categories and the number of labeled news articles, where NoA stands for "number of articles"

Category	NoA	Category	NoA
Politics	693	Health	479
Technology	444	History	295
Military	241	Real estate	347
Sports	549	Automobiles	500
Entertainment	929	Games	523

SLLDA-C, and SVM. In the third part, we compare the classification results on documents of different content complexity.

3.4.1.5.2   Accuracy measurements

In each round i, $i = 1, 2, \cdots, M$, let P_i denote the precision, R_i the recall, TP_i the number of true positives, FP_i the number of false positives, and NF_i the number of false negatives. That is,

$$
\begin{aligned}
P_i &= \frac{TP_i}{TP_i + FP_i} \\
R_i &= \frac{TP_i}{TP_i + FN_i}
\end{aligned}
$$

The overall precision $\mathcal{P}$ and the overall recall $\mathcal{R}$ are calculated by, respectively, the following formulas:

$$
\begin{aligned}
\mathcal{P} &= \frac{\sum_{i=1}^{M} P_i}{M} \\
\mathcal{R} &= \frac{\sum_{i=1}^{M} R_i}{M}
\end{aligned}
$$

3.4.1.5.3   Overall precision and recall comparisons

Table 3.4 lists the overall precision and recall results for LLDA-C, SLLDA-C, and SVM for the given data set of 5000 news articles.

Figure 3.17 and 3.18 show, respectively, the precision and recall rates for data sets of different sizes by using data sets of 100, 200, 300, 500, 1000, 2000, 3000, 4000, and 5000 news articles selected uniformly at random from the training data set, where the horizontal axis represents the volume of the data sets, and the vertical axis represents the overall precision.

From Figure 3.17 and 3.18, and Table 3.4 we can conclude the following:

1. For all classifiers, a larger training set will produce higher accuracy.

2. LLDA-C has higher precision and recall than SLLDA-C and SVM.

3. SLLDA-C has higher precision than SVM, but has lower recall than SVM.

4. For a small training set with less than 500 items, LLDA-C still produces high accuracy, much better than SVM. Thus, LLDA-C is a clear winner, particularly when we have new classifications for new types of data. We may use LLDA-C with a small set of training data to achieve classification results of over 75% precision.

Figure 3.19 shows the log scale of the running time for SLLDA-C, LLDA-C, and SVM on data sets of different sizes. The running time of LLDA-C depends on the number of iterations in Gibbs sampling, which we set to 100. From Figure 3.19, we can see that SLLDA-C is much more efficient than SVM, which is more efficient than LLDA-C.

TABLE 3.4:   Overall precision and recall of classifiers on the data set of 5000 news articles

	LLDA-C	SLLDA-C	SVM
Precision	0.905	0.894	0.884
Recall	0.881	0.867	0.875

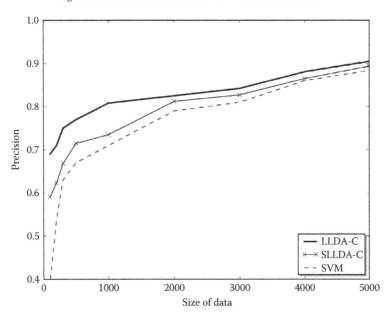

FIGURE 3.17: Comparison of overall precision of LLDA-C, SLLDA-C, and SVM.

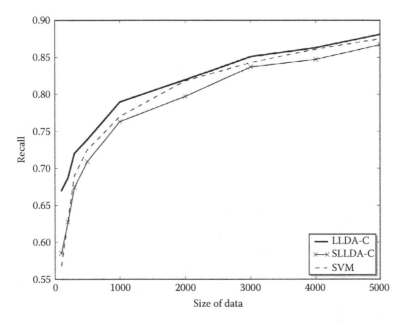

FIGURE 3.18: Comparison of overall recall of LLDA-C, SLLDA-C, and SVM.

3.4.1.6   Content complexity

We use the entire training set of 5000 news articles to run this experiment. In each round, we calculate $\boldsymbol{\theta}^{(d)}$ for each news article d, count the number of SCC and HCC documents, and record the number of SCC and HCC documents that are correctly classified. Finally, we calculate the percentage of SCC in the test data. The results are shown in Figures 3.20 and 3.21.

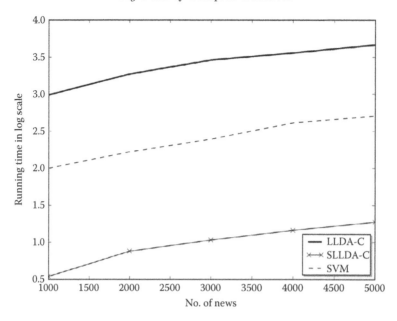

FIGURE 3.19: Log scale of running time for LLDA-C, SLLDA-C, and SVM.

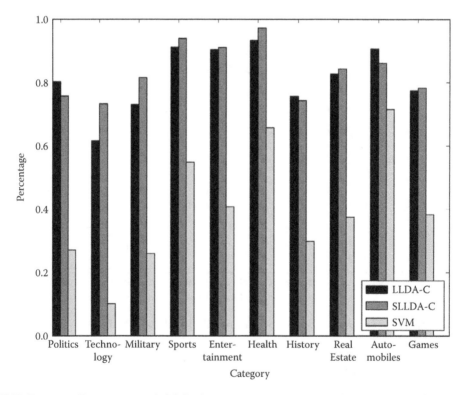

FIGURE 3.20: Percentage of SCC documents in news articles correctly classified by LLDA-C, SLLDA-C, and SVM in each category.

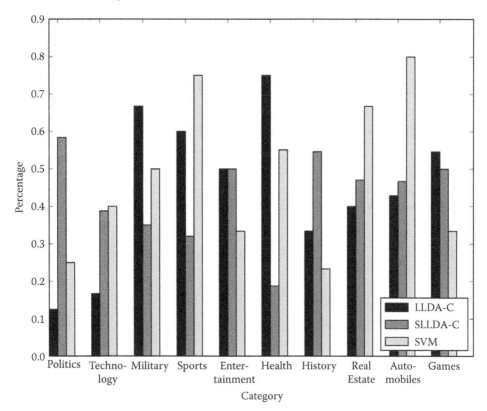

FIGURE 3.21: Percentage of SCC documents misclassified by LLDA-C, SLLDA-C, and SVM in each category.

We can see that for documents correctly classified by LLDA-C, SLLDA-C, and SVM, the percentage of SCC documents in each category are roughly the same for LLDA-C and SLLDA-C, and they are both larger than that of SVM. For documents incorrectly classified by LLDA-C, SLLDA-C, and SVM, the percentage of SCC documents in each category are much different and there is no clear pattern.

We note that for an HCC document, it is better for an LLDA classifier to give it multiple labels, instead of just one label as restricted by SVM.

3.4.2   PopuRank

KWB implements a Labeled-LDA classifier to classify all the news items stored in the central DB. To determine the popularity ranking of news items, KWB uses the notion of PopuRank defined by Bai et al [3].

We observe that breaking news will be reported and reposted online everywhere. In this case, the TF of certain words describing this news will increase sharply. Meanwhile, the document frequency (DF) of certain words describing the breaking news will also increase. KWB monitors each word (except stop words) in each time frame. By monitoring the TF and DF fluctuations of words, KWB calculates the PopuRank of the news items collected in each time unit u. A news item with a higher PopuRank is more popular. The time unit u may be changed according to the actual needs and user interests. For example, if we want to determine popular news items in each hour, then we may set u to be the unit of hour. The

PopuRank of each article remains valid for a fixed number ℓ of time frames. For example, we may let $\ell = 24$ or 48, when u is hour. The value of ℓ may also be changed.

Let t_v denote the current time frame. Let

$$\mathcal{D}_v = \{D_1, D_2, \cdots, D_N\}$$

denote the corpus of all news items collected in this time frame with duplicates removed, where D_i is a news article and D_i contains N_i words in the bag of words model, denoted by

$$D_i = (w_1, w_2, ..., w_{N_i})$$

where each word is a segment of two or more Chinese characters after segmentation.

We define the following terms:

1. **Term frequency (TF)**. The term frequency of word w_j in D_i in time frame t_v, denoted by $tf(w_j, D_i, t_v)$, is the number of times it appears in D_i, denoted by N_{ij}, divided by N_i. That is,

$$tf(w_j, D_i, t_v) = \frac{N_{ij}}{N_i}$$

   Note that if $w_j \notin D_i$, then $tf(w_j, D_i, t_v) = 0$.

2. **Document frequency (DF)**. The document frequency of word w_j in the corpus $\mathcal{D}_v$, denoted by $df(w_j, \mathcal{D}_v)$, is defined as the total number of documents in $\mathcal{D}_v$ that contain w_j, denoted by N_j, divided by the total number of words in $\mathcal{D}_v$, denoted by N. That is,

$$df(w_j, t_v) = \frac{N_j}{N}$$

3. **Average term frequency (ATF)**. Let $atf(w_j, \mathcal{D}_v)$ denote the average term frequency of word w_j in corpus $\mathcal{D}_v$. That is,

$$atf(w_j, t_v) = \frac{\sum_{i=1}^{N} tf(w_j, D_i, t_v)}{N}$$

4. **Term rank (TR)**. We define the term rank of word w_j in document D_i in time frame t_v, denoted by $tr(w_j, D_i, t_v)$, as follows:

$$\begin{aligned} tr(w_j, D_i, t_v) &= \alpha \cdot tf(w_j, D_i, t_v) + \\ &\quad \beta \cdot df(w_j, \mathcal{D}_v) \end{aligned}$$

   where $\alpha \geq 0$, $\beta \geq 0$, and $\alpha + \beta = 1$. For example, we may let $\alpha = 0.6$ and $\beta = 0.4$ to indicate that we place more weight on term frequency over document frequency.

For each word w_j appearing in $\mathcal{D}_v$, compute $df(w_j, \mathcal{D}_v)$ and $atf(w_j, \mathcal{D}_v)$, and keep them for ℓ time frames.

The PopuRank of a document is defined in the following. Assume that word w_j appears in the current time frame t_v. Let T denote the following sequence of consecutive time frames, called a window:

$$T = (t_{\ell-v+1}, t_{\ell-v+2}, \cdots, t_v)$$

At each time frame in this window, we monitor the DF and ATF values for each word. Let t_v be the current time frame. For each word w_j in $\mathcal{D}_v$, we have the following two cases:

Case 1: w_j is a new word, that is, it did not appear in the previous time frames in the window T. Then we compute the TF-IDF values of all the new words in this time frame and mark the top d percent of the new words as popular words.

Case 2: w_j is not a new word. Compute $atf(w_j, t_v)$ and $df(w_j, t_v)$. If the ATF and DF values of word w_j at time t_v suddenly increase k_1 and k_2 times over the previous average ATF and DF values, respectively, for word w_j, denoted by $avgATF(w_j, t_v)$ and $avgDF(w_j, t_v)$, then we will consider the word w_j a popular word, where

$$avgATF(w_j, t_v) = \frac{ATF(w_j, t_v)}{\ell - 1}$$

$$avgDF(w_j, t_v) = \frac{DF(t_v)}{\ell - 1}$$

$$ATF(w_j, t_v) = \sum_{t_i \in T - \{t_v\}} atf(w_j, t_i)$$

$$DF(w_j, t_v) = \sum_{t_i \in T - \{t_v\}} df(w_j, t_i)$$

To specify the values of k_1 and k_2, let

$$ratATF(w_j, t_v) = \frac{atf(w_j, t_v)}{avgATF(w_j, t_v)}$$

$$ratDF(w_j, t_v) = \frac{df(w_j, t_v)}{avgDF(w_j, t_v)}$$

If

$$ratATF(w_j, t_v) > \delta$$
$$ratDF(w_j, t_v) > \sigma$$

where δ and σ are threshold values, then we say that word w_j is popular in time frame t_v.

Let H_v denote the set of all popular words in time frame t_v. We define the PopuRank of news article $D_i \in \mathcal{D}_v$ to be the sum of the term rank of the popular words in D_i in time frame t_v. Namely,

$$\text{PopuRank}(D_i, t_v) = \sum_{w \in H_v \cup D_i} tr(w, D_i, t_v) \qquad (3.3)$$

Figure 3.22 depicts the top 20 news items in all categories within one time frame together with a time stamp when a news article becomes popular, while Figure 3.23 depicts the top 20 news items in the category of sports in the time frame. The values of parameters for our PopuRank calculation are u = hour, $\ell = 24$, $d = 20\%$, $\alpha = 0.6$, $\beta = 0.4$, $\delta = 1.5$, and $\sigma = 1.5$. The time stamp 1430445600 is the Unix epoch time, which is equal to the total number of seconds since 00:00:00, January 1, 1970 Greenwich time, corresponding to 22:00:00, April 30, 2015, Eastern time.

Parameters α and β is related to TR and PopuRank. The value of α and β are decided by which character, TF or DF, is regarded as more important. Figure 3.24 shows the TR of a particular word with different α. Meanwhile, since TR varies, PopuRank of the news also varies; Figure 3.25 shows the different PopuRank of one news with different α and β in same time frame.

	A	B	C
1	Title	Hot Time (timestamp)	PopuRank
2	南宁哪里交通堵刷手机就懂	1430445600	579
3	广西今年投520.7亿为民办实事	1430445600	546
4	春晚福娃邓鸣贺因白血病去世	1430445600	519
5	五月起个人篡乱发布天气预报	1430445600	483
6	巡视追回中粮2.4亿流失国资没见过这么乱的企业	1430445600	478
7	京津冀协同发展2020年北京人口不超2300万	1430445600	475
8	精购房留意政策新变化 买家迎购房"窗口期"	1430445600	465
9	南宁市区小学地段划分将公布	1430917200	458
10	甘肃省人民政府关于进一步加强新时期爱国卫生工作的实施意见	1430445600	457
11	美国炒作蒋介石"婚外情"内幕	1430445600	443
12	成都企业启动"蛛网"计划 民资投	1430445600	440
13	市窗促会副会长李焕亭网谈	1430445600	434
14	抗战期间蒋介石为何布重兵却守不住南京?	1430445600	424
15	海口养生胜地大盘点	1430445600	424
16	互联网+引领白领跨界流动	1430445600	421
17	让万隆精神绽放新的光彩	1430445600	421
18	习近平总书记向全国劳动群众致节日祝贺	1430445600	418
19	武长顺曾遇威胁接电话称某中央领导送书	1430445600	415
20	重磅! 中央政治局会议释放9大信号	1430445600	413
21	揭秘贪官为何偏爱现金	1430445600	411

FIGURE 3.22: The top 20 news items in all categories in a time frame.

	A	B	C
1	Title	Hot Time (timestamp)	PopuRank
2	成都将申办世预赛 长远目标已瞄准2026年世界杯	1430917200	265
3	成都某体育校大学生练后空翻 头部着地身亡	1430445600	256
4	青岛与5城市争办中国足首个主场 硬件条件不落下风	1430917200	244
5	谈球衣退役纪念和规定 奇葩的故事	1430445600	241
6	浙江马拉松推出积分赛规范赛事 办出特色	1430917200	230
7	四川草根足球缺啥? 缺教练缺裁判缺规范	1430917200	217
8	丁俊晖腹背受敌	1430917200	214
9	一代球王与中国足球的"黄金时代"	1430445600	211
10	CBA解析下赛季新政 续约外援细则限制薪资疯涨	1430445600	204
11	一周体坛论语福原爱自曝单身 卡卡主动请战	1430744400	190
12	LOL季中赛战队AHQ禁选解析	1430445600	187
13	中超前瞻:申花叫板恒大望抢分 京鲁争胜有难度	1430445600	184
14	国象男队夺冠凯旋 余泱漪爆料末轮遭"午夜惊铃"	1430445600	182
15	拳王争霸,赛前已收4亿	1430445600	176
16	刘国梁波尔入乡随俗能力强 他喝啤酒爱兑雪碧	1430917200	173
17	互联网巨头开价10亿绿城足球要改门庭?宋卫平暂未回应	1430917200	173
18	成都申办世预赛中卡之战承办权 综合实力有优势	1430917200	171
19	梅西和瓜迪奥拉没联系 伤病不能成拜仁借口	1430917200	169
20	前有上海上港后有多路追兵 恒大为何不再独大	1430445600	164
21	美媒曝世纪大战计分表弄错 帕奎奥被陷害了?	1430917200	160

FIGURE 3.23: The top 20 news items in the sports category in a time frame.

Word	Alpha	Term Rank
事故	0.9	0.107
	0.8	0.104
	0.7	0.101
	0.6	0.098
	0.5	0.095
	0.4	0.092
	0.3	0.089
	0.2	0.086
	0.1	0.083

FIGURE 3.24: Term rank of a word with different values of α.

Title	Alpha	PopuRank
决不放弃任何一丝生的希望——"东方之星"沉船水下搜救纪实	0.9	8
	0.8	16
	0.7	15
	0.6	63
	0.5	35
	0.4	34
	0.3	35
	0.2	37
	0.1	54

FIGURE 3.25: PopuRank of one news item with different values of α.

Thresholds δ and σ decide the number of popular words. Figure 3.26 shows that the number of popular words decrease when δ and σ increase; δ and σ have same value in Figure 3.26.

The running time of calculating PopuRank on news items in each time frame depends on the number of news items waiting to be processed. Table 3.5 shows the number of news items in each time frame on an average day and the time to compute PopuRank of all news items in each time frame on a server running QEMU Virtual CPU version 1.2.0 with 2.6 GHz and 16 GB RAM.

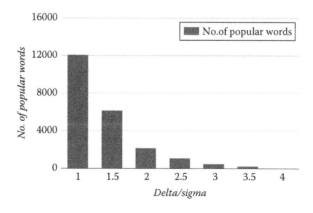

FIGURE 3.26: Number of popular words with different values of δ/σ.

TABLE 3.5: Running time (seconds) for computing PopuRank for news items on an average day

Time Frame	No. of News Items	Running Time
00:00	1238	13.671
01:00	11	0.119
02:00	16	0.116
03:00	5	0.088
04:00	4	0.076
05:00	2	0.070
06:00	3	0.082
07:00	15	0.249
08:00	3	0.196
09:00	7	0.203
10:00	841	6.343
11:00	602	4.735
12:00	6	0.283
13:00	1007	8.848
14:00	2089	38.700
15:00	1444	13.767
16:00	2100	25.918
17:00	2485	40.937
18:00	685	4.437
19:00	5	0.400
20:00	3	0.321
21:00	2	0.320
22:00	4	0.325
23:00	34	0.361

3.5 Conclusion

In this chapter we described two recently developed systems on Big Data text automation. One was targeted on data acquired from OSNs and the other on news media websites on the Internet. We described how text data are acquired, stored, and processed to generate information automatically, including fast and accurate content extractions, fast and accurate keyword and phrase searches on OSN data that are tuned for performing statistical analysis, and fast and accurate automatic classification of contents. We also introduced a method called PopuRank to measure timely popularity of keywords and news items.

Our systems have been found useful to process a large volume of text data accurately and efficiently on commodity computers in a small lab environment and generate useful information for users. For example, we have successfully applied the algorithms and techniques we developed in these systems to help a medium-size medical software company to collect, classify, summarize, rank, and search for new information on cancer treatments, including clinical trials, new treatments, new policies, DNA research, medical conferences, patient reactions on treatments, and comments by physicians on treatment plans, helping physicians and medical researchers acquire information in a timely manner for treating patients and conducing research.

References

1. Ming Jia and Jie Wang. Handling Big Data of online social networks on a small machine. In *Computing and Combinatorics*, pages 676–685. Springer, 2014.

2. Ming Jia, Hualiang Xu, Jingwen Wang, Yiqi Bai, Benyuan Liu, and Jie Wang. Handling Big Data of online social networks on a small machine. *Computational Social Networks*, 2(1):5, 2015.

3. Yiqi Bai, Wenjing Yang, Hao Zhang, Jingwen Wang, Ming Jia, Roland Tong, and Jie Wang. KWB: An automated quick news system for Chinese readers. *ACL-IJCNLP 2015*, page 110, 2015.

4. Open Document for Sina Micro-blog API. Project Website: http://open.weibo.com/wiki/2/statuses/publictimeline/en.

5. Reihaneh Emamdadi, Mohsen Kahani, and Fattane Zarrinkalam. A focused linked data crawler based on HTML link analysis. In *Computer and Knowledge Engineering (ICCKE), 2014 4th International eConference on*, pages 74–79. IEEE, 2014.

6. Hui Lin and Jeff Bilmes. A class of submodular functions for document summarization. In *Proceedings of the 49th Annual Meeting of the Association for Computational Linguistics: Human Language Technologies-Volume 1*, pages 510–520. Association for Computational Linguistics, 2011.

7. Xueming Li, Minling Xing, and Jiapei Zhang. A comprehensive prediction method of visit priority for focused crawler. In *Intelligence Information Processing and Trusted Computing (IPTC), 2011 2nd International Symposium on*, pages 27–30. IEEE, 2011.

8. Li Wei-jiang, Ru Hua-suo, Zhao Tie-jun, and Zang Wen-mao. A new algorithm of topical crawler. In *Computer Science and Engineering, 2009. WCSE'09. Second International Workshop on*, volume 1, pages 443–446. IEEE, 2009.

9. Li Wei-jiang, Ru Hua-suo, Hong Kun, and Luo Jia. A new algorithm of blog-oriented crawler. In *Computer Science-Technology and Applications, 2009. IFCSTA'09. International Forum on*, volume 1, pages 428–431. IEEE, 2009.

10. Li Peng and Teng Wen-Da. A focused web crawler face stock information of financial field. In *2010 IEEE International Conference on Intelligent Computing and Intelligent Systems*, volume 2, pages 512–516, 2010.

11. Xiaolin Zheng, Tao Zhou, Zukun Yu, and Deren Chen. URL rule based focused crawler. In *e-Business Engineering, 2008. ICEBE'08. IEEE International Conference on*, pages 147–154. IEEE, 2008.

12. Ziv Bar-Yossef and Sridhar Rajagopalan. Template detection via data mining and its applications. In *Proceedings of the 11th International Conference on World Wide Web*, pages 580–591. ACM, 2002.

13. Parag Mulendra Joshi and Sam Liu. Web document text and images extraction using DOM analysis and natural language processing. In *Proceedings of the 9th ACM Symposium on Document Engineering*, pages 218–221. ACM, 2009.

14. Christian Kohlschütter, Peter Fankhauser, and Wolfgang Nejdl. Boilerplate detection using shallow text features. In *Proceedings of the Third ACM International Conference on Web Search and Data Mining*, pages 441–450. ACM, 2010.

15. Erdinç Uzun, Hayri Volkan Agun, and Tarık Yerlikaya. A hybrid approach for extracting informative content from web pages. *Information Processing & Management*, 49(4):928–944, 2013.

16. Jyotika Prasad and Andreas Paepcke. Coreex: Content extraction from online news articles. In *Proceedings of the 17th ACM Conference on Information and Knowledge Management*, pages 1391–1392. ACM, 2008.

17. David Gibson, Kunal Punera, and Andrew Tomkins. The volume and evolution of web page templates. In *Special Interest Tracks and Posters of the 14th International Conference on World Wide Web*, pages 830–839. ACM, 2005.

18. Hung-Yu Kao, Ming-Syan Chen, Shian-Hua Lin, and Jan-Ming Ho. Entropy-based link analysis for mining web informative structures. In *Proceedings of the Eleventh International Conference on Information and Knowledge Management*, pages 574–581. ACM, 2002.

19. Hung-Yu Kao, Shian-Hua Lin, Jan-Ming Ho, and Ming-Syan Chen. Mining web informative structures and contents based on entropy analysis. *Knowledge and Data Engineering, IEEE Transactions on*, 16(1):41–55, 2004.

20. Cai-Nicolas Ziegler and Michal Skubacz. Content extraction from news pages using particle swarm optimization on linguistic and structural features. In *Proceedings of the IEEE/WIC/ACM International Conference on Web Intelligence*, pages 242–249. IEEE Computer Society, 2007.

21. Lan Yi, Bing Liu, and Xiaoli Li. Eliminating noisy information in web pages for data mining. In *Proceedings of the Ninth ACM SIGKDD International Conference on Knowledge Discovery and Data Mining*, pages 296–305. ACM, 2003.

22. Yan Guo, Huifeng Tang, Linhai Song, Yu Wang, and Guodong Ding. Econ: An approach to extract content from web news page. In *Web Conference (APWEB), 2010 12th International Asia-Pacific*, pages 314–320. IEEE, 2010.

23. Tim Weninger and William H Hsu. Text extraction from the web via text-to-tag ratio. In *Database and Expert Systems Application, 2008. DEXA'08. 19th International Workshop on*, pages 23–28. IEEE, 2008.

24. Shumeet Baluja. Browsing on small screens: Recasting web-page segmentation into an efficient machine learning framework. In *Proceedings of the 15th International Conference on World Wide Web*, pages 33–42. ACM, 2006.

25. John Gibson, Ben Wellner, and Susan Lubar. Adaptive web-page content identification. In *Proceedings of the 9th Annual ACM International Workshop on Web Information and Data Management*, pages 105–112. ACM, 2007.

26. Deepayan Chakrabarti, Ravi Kumar, and Kunal Punera. Page-level template detection via isotonic smoothing. In *Proceedings of the 16th International Conference on World Wide Web*, pages 61–70. ACM, 2007.

27. Jeff Pasternack and Dan Roth. Extracting article text from the web with maximum subsequence segmentation. In *Proceedings of the 18th International Conference on World Wide Web*, pages 971–980. ACM, 2009.

28. Jingwen Wang and Jie Wang. Fast and accurate article extractions from web pages using partition features optimizations. In *Proceedings of the 7th International Conference on Knowledge Discovery and Information Retrieval (KDIR)*, 2015.

29. MongoDB. Project Website: http://www.mongodb.org/.

30. Dirk Bahle, Hughe Williams, and Justin Zobel. Compaction techniques for nextword indexes. In *Spire*, page 0033. IEEE, 2001.

31. Xingyuan Chen, Yunqing Xia, Peng Jin, and John Carroll. Dataless text classification with descriptive LDA. In *Twenty-Ninth AAAI Conference on Artificial Intelligence*, 2015.

32. Seonggyu Lee, Jinho Kim, and Sung-Hyon Myaeng. An extension of topic models for text classification: A term weighting approach. In *Big Data and Smart Computing (BigComp), 2015 International Conference on*, pages 217–224. IEEE, 2015.

33. Yung-Shen Lin, Jung-Yi Jiang, and Shie-Jue Lee. A similarity measure for text classification and clustering. *Knowledge and Data Engineering, IEEE Transactions on*, 26(7):1575–1590, 2014.

34. Fabrizio Sebastiani. Machine learning in automated text categorization. *ACM Computing Surveys (CSUR)*, 34(1):1–47, 2002.

35. Jingnian Chen, Houkuan Huang, Shengfeng Tian, and Youli Qu. Feature selection for text classification with naïve Bayes. *Expert Systems with Applications*, 36(3):5432–5435, 2009.

36. Simon Tong and Daphne Koller. Support vector machine active learning with applications to text classification. *The Journal of Machine Learning Research*, 2:45–66, 2002.

37. David M Blei, Andrew Y Ng, and Michael I Jordan. Latent Dirichlet allocation. *The Journal of Machine Learning Research*, 3:993–1022, 2003.

38. Daniel Ramage, David Hall, Ramesh Nallapati, and Christopher D Manning. Labeled LDA: A supervised topic model for credit attribution in multi-labeled corpora. In *Proceedings of the 2009 Conference on Empirical Methods in Natural Language Processing: Volume 1-Volume 1*, pages 248–256. Association for Computational Linguistics, 2009.

39. Jon D Mcauliffe and David M Blei. Supervised topic models. In *Advances in Neural Information Processing Systems*, pages 121–128, 2008.

40. Simon Lacoste-Julien, Fei Sha, and Michael I Jordan. DiscLDA: Discriminative learning for dimensionality reduction and classification. In *Advances in Neural Information Processing Systems*, pages 897–904, 2009.

41. Daniel Ramage, Paul Heymann, Christopher D Manning, and Hector Garcia-Molina. Clustering the tagged web. In *Proceedings of the Second ACM International Conference on Web Search and Data Mining*, pages 54–63. ACM, 2009.

42. Yiqi Bai and Jie Wang. News classifications with labeled LDA. In *Proceedings of the 7th International Conference on Knowledge Discovery and Information Retrieval (KDIR)*, 2015.

43. Balaji Lakshminarayanan and Raviv Raich. Inference in supervised latent Dirichlet allocation. In *Machine Learning for Signal Processing (MLSP), 2011 IEEE International Workshop on*, pages 1–6. IEEE, 2011.

44. Thomas L Griffiths and Mark Steyvers. Finding scientific topics. *Proceedings of the National Academy of Sciences*, 101(suppl 1), pages 5228–5235. 2004.

45. William M Darling. A theoretical and practical implementation tutorial on topic modeling and Gibbs sampling. In *Proceedings of the 49th Annual Meeting of the Association for Computational Linguistics: Human Language Technologies*, pages 642–647, 2011.

Chapter 4

Big Data Visualization for Large Complex Networks

Lei Shi, Yifan Hu, and Qi Liao

4.1   Introduction

Visualization methods are essential to the analysis of Big Data. In this chapter, we focus on large complex networks (interchangeably called the big network) that may have millions of nodes, multiple node/edge attributes and nontrivial topologies beyond the random graph. We present both current challenges and state-of-the-art methods to visually analyze these networks. This should help practitioners and researchers in Big Data-related areas handle network data.

We start by introducing the concept of networks in the visualization research. In general, the network discussed here is composed of a set of nodes representing various types of entities or items, and a set of edges representing the complex relationship between these nodes. For example, social networks are typical instances of such a concept. The nodes can be registered users of an online application, or students in an offline teaching class. Both kinds of users can have rich demographics, for example, gender, age or interests. The edges in these social networks can be friendship ties with various degrees of familiarity, short-message communications with informative sending/receiving patterns, or even mutual hostile behaviors at school, indicating the negative relationship. On these networks, the problem of visualization arises as ordinary people fail to interpret the structure and useful patterns from the network only by looking at the list of nodes and edges. Researchers seek to design automatic algorithms to draw the networks into a virtual space, namely world coordinates, which can be further displayed in computer screens or printed on paper to amplify people's cognition on networks.

The fundamental challenge in network visualization is how to assign appropriate locations for nodes and how to connect edges between these nodes, known as the node and edge layout problem. This is referred to as the graph drawing problem where the term graph is often used interchangeably with the term network. Common graph drawing frameworks identify general constraints (called aesthetics) in the desired network layout, develop fast algorithms to satisfy these constraints, and then design visual representations to draw the network nodes and edges for better interpretation. Among known drawing aesthetics, the objective to minimize the edge crossings is often considered to be the most critical. While theoretical methods to avoid any or most edge crossings in the graph layout have been developed, known as the planarity-based approach, another class of methods called the force-directed approach is more popular in drawing real-life graphs that can have larger size and complexity. The force-directed method was first invented by Eades in his spring-electrical model [1] and later improved by Fruchterman and Reingold on the force formula [2]. The main idea is to translate the graph layout problem into the energy minimization of a physical system. Similar models include the stress model introduced by Kamada and Kawai [3], on which fast solvers have been available [4]. In visualizing networks given the node and the edge layouts, the node-link representation is the most popular, which draws nodes by enclosures and edges by lines connecting them. This representation exploits people's perception-level characteristics that tend to relate visual elements connected by uniform visual properties, also known as the connectedness principle of the Gestalt laws [5]. Other visual metaphors for networks, such as the adjacency matrix [6] and statistical charts [7], have also been studied and shown to outperform the node-link representation under certain conditions. The full technical details of the classical graph drawing algorithms and frameworks are not the focus of this chapter. Readers are referred to the book in [8] for a comprehensive understanding of the field. In particular, for ease of discussion, we will limit the scope in this chapter to the node-link visualization of simple and undirected graphs, that is, there are no loops (edges that connect one node to itself), no duplicate edges (more than one edge that connects the same pair of nodes together), and no hyperedges (edges that connect to more than two nodes) on the graph.

Over the past few decades, we have witnessed the development of network visualization techniques in both theories and applications. For example, the Graphviz open source graph layout and rendering engine [9] developed at AT&T Labs has been widely applied in various domains as the standard graph drawing toolkit. The popular D3.js library in JavaScript and the Prefuse library in Java both carry implementations of the force-directed graph layout algorithm and example routines for network visualization. Other relevant software packages developed in academia include Gephi [10], OGDF [11], and Tulip [12]. Nowadays, the visualization of small commodity networks have grown to an extent that beginners can draw an elegant graph within 100 lines of code in a short amount of time with the help of well-developed libraries. However, on the large complex networks collected in the current Big Data era, such as the massive online social networks and the semantic-rich knowledge graphs, established graph drawing methods often face new and unsolved challenges due to the unique properties of the big network. We shall briefly discuss these challenges by looking at the four Vs of Big Data.

On data volume, Facebook now hosts more than one billion registered users, and Twitter has 288 million monthly active users. Their social graphs are millions of times larger than the classical network data set measured in Zachary's Karate Club (34 nodes). It is difficult, if not impossible, to store the entire graph ($\sim 10^9$ nodes with attributes) for visualization in a commodity desktop. When the network is displayed in browser through cloud-based services, on a small portion of such graphs ($\sim 10^7$ nodes), the time to load the data can exceed the amount that can be tolerated by users (> 1 min). Furthermore, if the full layout is to be computed in-situ at the client side, a moderately-sized graph ($\sim 10^5$ nodes) can take an unacceptable time to process. Finally, in cases where all earlier mentioned difficulties can be settled, normal people can hardly interpret the network structure with more than 10^3 nodes visually, known as the perception constraint. In summary, the sheer volume of big networks brings huge challenges to network visualization due to its growing space, time, and visual complexities.

On data variety, the social graphs on Facebook and Twitter have more than a hundred attributes on each account, including their user types, demographics, and various operational/security settings; the gene regulatory network can have tens of interaction types modeled on edges. Understanding the distribution of these attributes on big networks can be critical to many real-world problems, such as the network diffusion, evolution, and so on. While the straightforward method to encode node/edge attributes through visual channels can serve a limited number of such dimensions, the navigation and visual analysis of the entire high-dimensional network remains an open challenge. New methods need to be developed to visually represent the correlation between the topology and node/edge attributes of big networks.

On data velocity, all networks are live, that is, they change over time. This first duplicates the network size by the number of time slots under study, and makes the challenge on data volume even greater. On the other hand, the visualization of network dynamics brings the new requirement to stabilize the network layout. Furthermore, how to appropriately represent the time dimension on the network visualization remains an open problem, when the classical 2D network layout has been assigned to illustrate the topology information. This thread of research has attracted significant attention recently on the topic of dynamic network visualization. Due to space limits, we will not cover more details on this part. Readers can refer to the latest survey [13] for more information.

In a few reports, the fourth V of the Big Data is known as the value. On the visualization of large complex networks, the value for end users is greatly diminished by the increasing size and dimensionality of the network data. Users find it hard to understand the hairballs created on large networks and the information-intensive drawing of most multivariate network visualizations. The key to this challenge often depends on managing the complexity in the first three facets of big networks.

Existing literature on large complex network visualization can be roughly categorized into four classes of approaches, as shown in Figure 4.1: graph layout–based, graph mining–based, graph and view transformation–based, and interaction-based. In this chapter, we

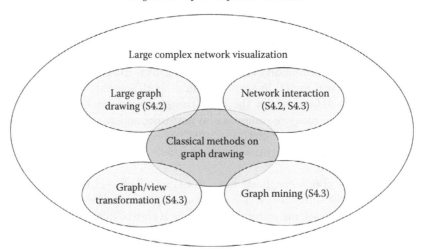

FIGURE 4.1: Classifications of large complex network visualization methods and the organization of this chapter.

will focus on the first three classes and their methodologies to tackle the data volume and variety challenges of large complex network visualizations. The fourth class related to network interactions is also involved in these discussions. The rest of this chapter is organized as follows:

- In Section 4.2, we talk about large graph drawing methods, which extend the classical drawing algorithms to support large and complex networks that may exhibit small world tendencies.

- In Section 4.3, we describe the visual abstraction methods for large and multivariate networks, which can employ both graph mining–based and graph transformation–based approaches.

- In Section 4.4, we give a few examples on the application of the proposed methods in real-world visualization problems.

- Finally in Section 4.5, we summarize existing approaches and potential future directions on large complex network visualization.

4.2 Algorithms for Large Network Visualization

In this section we look at algorithms and techniques for laying out large graphs. One important consideration when working on large graphs is that the scalability of the algorithm becomes very important. An algorithm that has, say, a computational complexity quadratic to the number of nodes may work well for graphs of a few hundred nodes. But such an algorithm would take a very long time to find a layout for a graph of a million nodes. Furthermore, an algorithm with quadratic memory complexity would not even run on a graph of a million nodes, because no single computer would be able to store 10^{12} real numbers in memory. In addition, many algorithms that can scale to a large size are iterative methods that attempt to find an optimal solution to a cost function. For large graphs, such a cost function tends to have many local minima. So the ability of the algorithm in escaping from a local minimum and achieving a globally optimal drawing is also very important on large graphs.

4.2.1 Layout algorithms for large graphs

We first define some notations. An undirected graph G consists of a set of nodes (vertices) V, and a set of edges E, which are tuples of nodes. Denote by $|V|$ and $|E|$ the number of vertices and edges, respectively. If vertices i and j form an edge, we denote that $i \leftrightarrow j$, and call i and j neighboring vertices.

A graph is traditionally visualized as a node-link diagram like Figure 4.2, which displays a 14-node social network. To get this diagram, we must lay out the graph, that is, assign a location to each node. We denote by x_i the location of vertex i in the layout. Here x_i is in d-dimensional Euclidean space. Typically $d = 2$ or 3. The aim of laying out a graph is that the resulting drawing gives an aesthetic visual representation of the connectivity information among vertices.

The most widely used techniques turn the problem of graph layout into one of finding a minimal energy configuration of a physical system. The two most popular methods in this category are the spring-electrical model [1,2], and the stress model [3]. These are among the algorithms we will discuss next.

4.2.1.1 Spring-electrical model–based algorithms

The spring-electrical model was first introduced by Peter Eades in 1984 [1]. He suggested embedding a graph by replacing the vertices by steel rings and each edge with a spring to form a mechanical system. The vertices are placed in some initial layout and let go so that the spring forces on the rings move the system to a minimal energy state. At the same time he made nonadjacent vertices repel each other. This intuitive idea describes the spring-electrical model, as well as the force-directed solution framework now widely used for solving this and other virtual physical models. The version of the spring-electrical model that is used today is a modification of Eades's formulation, due to Fruchterman and Reingold [2]. In this version, attractive spring force exerted on vertex i from its neighbor j is proportional to the squared distance between these two vertices,

$$F_a(i,j) = -\frac{\|x_i - x_j\|^2}{K} \frac{x_i - x_j}{\|x_i - x_j\|}, \ i \leftrightarrow j \tag{4.1}$$

where K is a parameter related to the nominal edge length of the final layout. The repulsive electrical force exerted on vertex i from any vertex j is inversely proportional to the distance between these two vertices,

$$F_r(i,j) = \frac{K^2}{\|x_i - x_j\|} \frac{x_i - x_j}{\|x_i - x_j\|}, \ i \neq j \tag{4.2}$$

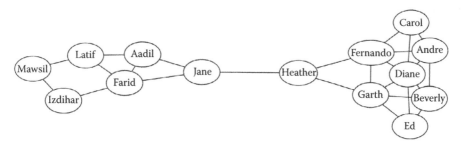

FIGURE 4.2: Visualization of a small social network.

The energy of this physical system [14] is

$$E(x) = \sum_{i \leftrightarrow j} \|x_i - x_j\|^3/(3K) - \sum_{i \neq j} K^2 \ln\left(\|x_i - x_j\|\right)$$

with its derivatives a combination of the attractive and repulsive forces.

The spring-electrical model can be solved with a force-directed procedure by starting from an initial (e.g., random) layout, calculating the combined attractive and repulsive forces on each vertex, and moving the vertex along the direction of the force for a certain step length. This process is repeated for every vertex, with the step length decreasing after a complete pass over all the vertices. The process continues until the layout stabilizes.

This procedure can be enhanced by an adaptive step length updating scheme [15,16]. It usually works well for small graphs. For large graphs, this simple iterative procedure is prone to be trapped in one of the many local energy minima that exists in the space of all possible layouts. Furthermore, the computational complexity of the spring-electrical model is $O(|V|^2)$, making it unsuitable for very large graphs. We discuss two techniques next to make the spring-electrical model scalable and better at finding a global optimal solution.

Fast force approximation. The force-directed algorithm involves computing the repulsion force, Equation 4.2, of all other nodes j to one node i, and repeats this for every node i. Thus to compute the repulsion force once for every node requires a quadratic amount $(|V|^2)$ of force calculations.

One way to reduce this complexity is to use one of a few possible force approximation schemes. The scheme proposed by Barnes and Hut [17] approximates the repulsive forces in $O(n \log n)$ time with good accuracy, but does so without ignoring long range forces. It works by treating groups of far away vertices as supernodes, using a nested space decomposing data structure. This idea was adopted by Tunkelang [18] and Quigley [19]. They both used a quadtree (or octree in 3D) data structure.

A quadtree forms a recursive grouping of vertices (Figure 4.3), and can be used to efficiently approximate the repulsive force. When calculating the repulsive force on a vertex i, if a group of vertices, S, lies in a square that is sufficiently "far" from i, the whole group can be treated as a supernode. Otherwise we traverse down the tree and examine the four sibling squares.

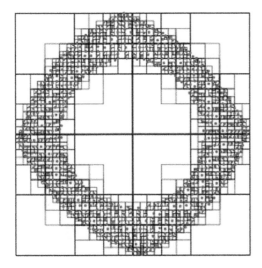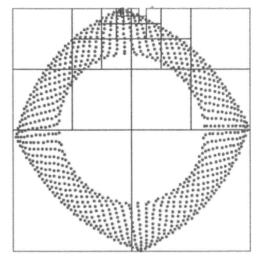

FIGURE 4.3: An illustration of the quadtree data structure. Left: the overall quadtree. Right: supernodes with reference to a vertex at the top-middle part of the graph, with $\theta = 1$.

Figure 4.3 (right) shows all the super-nodes (the squares) and the vertices these super-nodes consist of, with reference to vertex i located at the top-middle part of the graph. In this case there are 936 vertices, and 32 supernodes.

Under a reasonable assumption [17,20] that the vertices do not distribute pathalogically (in practice this assumption is satisfied when the initial layout is not highly skewed in some areas), it can be proved that building the quadtree takes a time complexity of $O(|V|\log|V|)$. Finding all the supernodes with reference to a vertex i can be done in a time complexity of $O(log|V|)$. Overall, using a quadtree structure to approximate the repulsive force, the complexity for each iteration of the force-directed algorithm using the spring-electrical model is $O(|V|\log|V|)$. This force approximation scheme can be further improved by considering force approximation at the supernode–supernode level instead of the vertex–supernode level [21].

The multipole method [22] is another force approximation algorithm with the same $O(|V|\log|V|)$ complexity. The advantage of this approach is that its complexity is independent of the distribution of vertex positions. It also uses on a quad-tree space decomposition. Hachul and Jünger applied this force approximation to graph drawing [23] in the FM^3 code.

Multilevel approach. The force-directed algorithm with the spring-electrical model repositions one vertex at a time to minimize the energy locally. On a graph, this system of springs and electrical charges can have many local minimum configurations. Therefore, starting with a random initial layout, the algorithm often cannot converge to a global optimal final layout on large graphs. A multilevel approach can overcome this limitation. In this approach, a sequence of smaller and smaller graphs is generated from the original graph, each capturing the essential connectivity information of its parent. The global optimal layout can be found much more easily on small graphs. A good layout for a coarser graph is thus always used as a starting layout for its parent. From this initial layout, further refinement is carried out to achieve the optimal layout of the parent.

Multilevel approaches have been used in many large-scale combinatorial optimization problems, such as graph partitioning [24–26], matrix ordering [27,28], and the traveling salesman problem [29], and were proved to be a useful meta-heuristic tool [30]. They were later used in graph drawing algorithms [31–34], sometimes under the name "multiscale." Note that a multilevel approach is not limited to the spring-electrical model, but for convenience it is discussed in the context of this model.

A multilevel approach has three distinctive phases: coarsening, coarsest graph layout, and prolongation and refinement. In the coarsening phase, a series of coarser and coarser graphs, $G^0 = G, G^1, \ldots, G^l$, is generated, each coarser graph G^{k+1} encapsulating the information needed to lay out its parent G^k, while containing fewer vertices and edges. The coarsening continues until a graph with only a small number of vertices is reached. The optimal layout for the coarsest graph can be found cheaply. The layouts on the coarser graphs are recursively prolonged to the finer graphs, with further refinement at each level.

Graph coarsening and initial layout is the first phase in the multilevel approach. There are a number of ways to coarsen an undirected graph. One often-used method is based on edge collapsing [24–26]. In this scheme, a maximal independent edge set (MIES) is selected. This is a maximal set of edges, with no edges incident to the same vertex. The vertices corresponding to this edge set form a maximal matching. Each edge, and its corresponding pair of vertices, are coalesced into a new vertex. Each vertex of the resulting coarser graph has an associated weight, equal to the number of original vertices it represents. Each edge of the coarser graph also has a weight associated with it. Initially, all edge weights are set to one. During coarsening, edge weights are unchanged unless both merged vertices are adjacent to the same neighbor. In this case, the new edge is given a weight equal to the sum of the weights of the edges it replaces. Figure 4.4 illustrates MIES and the result of coarsening using edge collapsing.

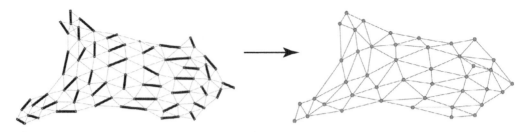

FIGURE 4.4: An illustration of edge collapsing–based graph coarsening. Left: original graph with 85 vertices. Edges in a maximal independent edge set are thickened. Right: a coarser graph with 43 vertices resulting from coalescing thickened edges.

The coarsest graph layout is carried out at the end of the recursive coarsening process. Coarsening is performed repeatedly until the graph is very small; at that point because the graph on the coarsest level is very small, many algorithms, including the force-directed algorithm with the spring-electrical mode, can find a globally optimal layout.

The prolongation and refinement step is the third phase in a multilevel procedure. The layout on the coarser graphs are recursively interpolated to the finer graphs, with further refinement at each level. Because the layout is derived from the layout of a coarser graph, it is typically already well placed globally and what is required is just some local adjustment, therefore a conservative step-length update scheme should be used.

Putting it all together. When combining the fast force approximation scheme with the multilevel approach, the resulting spring-electrical algorithm can scale to graphs with millions of nodes and up to a billion edges [35]. Row "Spring electrical" of Figure 4.5 shows drawings of two graphs using such a multilevel force-directed algorithm [16]. The drawings are of good quality for both `dw256A`, a mesh-like graph, and `qh882`, a sparser graph. Two implementations of the multilevel force-directed algorithm are `sfdp` [16] and `fm`^{3} [23]. Figure 4.6 gives some examples using sfdp.

4.2.1.2   Stress and strain model–based algorithms

The spring-electrical model assumes a constant ideal edge length. But in many applications, edges of a graph often have a variable ideal length. For example, the length of an edge may represent the dis-similarity of the end nodes. It is useful to lay out such graphs so that similar nodes are closer while dissimilar nodes are further apart. While the spring-electrical model can be modified to take into account the edge length in a heuristic fashion, this kind of information is best encoded with the stress model.

The stress model. The stress model assumes that there are springs connecting all pairs of vertices in the graph, with the ideal spring length equal to the predefined edge length. The stress energy of this spring system is

$$\sum_{i \neq j} w_{\mathrm{ij}} \left(\|x_i - x_j\| - d_{\mathrm{ij}} \right)^2 \tag{4.3}$$

where d_{ij} is the ideal distance between vertices i and j. The layout that minimizes the earlier mentioned stress energy is an optimal layout of the graph according to this model. In the stress model w_{ij} is a weight factor. A typical choice is $w_{\mathrm{ij}} = 1/d_{\mathrm{ij}}{}^2$. With this choice, Equation 4.3 can be written as $\sum_{i \neq j} \left(\|x_i - x_j\|/d_{\mathrm{ij}} - 1 \right)^2$, thus this stress energy measures the relative difference between the actual edge length and ideal edge length. If $w_{\mathrm{ij}} = 1/d_{\mathrm{ij}}$, the resulting embedding is known as Sammon mapping.

The stress model has its root in multidimensional scaling (MDS) [37,38], and the term MDS is sometimes used to describe the embedding process based on the stress model.

Algorithms	dw256A	qh882
Spring electrical		
Stress		
Classical MDS		
Pivot MDS		
HDE		

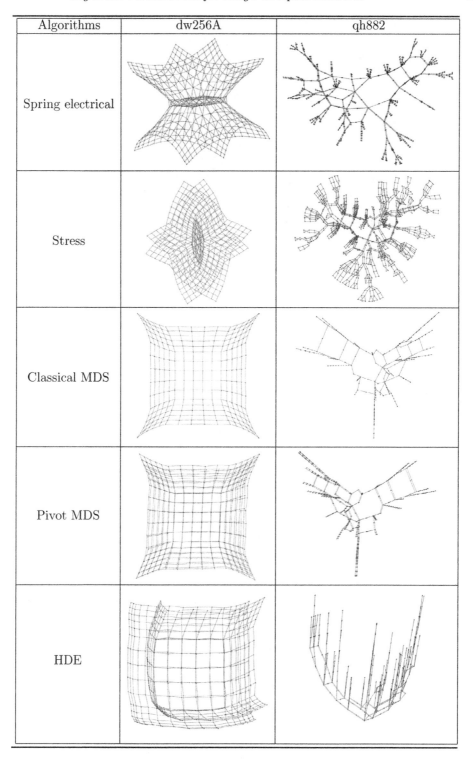

FIGURE 4.5: Result of some of the algorithms applied to two graphs, dw256A and qh882.

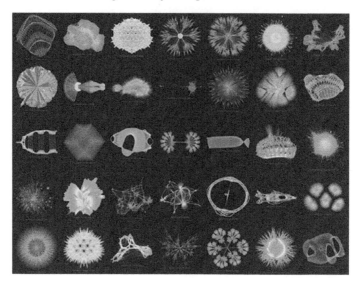

FIGURE 4.6: Example drawings of some graphs from the University of Florida Matrix Collection. (From T. A. Davis and Y. Hu. University of Florida Sparse Matrix Collection. *ACM Transaction on Mathematical Software*, 38:1–18, 2011 [36].)

When used to embed multidimensional data, distance among any pairs of item can be easily calculated using the appropriate metric defined in that space. However when used to embed graphs, typically only the lengths of the edges are known. In this case the distance between two non-adjacent nodes are assumed to be their shortest path distance. This practice dates back to at least 1980 in social network layout [39], and in graph drawing using classical MDS [38], even though it is often attributed to Kamada and Kawai [3] in the graph drawing literature.

There are several ways to minimize Equation 4.3. A simple (though not always robust) approach is to place each node iteratively [4],

$$x_i = \frac{1}{\sum_{j \neq i} w_{ij}} \sum_{j \neq i} w_{ij} \left(x_j + d_{ij} \frac{x_i - x_j}{\|x_i - x_j\|} \right)$$

Note that x_i and $x_j + d_{ij} \frac{x_i - x_j}{\|x_i - x_j\|}$ are of distance d_{ij} from each other. So the earlier mentioned simply places node i at the average of the ideal positions with regard to the other nodes. Here we treat the right-hand side as known by using the current best guess of node positions. A more robust and preferred approach is to employ the majorization technique [4]. While the technique is typically derived by a careful use of Cauchy–Schwarz inequality on the stress function, it can be derived simply by rearranging the earlier mentioned equation as

$$\sum_{j \neq i} w_{ij}(x_i - x_j) = \sum_{j \neq i} w_{ij} d_{ij} \frac{x_i - x_j}{\|x_i - x_j\|}$$

The left-hand side can be written as the product of a $|V| \times |V|$ weighted Laplacian matrix and a $|V| \times d$ tall matrix of node positions. The right-hand side is seen as a weighted sum of unit vectors, and is assumed to be known by using the current best guess of node positions. The stress majorization process thus repeatedly solves the earlier mentioned linear system, and uses the resulting solution as the new node position for the right-hand side. We iterate until the node positions no longer change. Row "Stress" of Figure 4.5 displays drawings given by the stress majorization algorithm. It performed very well on both graphs.

On large graphs, the stress model in Equation 4.3 is not scalable. Formulating the model requires computing the all-pairs shortest path distances, and a quadratic amount of memory to store the distance matrix. Furthermore the solution process has a computational complexity at least quadratic to the number of nodes. In recent years there have been attempts at developing more scalable ways of drawing graphs that still satisfy the user-specified edge length as much as possible. We will discuss some of these after we introduce the strain model.

The strain model (classical MDS). The strain model, also known as classical MDS [40], predates the stress model. Classical MDS tries to fit the inner product of positions, instead of the distance between points. Specifically, assume that the final embedding is centered around the origin. Furthermore, assume that in the ideal case, the embedding fits the ideal distance exactly: $\|x_i - x_j\| = d_{ij}$. It is then easy to prove [41] that the product of the positions, $x_i^T x_j$, can be expressed as the squared and double centered distance,

$$x_i^T x_j = -1/2 \left(d_{ij}^2 - \frac{1}{|V|} \sum_{k=1}^{|V|} d_{kj}^2 - \frac{1}{|V|} \sum_{k=1}^{|V|} d_{ik}^2 + \frac{1}{|V|^2} \sum_{k=1}^{|V|} \sum_{l=1}^{|V|} d_{kl}^2 \right) = b_{ij} \qquad (4.4)$$

In real data, it is unlikely that we can find an embedding that fits the distances perfectly, but we would still expect b_{ij} to be a good approximation of $x_i^T x_j$. Therefore we try to find an embedding that minimizes the difference between the two,

$$\min_X \ \|X^T X - B\|_F \qquad (4.5)$$

where X is the $|V| \times d$ dimensional matrix of x_i's, B is the $|V| \times |V|$ symmetric matrix of b_{ij}'s, and $\|.\|_F$ is the Frobenius norm. Denote the eigen-decomposition of B as $B = Q^T \Lambda Q$, then the solution to Equation 4.5 becomes $X = \Lambda_d^{1/2} Q$, where Λ_d is the diagonal matrix of Λ, with all but the d-largest eigenvalues on the diagonal set to zero. In other words, the strain model works by finding the top d eigenvectors of B, and uses that, scaled by the eigenvalues, as the coordinates. Because the strain model does not fit the distance directly, edge length in the layout may not fit the length specified by the user as well as the stress model. Nevertheless, the layout from the strain model can be used as a good starting point for the stress model. Row "Classical MDS" of Figure 4.5 gives drawings using classical MDS. It performed better on the mesh-like `dw256A` graph than on the sparser `qh882` graph. On the latter it gives a drawing with many vertices close to each other, making details of the graph unclear, even though it captures the overall structure well.

MDS for large graphs. In the stress model (as well as in the strain model), the graph distances between all pairs of vertices have to be calculated, which necessitates an all-pairs shortest path calculation. Using Johnson's algorithm, this needs $O(|V|^2 \log |V| + |V||E|)$ computation time, and a storage of $O(|V|^2)$. Solution of the dense linear systems takes even more time than the formation of the distance matrix. Therefore for very large graphs, the stress model is computationally expensive and prohibitive from a memory standpoint.

A number of attempts have been made to approximately minimize the stress energy or to solve the strain model.

The multiscale algorithm of Hadany and Harel [32] improved the Kamada and Kwai [3] solution process by speeding up its convergence; Harel and Koren[33] improved that further by coarsening with k-centers.

A multiscale algorithm of Gajer et al. [31] applies the multilevel approach in solving the stress model. In the GRIP algorithm [31], graph coarsening is carried out through vertex filtration, an idea similar to the process of finding a maximal independent vertex set. A sequence of vertex sets, $V^0 = V \subset V^1 \subset V^2, \ldots, \subset V_L$, is generated. However, coarser graphs are not constructed explicitly. Instead, a vertex set V^k at level k of the vertex set hierarchy is constructed so that distance between vertices is at least $2^{k-1} + 1$. On each level k, the stress model is solved by a force-directed procedure. The spring force on each vertex $i \in V^k$ is calculated by considering a neighborhood $N^k(i)$ of this vertex, with $N^k(i)$ the set of vertices in level k—chosen so that the total number of vertices in this set is $O(|E|/|V^k|)$. Thus the force calculation on each level can be done in time $O(|E|)$. It was proved that with this multilevel procedure and the localized force calculation algorithm, for a graph of bounded degree, the algorithm has close to linear computational and memory complexity. However in the actual implementation, at the finest level, GRIP reverts back to the spring-electrical model of Fruchterman and Reingold [2], making it difficult to assess whether GRIP can indeed solve the stress model well.

LandmarkMDS [42] approximates the result of classical MDS by choosing $k << |V|$ vertices as landmarks, and calculating a layout of these vertices using classical MDS, based on distances among these vertices. The positions for the rest of vertices are then calculated by placing them at the weighted barycenter, with weights based on distances to the landmarks. So essentially classical MDS is applied to a $k \times k$ submatrix of the $|V| \times |V|$ matrix B. The complexity of this algorithm is $O(k|E| + k^2)$, and only $O(k|V|)$ distances need to be stored.

PivotMDS [43], on the other hand, takes a $|V| \times k$ submatrix C of B. It then uses the eigenvectors of CC^T as an approximation to those of B. The eigenvectors of CC^T are found via those of the $k \times k$ matrix $C^T C$. By an algebraic argument, if v is an eigenvector of $C^T C$, then

$$CC^T(Cv) = C(C^T Cv) = \lambda(Cv)$$

hence Cv is an eigenvector of CC^T. Therefore PivotMDS proceeds by finding the largest eigenvectors of this smaller $k \times k$ matrix $C^T C$, then projects back to $|V|$-dimensional space by multiplying them with C. It uses these projected eigenvectors, scaled by the inverse of the quartic root of the eigenvalues, as coordinates. Using this technique, the overall complexity is similar to LandmarkMDS, but unlike LandmarkMDS, PivotMDS utilizes distances between landmark vertices (called pivots) and all vertices in forming the C matrix. In practice, PivotMDS was found to give drawings that are closer to the classical MDS than LandmarkMDS. It is extremely fast when used with a small number (e.g., $k \approx 50$) of pivots.

It is worth pointing out that, for sparse graphs, there is a limitation in both algorithms. For example, if the graph is a tree, and pivots/landmarks are chosen to be non-leaves, then two leaf nodes that have the same parent will have exactly the same distances to any of the pivots/landmarks; consequently their final positions based on these algorithms will also be the same. This problem may be alleviated to some extent by utilizing the layout given by these algorithms as an initial placement for a sparse stress model [4,39].

The MaxEnt algorithm [44] is based on the following argument. The user only specifies the length of edges. The stress model, Equation 4.3, assumes that the unknown distance between nonneighboring vertices should be the shortest graph-theoretic distance. This is reasonable, but does add artificial information that is not given in the input. In addition, calculating all-pairs shortest distances for large graphs is costly anyway. On the other hand, specifying only the length of edges leaves too many degrees of freedom in possible node placement. A reasonable yet efficient way to satisfy these degrees of freedom is thus needed. MaxEnt proposed to resolve the extra degrees of freedom in the node placement by maximizing a notion of entropy that is maximized when vertices are uniformly spread out, subject to the edge length constrains. In the final implementation, MaxEnt minimized the following sparse stress function defined on the edges, together with a penalty term that helps to spread out vertices:

$$\min \sum_{\{i,j\}\in E} w_{ij} \left(\|x_i - x_j\| - d_{ij} \right)^2 - \alpha \sum_{\{i,j\}\notin E} ln\,\|x_i - x_j\| \qquad (4.6)$$

It turns out that a solution technique similar to stress-majorization can be employed, except that the matrices involved are all sparse, and the right-hand side can be approximated efficiently using the fast force approximation technique discussed earlier. MaxEnt was found to be more efficient than the full stress model. Although it is not as fast as PivotMDS, MaxEnt does not suffer from the same limitation as PivotMDS on sparse graphs.

There are at least two more recent attempts at a scalable stress model. Very briefly, the MARS algorithm [45] tried to approximate the full stress model by forming an approximate SVD of the off-diagonal matrix of the Laplacian, using a small set of selected columns of the matrix. The COAST [46] algorithms reformulate the stress model so that fast convex optimization techniques can be applied.

4.2.1.3 Other graph layout algorithms for large graphs

ACE. Koren et al. [47] proposed an extremely fast algorithm for calculating the two extreme eigenvectors using a multilevel algorithm. The algorithm is called algebraic multigrid computation of eigenvectors (ACE). Using this algorithm, they were able to lay out graphs of millions of nodes in less than a minute. However, such eigenvector-based algorithms tend to give relatively poor graph layout.

High-dimensional embedding. The high-dimensional embedding (HDE) algorithm [48] finds coordinates of vertices in a k-dimensional space, then projects back to two- or three-dimensional space. First, a k-dimensional coordinate system is created based on k-centers, where the k-centers are chosen as in LandmarkMDS/PivotMDS. The graph distances from each vertex to the k-centers form a k-dimensional coordinate system. The $|V|$ coordinate vectors form an $|V| \times k$ matrix Y, where the ith row of Y is the k-dimensional coordinates for vertex i. The dimension of this coordinate system is then reduced from k to d ($d = 2$ or 3) by principal component analysis, which finds the d largest eigenvectors v_i of $Y^T Y$ (Y is first normalized so that the sum of each column is zero), and uses $Y v_i$ as the coordinates of the final embedding.

HDE has many commonalities to PivotMDS. Due to its reliance on distances from the k-centers, HDE may suffer from the same issue as PivotMDS and LandmarkMDS on sparse graphs. In practice it tends to do worse than PivotMDS on such graphs. Row "HDE" of Figure 4.5 gives drawings using HDE. It performs particularly badly on the qh882 graph, with vertices close to each other, obscuring many details.

4.2.2 Software for large graph layout

While it is not difficult to write simple code to lay out small graphs, for large graphs, a lot of techniques must be used to make the code efficient and effective. Fortunately there are many free noncommercial software programs and frameworks for visualizing and drawing large graphs. The following is a non-exhaustive list that we are aware of:

- Cytoscape [49] is a Java-based software platform particularly popular with the biological community for visualizing molecular interaction networks and biological pathways.

- Gephi [10] is a Java-based network analysis and visualization software package which is able to handle static and dynamic graphs. It is supported by a consortium of French organizations.

- Graphviz [9] is one of the oldest open source graph layout and rendering engines, developed at AT&T Labs. It is written in C and C++ and hosts layout algorithms for both undirected (multilevel spring-electrical and stress models) and directed graphs (layered layout), as well as various graph theory algorithms. Graphviz is incorporated in Doxygen, and available via R and Python.

- NetBioV [50] is an R package that allows the visualization of large network data in biology and medicine. The purpose of NetBioV is to enable an organized and reproducible visualization of networks by emphasizing or highlighting specific structural properties that are of biological relevance.

- OGDF [11] is a C++ class library for automatic layout of diagrams. Developed and supported by German and Australian researchers, it contains a spring-electrical model–based algorithm with fast multipole force approximation, as well as layered, orthogonal, and planar layout algorithms.

- Tulip [12] is a C++ framework from the University of Bordeaux I for development of interactive information visualization applications. One of the goals of Tulip is to facilitate the reuse of components; it integrates OGDF graph layout algorithms as plugins.

There is other free software each with its own unique merits, but they may not be designed to work on very large graphs. For example,

- D3.js [51] is a JavaScript library for manipulating web documents based on data. Within D3 are spring-electrical model–based layout modules solved by a force-directed algorithm. D3 makes it easy to take advantage of the full capabilities of modern browsers, making the resulting visualization highly portable. D3 is developed by Michael Bostock, Jeffrey Heer, and Vadim Ogievetsky at Stanford University.

- Dunnart [52] is a C++ diagram editor. Its unique feature is the ability to lay out diagrams with constraints, and it has features such as constraint-based geometric placement tools, automatic object-avoiding polyline connector routing, and continuous network layout. It is developed at Monash University, Australia.

An important resource for testing algorithms and systems is real-world network data. There are a number of websites that host large graph data from real-world applications. For example, the University of Florida Matrix Collection [36] contains over 2700 sparse matrices (which can be considered as the adjacency matrix of graphs) contributed by practitioners from a wide range of fields, with the oldest matrices dating as far back as 1970. The Stanford Large Network Dataset [53] hosts about 50 networks from social networks, web crawls, and others. The Koblenz Network Collection [54] has about 150 large networks. GraphArchive [55] contains many thousands of smaller graphs collected by the graph drawing community. House of Graphs [56] offers graphs that are considered interesting and relevant in the study of graph-theoretic problems.

4.3 Visual Abstraction of Large Multivariate Networks

In this section, we look at another class of large complex network visualization methods that do not display the entire graph in full detail like the graph drawing methods. Instead, the graph data has first undergone some kind of transformations into smaller and simpler abstractions. These graph abstractions generally fit into the feasible set for classical graph drawing methods and can be displayed in readable visual forms. To study particular details on the visual abstraction, network interactions, such as various types of navigation methods, are often integrated to support the user-driven visual analysis.

The idea of assembling visual abstractions on networks is not new to the visualization community, and there are quite a few studies on this topic. Here we describe three of them based on graph clustering, graph simplification and multivariate network approaches, respectively. Interested readers are pointed to the related surveys [57–59] for more information.

4.3.1 Hierarchical clustering–based large network visualization

Clustering is a popular unsupervised data mining method to learn a partition of the data set. On graphs and networks, it generally works by first identifying a similarity measure over graph nodes, such as the distance-based [60] and topology-based measures [61]. Then these nodes are divided into a few groups by clustering algorithms. The visualization of these clustered graphs is promising for users in that they can reveal the high-level structure of the graph data. However, classical graph clustering algorithms (e.g., k-means [62]) create only one level of detail and do not allow the fine-grained navigation of these graphs. We describe the hierarchical interactive map HiMap [63] for large networks that builds over the hierarchical graph clustering algorithms and supports both flexible visual complexity control and interactive navigation over the visualization.

Framework and algorithm. Figure 4.7a shows the visualization pipeline for HiMap. It could be divided into two separate stages: offline data manipulation and online

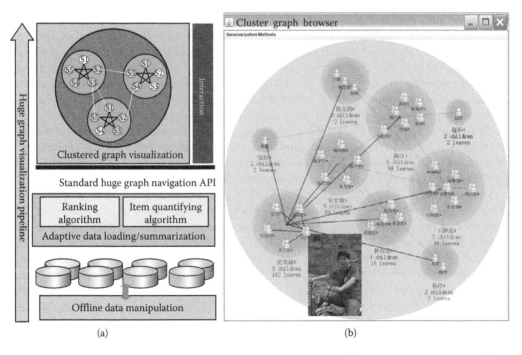

(a)   (b)

FIGURE 4.7: HiMap large network visualization approach: (a) the system pipeline; (b) the visual design in displaying the online social network of students in a university's department. (From Lei Shi et al., *Proceedings of the IEEE Pacific Visualization Symposium*, pages 41–48, 2009 [63].)

adaptive visualization. The offline data manipulation stage involves data collection, cleaning, and the hierarchical graph clustering. This stage prepares the network data required for the on-line visualization of large networks. The adaptive visualization stage involves data loading, graph layout, projection, and rendering. Different from the classical visualization methods that show all nodes of the large network in one view, the HiMap approach clusters the entire network into a hierarchical tree and only presents the nodes within a certain depth from the root node of the current view. This strategy greatly reduces the visual clutter commonly found in large network visualization design. Moreover, through a novel adaptive data loading method, the visual complexity of each network view is maintained within the capability of human perception. In this way, the amount of data required to load upon each view transition is also kept small by the screen's constraint, so that users do not find significant lag in the experience, but only the smooth animation designed to assist in network comprehension. Data caching is also maintained to store all the abstracted network data that users have accessed in their navigation history. When users switch back to one of the previous views, no extra time and cost is required for the repeated loading.

In particular, HiMap adopted the hierarchical graph clustering algorithm by Newman and Girvan [64], which can group all nodes in a graph into a binary tree in an agglomerative manner. To obtain the balanced hierarchical clustering structure, this algorithm is invoked recursively: (1) All the nodes are grouped into the binary tree until the termination condition of the algorithm is triggered (i.e., the delta Q value decreases below zero). The remaining groups become the clusters in the top hierarchy. (2) The subgraph of each such cluster is processed separately to obtain the subclusters in the next hierarchy. (3) This process works iteratively until the predefined maximal tree depth is reached. Finally, a static tree structure is generated over the large network.

On the clustering tree of large networks, we further apply summarization algorithms to generate the view for visualization. The objective is to maintain an interpretable visual density and readability in the final view based on the given screen real estate. The algorithm involves three steps. First, the nodes/clusters under the same father node are ranked by their customized importance. Two ranking algorithms are supported: one seeks to maximize the coverage on the subgraph given the top-ranked nodes/clusters. The second algorithm ranks the nodes/clusters according to their clustered betweenness centrality, so that the brokers connecting clusters together are displayed with higher priorities for network diagnosis purposes. The ranks with both algorithms are precomputed so that users could switch to a new configuration without any delay. Second, based on the current screen real estate and the user specification, the number of visual items to display in the entire view is computed. Third, a recursive allocation algorithm is run to determine the number and the set of displayed clusters/nodes in each hierarchy.

Visual design. Figure 4.7b shows the HiMap visualization of an online social network with more than 2000 users. These users are from the same department of a university. The entire view is composed of multiple filled circles enclosing each other. The largest circle in the view is drawn as the background canvas, which denotes the root node of the clustering tree, and the smaller circles indicate the first-level clusters in this tree. The fill color is selected in an ascending saturation from the clusters in the top hierarchy to the bottom hierarchy, so that the lower hierarchies of the graph are drawn in the foreground to display the fine details. Within the lowest-level hierarchy in the view, there are icons to represent the node/cluster within it. The group icon indicates the intermediate cluster in the tree, with the "+" sign following their node labels. The single icon indicates the leaf node in the tree. In the current figure, the view is set to a maximal visible depth of two.

By default, the edges between any two leaf nodes are drawn in the view by straight lines. To reduce the visual clutters common in the densely connected graph, two edge bundling methods are applied: geometric edge bundling and hierarchical edge bundling. Geometric bundling implements the solution in [65]. Some control points are selected in the graph according to the topology, then all the edges are forced to traverse these points. The other method bundles all the edges between any two upper-hierarchy clusters together into one aggregated edge. The intra-cluster edges inside each cluster are not changed.

Interaction. HiMap supports a suite of interactions to inspect the details of the large network. For example, as shown in Figure 4.7b, users can mouse over a node/cluster in the view to show the corresponding profile of the representative leaf node (in this data set, the photo of the user). All the neighboring nodes and edges are highlighted to facilitate the user's analysis. More importantly, HiMap provides navigation interactions to visually traverse the large complex network based on its clustering tree. The basic interaction method is the hierarchical drill-in/roll-up. By double-clicking on one cluster in the view, the subgraph under this cluster is summarized again and shown in the entire view, with more visual items for the detail. This is demonstrated in Figure 4.8. In the first stage of the interaction, the selected cluster in the blue frame is expanded geometrically, while the other clusters in the same hierarchy fade out gradually. In the second stage, this selected cluster becomes the root node and fills the entire canvas, while new clusters start to emerge under the selected cluster. In the third stage, subclusters/nodes appear within the new clusters to fill up the view and ensure the visual complexity requirement. Similarly, the view can be rolled up to go back to the upper hierarchy by double-clicking on the current canvas. Animated transitions are used during these interaction stages to provide a smooth viewing experience.

4.3.2 CompressedGraph: Large network visualization by graph simplifications

While graph clustering–based visualization is a good fit for large-scale social networks, the same method can be inadequate on other domains when the network does not necessarily

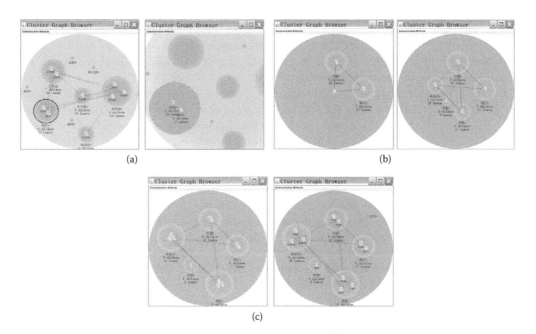

(a) (b)

(c)

FIGURE 4.8: The drill-in interaction in HiMap; the circled frame in (a) indicates the selected cluster to drill-in: (a) Stage I; (b) Stage II; (c) Stage III. (From Lei Shi et al., *Proceedings of the IEEE Pacific Visualization Symposium*, pages 41–48, 2009 [63].)

have community structures. A typical case is the network traffic graph in a lab setting. Each host in the lab (the network node) can have more connections to the external network, for example, by surfing the web, than the internal connections. Figure 4.9b illustrates the visualization result using the clustering–based method over the original traffic network in Figure 4.9a. The clustered view does not reduce the visual complexity for user analysis, and even worse, the original network topology is destroyed to some extent. In this part, we describe a new graph simplification method, called the CompressedGraph [66], based on the concept of structural equivalence in social network analysis [67]. Rather than detecting communities, the structural equivalence method classifies the network nodes into categories by their positions taken in the network, depending only on the network topology. On large networks in domains such as the computer network, the visualization result can be much effective than the clustering–based method. As shown in Figure 4.9c, the 3460-node network traffic graph is reduced to 18 grouped nodes and 28 edges, while the same graph using modularity-based clustering [64] generates 50 intermediate clusters (with a maximal depth setting of four), 939 leaf nodes and 15, 438 edges in the top view.

We should note that the similar idea of visually compressing graphs by exploiting the topology redundancy has also been studied in a few other studies [68–70]. Similar visual effects can be achieved; however, they vary in the compression rate and the applicability to different domains.

Framework and algorithm. The main algorithm in the CompressedGraph is called structural equivalence grouping (SEG). SEG has a time complexity linear to the number of edges in graphs. Also, SEG enjoys an advantage in that it can support not only the undirected graphs, but also the directed, weighted graphs, and provide a fuzzy version of the algorithm to control the visual complexity after the compression.

Intuitively, SEG aggregates the graph nodes with the same neighbor set together into groups and constructs a new graph for visualization. For example, in the traffic graph of Figure 4.10a, the host "192.168.2.23" can be combined with the other three surrounding hosts with exactly the same connection pattern. The new graph after SEG processing (Figure 4.10b) is called the *compressed graph*. The compressed graph has two kinds of nodes: the *single-node* remaining from the original graph (drawn as hollow) and the *mega-node* grouped from multiple *subnodes* in the original graph (drawn as filled colors). In the following, the SEG algorithm is formally defined.

Let $G = (V, E)$ be a directed, weighted, and connected graph where $V = \{v_1, ..., v_n\}$ and $E = \{e_1, ..., e_m\}$ denote the node set and the link set. Let W be the graph adjacency matrix where $w_{ij} > 0$ indicates a link from v_i to v_j, with w_{ij} denoting the link weight. In each row of W, $R_i = \{w_{i1}, ..., w_{in}\}$ denotes the row vector for node v_i, representing its connection pattern. The compressed graph after SEG is denoted as $G^* = (V^*, E^*)$. The compression rate is defined by $\Gamma = 1 - |V^*|/|V|$.

The basic SEG algorithm takes the graph as a simple, undirected, and unweighted one by setting $w_{ii} = 0$ and $w_{ij} = w_{ji} = 1$ for any $w_{ij} > 0$. On a graph G, for any collection of nodes with the same row vector (including the single outstanding node), SEG aggregates them into a new mega-node/single-node $Gv_i = \{v_{i_1}, ..., v_{i_k}\}$. All Gv_i form the node set V^* for the compressed graph G^*. Let $fv_i = v_{i_1}$ denote the first subnode in Gv_i. The link set E^* in G^* is generated by replacing all fv_i with Gv_i in the original link set, and removing all the links not incident to any fv_i. For directed graphs, the adjacency matrix W is transformed to encode the two connection directions for each node. Each row vector $R_i(i = 1, ..., n)$ becomes $R_i = \{w_{i(-n)}, ..., w_{i(-1)}, w_{i1}, ..., w_{in}\}$ with $w_{i(-j)} = w_{ji}$ for $j = 1, ..., n$. For weighted graphs, the adjacency matrix W is switched to the weighted one by mapping a numeric data attribute of the link (i, j) (e.g., flow count in the traffic graph) to w_{ij} in the matrix. To further increase the compression rate, the discretization of the link weight

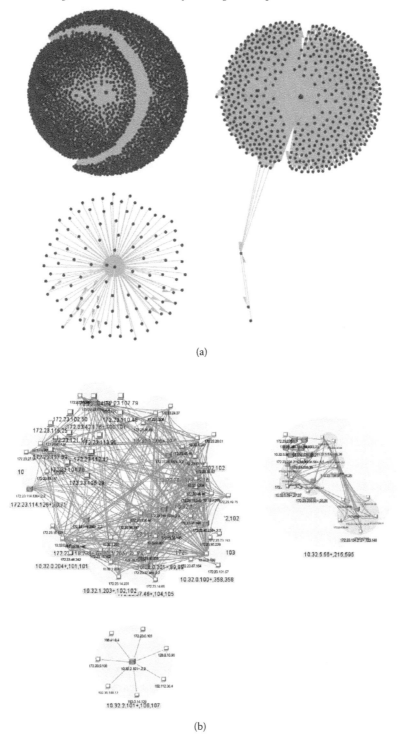

(a)

(b)

FIGURE 4.9: A comparison of large network visualization methods on the VAST 2012 Mini Challenge-II network traffic graph (3460 nodes, 48,599 edges): (a) Original view; (b) clustered view. (From Lei Shi et al., *IEEE International Conference on Big Data*, pages 606–612, 2013 [66].) *(Continued)*

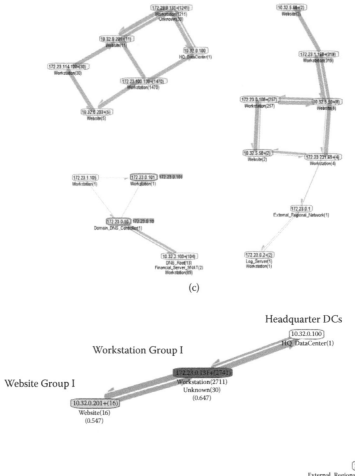

(c)

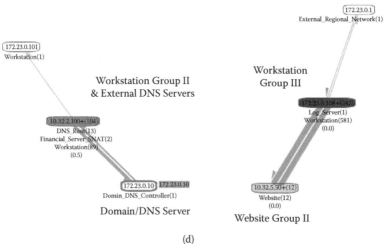

(d)

FIGURE 4.9: (*Continued*) A comparison of large network visualization methods on the VAST 2012 Mini Challenge-II network traffic graph (3460 nodes, 48,599 edges): (c) SEG-compressed view; (d) manually-compressed view over SEG. (From Lei Shi et al., *IEEE International Conference on Big Data*, pages 606–612, 2013 [66].)

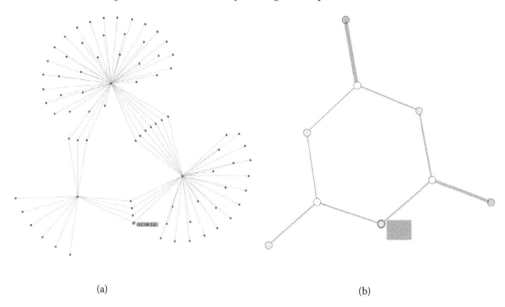

(a)   (b)

FIGURE 4.10: The result of the SEG algorithm on an undirected graph (a) original graph and (b) compressed graph. (From Lei Shi et al., *IEEE International Conference on Big Data*, pages 606–612, 2013 [66].)

is allowed: first, all link weights are transformed into $w_{ij} \in (0,1]$ by either the linear or nonlinear normalization; second, a bin count $B(B \geq 1)$ is picked to regenerate link weights by $w_{ij} = \lceil w_{ij} \times B \rceil$.

SEG is a single-pass, deterministic algorithm in that for the same original graph, it always produces the same compressed graph. In real usage, the user would like to flexibly control the visual complexity after the compression. This can be achieved through the fuzzy SEG algorithm. The basic idea is to group nodes with not only the same, but also similar neighbor sets. The compression rate can be increased with a bounded compensation on the accuracy. The key is to define the pairwise similarity score between graph nodes. The standard Jaccard similarity can be adopted, that is, between two sample sets A and B we have $J(A, B) = \frac{|A \bigcap B|}{|A \bigcup B|}$. For the directed and weighted graphs, we introduce a unified Jaccard similarity computation between node v_i and v_j in the graph G by $\rho = \frac{\sum_{\forall k} \min(w_{ik}, w_{jk})}{\sum_{\forall k} \max(w_{ik}, w_{jk})}$. Note that for the directed graph, $k = -n, ..., -1, 1, ..., n$. The fuzzy SEG is achieved by setting a similarity threshold ξ. The pair of nodes with $\rho \geq \xi$ is grouped together iteratively.

Visual design. The right panel of Figure 4.11 gives an example of the SEG-compressed graph visualization. As shown in the figure, the single-nodes have no fill color and the mega-nodes have solid fill colors, with the color saturation mapped to the number of subnodes in the group. The larger group is filled with the more saturated color. By default, the fill color hue is blue, for example, the node "192.168.1.10+" in the top-right of Figure 4.11. For the mega-nodes created by the fuzzy SEG, for example, the node "192.168.2.11+" on the right part of of Figure 4.11, the fill color is blue-green, indicating a pairwise similarity score below one. The fill color will gradually change from blue to green and finally to brown, as the similarity score drops from one to zero.

The labels on the mega-node are created by aggregating the labels of the subnodes in the original graph. Due to the space limitations of the graph view, an abstracted label is displayed on each mega-node as the node identifier. The full label will pop up upon a mouse hovering or a click action, for example, the one on the node "192.168.1.10+." The

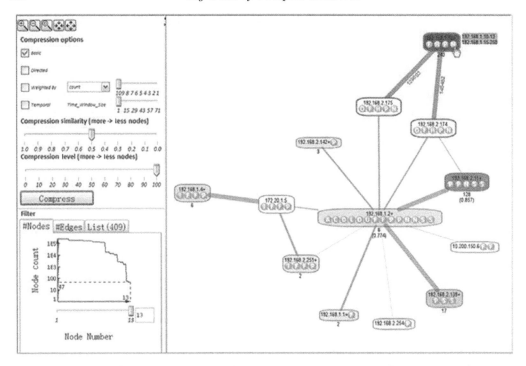

FIGURE 4.11: The compressed graph visualization interface. The main graph view is on the right, and the algorithm controller is on the left. The data set is from the VAST Challenge in 2011. (From Lei Shi et al., *IEEE International Conference on Big Data*, pages 606–612, 2013 [66].)

group size of each mega-node is drawn below the node, together with the within-group similarity score when it is below one. By default, straight lines are used to represent the links in the compressed graph, with the line thickness mapped to the sum of counts of all the corresponding links in the original graph. The node/edge attributes on compressed graphs can also be mapped onto the visuals. For example, in the case of the traffic graph in the security domain, the node label can be the host types (Figure 4.9c) or the alphabetical anomaly icons (Figure 4.11) indicating the detected type of anomalies on the host.

Interaction. The CompressedGraph visualization supports most basic graph interactions. In addition, it provides user control over the SEG algorithm settings. In the control panel of Figure 4.11, the "Compression level" section lists multiple checkboxes as switches for the undirected, directed, and weighted SEG. For the weighted SEG, the link weight mapped from the edge attribute of the graph can be specified. To use the normalized link weight, a bin number can be specified to discretize the weight. In the "Compression level" and "Compression similarity" sections, the compression rate can be tuned to generate smaller or larger graph abstractions through the fuzzy SEG algorithm.

Complementary to the automatic SEG operation, CompressedGraph introduces manual node grouping/splitting interaction. In many cases, the users have their own criteria towards a best graph abstraction. The user can either select a collection of nodes and click the "group" button in the navigation panel, or use drag-and-drop to group one node into another one at a time. In the drag-and-drop process, the pairwise similarities of the selected node with all the other subnodes in the candidate group are shown as the visual hint. On the other hand, the user can either select the mega-nodes and click the "split" button, or simply double-click on one mega-node. The mega-node grouped by the fuzzy SEG will first collapse to mega-nodes by the deterministic SEG, and upon another double-click, collapses to the original subnodes. Figure 4.10a gives an example of the manually created abstraction over the SEG algorithm, which further reduces the network size to 9 grouped nodes and 11 edges.

The proposed SEG-based compressed graph visualization method has been integrated into a tool called network security and anomaly visualization (NSAV) to address the problem in the network security domain, which will be detailed in Section 4.4.

4.3.3 OnionGraph: Multivariate network visualization by topology+attribute abstractions

The aforementioned visual abstraction methods can be effective in displaying large-scale networks. However, the modern networks are often multivariate in as there are multiple attributes on the network nodes and edges. Some networks are even heterogeneous as their nodes and edges are of different types. For example, in a bibliographic information network, a node can be an author with an affiliation, a paper with a certain topic, or a venue (i.e., conference/journal) held in some location. An edge can represent the relationship of citations, authorizations, presentations, and so on. On these multivariate networks, except for visualizing the network topology, how to display the network node/edge attributes and the correlations between these attributes and the network topology emerges as a new problem. This is particularly challenging because in the graph drawing approaches, the node locations have traditionally been used to represent the network topology. While a few visual channels, for example, the node shape, fill color, and edge thickness are available to display the node/edge attributes, the constraints on human perception and the nature of high-dimensional information on networks make direct encoding methods infeasible for many scenarios. In this part, we describe OnionGraph [71], an integrated framework for the exploratory visual analysis of large multivariate networks. OnionGraph creates several hierarchies on the multivariate network by combining the topology and attribute information on these networks. Through visual abstractions based on the network hierarchies and interactions similar to the compressed graph, OnionGraph supports level-of-detail viewing on multivariate networks.

Framework and algorithm. OnionGraph introduced a stratified semantic+topological principle. At high levels, the large network is aggregated by a combination of semantic information (the node type and attributes). Interesting portions of the abstraction can be drilled down in situ by exploiting topological features. After steps of user navigation, a stratified network view with different levels of abstraction is created on demand to serve the complicated heterogeneous network analysis tasks. Sketched examples are given in Figure 4.12, which shows the five-level hierarchical structure of OnionGraph. The design features the semantic+topological principle: the semantic aggregations (SA) in level-I and level-II are purely semantic, the level-III RRE combines semantic and topological information, and the level-IV (SSE) is mostly topological. In the following, we formally present the suite of OnionGraph algorithms to create the hierarchical abstraction over multivariate networks.

Heterogenous network. Let $G = (V, E)$ be a directed and weighted multivariate network. $V = \{v_1, ..., v_n\}$ and $E = \{e_1, ..., e_m\}$ denote the node and link sets. W denotes the adjacency matrix where $w_{ij} > 0$ indicates a link from v_i to v_j, with w_{ij} denoting the link weight. On each node v_i, $N^+(v_i) = \{j|w_{ij} > 0\}$ and $N^-(v_i) = \{j|w_{ji} > 0\}$ indicate the outbound and inbound neighborhood set, both representing its connection pattern. Let $D = \{d_1, ..., d_s\}$ be the type and attribute of network nodes in G, with s dimensions in total. $D(v_i) = \{d_1(v_i), ..., d_s(v_i)\}$ denotes the type/attribute values of the node v_i, with $d_k(v_i)$ indicating the value in the kth dimension.

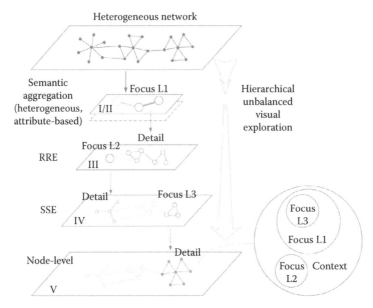

FIGURE 4.12: OnionGraph structure featuring five hierarchies below the original network: networks by semantic aggregations (SA) on the node type (heterogeneous abstraction) and the node attributes, relative regular equivalence (RRE), strong structural equivalence (SSE), and the single-node-level network in the finest granularity. In each hierarchy, the network can be expanded at certain points into their lower-hierarchy details. (From Lei Shi et al. *IEEE Pacific Visualization Symposium*, pages 89–96, 2014 [71].)

Network partition. Let $P : V \rightarrow \{1, 2, ..., t\}$ be a partition (role assignment, coloration, grouping, interchangeably) of the network G into t subgroups of nodes. $P(v_i)$ indicates the partition index of node v_i.

The OnionGraph algorithms to create a network abstraction is equivalent to finding a partition of the network according to the abstraction settings.

SA creates a partition of the network using a selected set of node types or attributes. Formally, for any nodes v_i and v_j in a network G, given the selected attribute set $\overline{D} \subseteq D$, the SA network partition P satisfies:

$$\overline{D}(v_i) = \overline{D}(v_j) \Leftrightarrow P(v_i) = P(v_j) \tag{4.7}$$

Figure 4.13a illustrates a partition based on the node attribute having values "I" or "II."

RRE [72] is defined recursively on the network node by the same set of neighborhood roles. For any nodes v_i and v_j in a network G, a regular equivalence network partition P satisfies:

$$P(v_i) = P(v_j) \Rightarrow P(N^+(v_i)) = P(N^+(v_j)) \;\; and \;\; P(N^-(v_i)) = P(N^-(v_j)) \tag{4.8}$$

However, directly applying the regular equivalence on a network will lead to many possible partitions. Figure 4.13b gives a particular case. In the extreme case, the identity partition (every node serves a different role) and the complete partition (every node serves the same role) are both regular. It is hard to find an appropriate regular equivalence partition in the real usage. OnionGraph introduces a practical solution to apply RRE, which can work on top of the existing SA partition. RRE explicitly derives the maximal regular equivalence

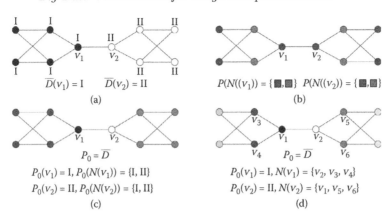

FIGURE 4.13: OnionGraph network partitions (undirected case). In each subfigure, the node fill color indicates the partition index: (a) semantic aggregation (the selected attribute value is labeled on the node); (b) a regular equivalence partition; (c) the regular equivalence relative to the semantic aggregation in (a); (d) SSE. (From Lei Shi et al. *IEEE Pacific Visualization Symposium*, pages 89–96, 2014 [71].)

partition by refining the semantic partition. Mathematically, the RRE partition P over a semantic partition P_0 satisfies:

$$P(v_i) = P(v_j) \Leftrightarrow P_0(v_i) = P_0(v_j) \quad and$$
$$P_0(N^+(v_i)) = P_0(N^+(v_j)) \quad and \quad P_0(N^-(v_i)) = P_0(N^-(v_j)) \tag{4.9}$$

Figure 4.13c gives an example of the RRE partition relative to the semantic partition in Figure 4.13a.

SSE partition [67] requires the network nodes to have exactly the same neighborhood set. For any nodes v_i and v_j in a network G, the SSE network partition P over a RRE partition P_0 satisfies:

$$P(v_i) = P(v_j) \Leftrightarrow P_0(v_i) = P_0(v_j) \quad and$$
$$N^+(v_i) = N^+(v_j) \quad and \quad N^-(v_i) = N^-(v_j) \tag{4.10}$$

Besides the standard definition, there are a few variations of the RRE/SSE partitions. The undirected RRE/SSE (Figure 4.13) considers the union of both inbound and outbound neighborhood sets, and the weighted RRE/SSE considers the number of role occurrences in the neighborhood (RRE) and the weight of the connecting edges (SSE). These options are included in the OnionGraph design and can be configured by users.

Visual design. A typical OnionGraph visualization is shown in Figure 4.14. Each colored node represents a group of the original nodes in the large network. The node size encodes the number of individual nodes in each group. The initial abstraction aggregates all the nodes by their types ("author," "paper," and "venue" in this figure), as indicated by the icon on the top-right part of each node. The node group in the top-level heterogeneous abstraction is displayed by a filled node, for example, the spring-green node in the center of Figure 4.14, representing all 9557 papers. All the other nodes in this figure are expanded from the top level. They are drawn by the "onion" metaphor composed of several concentric circles. The number of circles indicates the abstraction hierarchy: the SA has three circles (venue nodes in Figure 4.14), RRE has two circles (author nodes in Figure 4.14), and SSE has one circle. The individual node only remains a solid dot. Upon the top-down exploratory analysis, the visual complexity of each node group is reduced as the number of node groups increases, so as to balance the overall complexity of the OnionGraph view.

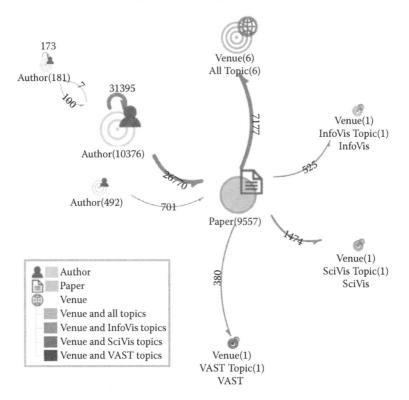

FIGURE 4.14: An OnionGraph network visualization of the author-paper-venue bibliographic network in the visualization community. Three author groups indicate the different connection patterns: normal authors with co-authors and publications, special authors who only write single-authored papers, and anomalous authors without a publication (potential errors in the data set). Four venue groups indicate the conferences/journals on different topics. (From Lei Shi et al. *IEEE Pacific Visualization Symposium*, pages 89–96, 2014 [71].)

Node color in OnionGraph is determined by the type/attribute values of each node group. In Figure 4.14, initially three colors (yellow, spring-green, indigo) are picked uniformly on the color hue. After the expansion of the venue node into subnodes, four new colors are assigned with the linear hue and saturation offsets from the indigo color, as shown by the legend in the bottom-left part of Figure 4.14. The node labels by default show the value of the currently selected node type/attributes. When a node group contains only one node, the detailed node title is also shown in the label. The selected nodes are drawn with dark-red outlines and labels, coupled with a "+/-" sign upon hovering to indicate the lower/upper level to explore. The neighborhood of the selected nodes and their connecting links are drawn in dark orange. The link thickness and the link label encode the number of individual links between the groups. Different from the simple graphs, OnionGraph usually has a loopback link on each node group, which indicates the internal connection. This is displayed by the arc above the node.

Interaction. The typical OnionGraph interaction works like this: the user selects a portion of interesting subgraphs in the network, specifies an abstraction profile, and finally executes the OnionGraph algorithm to display the finer/coarser-grained visual representation on these selected subgraphs. The network selection is supported in multiple ways, through single-clicks on network nodes, rubber band selection/deselection, and Ctrl plus single-click to select a node set with the same abstraction profile. The abstraction profile is configured in the control panel on the left side of the interface. The control panel includes the abstraction level control, the attribute multiselection, and the switches indicating the abstraction settings, for example, directed, weighted, and fuzzy RRE/SSE algorithms. The selected subgraph is processed by clicking the abstract button. In another usage, users can double-click on the selected node set in the main network view to expand to the next level abstraction. When the context is set to the "collapse" mode, the selected node set is regrouped into the upper-level abstraction.

From the menu of the OnionGraph interface on the top, a few configurations such as the layout algorithm, and the network node/link visual encoding, can be specified. OnionGraph also allows the neighborhood charting mode. Each node group aggregated by RRE is drawn by a chart instead of the onion metaphor, showing the distribution of attribute values in the node's neighborhood. The bottom-left filter panel allows the user to plug in node attribute filters on the network. The OnionGraph abstraction is executed over the filtered network. On the network link, a simplified filtering mechanism is supported by a multi-checkbox interface. The rightmost part of the OnionGraph interface shows the network details upon visualizations and interactions. The top-right panel displays the node legend indicating the icon/color coding of each node group in the network. The center-right panel displays the list of nodes currently selected in the main view. Upon choosing one node in the list, the node attributes are displayed in a key-value table in the bottom-right panel.

4.4 Applications of Large Complex Network Visualization in Security

4.4.1 Compressed graph for identification of attacks

Large-scale network graph analysis and visualization has been applied and shown to be effective in computer security and network management. One example is NSAV [66]. It is challenging to visualize network traffic graph due to the sheer volume and large number of hosts and network connections among them. NSAV looks to reduce the number of nodes (and their connections), thus reducing the visual complexity for network administrators and operators to analyze their networks.

4.4.1.1 Situation awareness

Consider John, the network operation lead of a large enterprise, who collects the Netflow firewall log on a daily basis; he is checking the corporate network status of the recent three days for noteworthy events. He starts by loading all of the network traffic in this period, as shown in Figure 4.15a. Because the view is too messy, he continues by creating a compressed traffic graph, as shown in Figure 4.15b. The comparison of the two graph visualizations is quite significant as the second graph is much smaller and easier to visualize.

4.4.1.2 Port scan and OS security holes

Based on his network structure, John bypasses three server machines (1.2, 1.6, 1.14), which routinely communicate with all the hosts for DNS/data services. Also according to his knowledge, the suspicious behaviors of a hub node, for example, port scan, is often associated with OS security holes. Therefore, he clicks on this anomaly type and highlights all the hosts with such an anomaly on the graph. To drill-down to individual hosts, he splits the fuzzy group and locates 2.171–2.175 as the threats. He finds that 2.174/175 are more

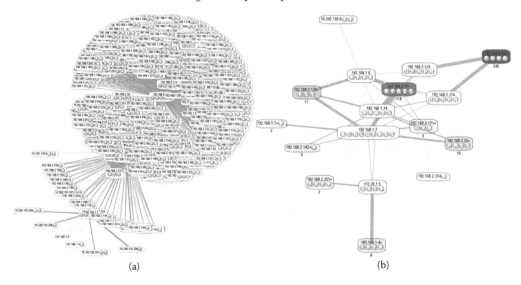

(a)   (b)

FIGURE 4.15: Corporate network traffic graphs using the NSAV compression tool: (a) Original and (b) compressed. (From Lei Shi et al., *IEEE International Conference on Big Data*, pages 606–612, 2013 [66].)

dangerous due to the higher port-scan rate (by the link thickness) and the cross-subnet floods to 1.10-250 where many hosts do not exist. The screenshot with anomaly views of 2.174/175 is given in Figure 4.16.

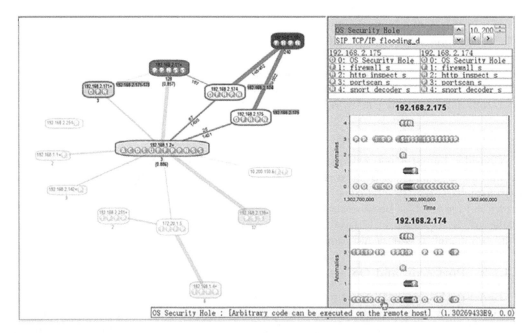

FIGURE 4.16: The hosts with security holes and the cross-subnet port scans from 192.168.2.174/175. (From Lei Shi et al., *IEEE International Conference on Big Data*, pages 606–612, 2013 [66].)

4.4.1.3 DoS attacks

One critical machine that John examines is the corporate external web server (172.20.1.5). With the compressed graph, the server is easily located by its unique connection patterns. A single click on the host shows up a noteworthy anomaly icon (I) on the morning of the first day, suggesting that there could be Denial-of-Service (DoS) attacks through SIP. John further drills down to that period with the time range selector and highlights the web server's egocentric traffic graph. Figure 4.17 confirms the potential DoS attacks from the external hosts 10.200.150.201, 206–209, through the anomalies happening simultaneously with the web server.

4.4.1.4 Botnet detection

In this example, NSAV is able to detect malware infection and botnet behavior via Internet Relay Chat (IRC) communication channels. In the first subgraph of the network traffic, as in Figure 4.18, it is identified that the I and M icons appeared frequently and almost in couples in reverse directions. A selection of the $IRC-Malware-Infection\_s$ anomaly (icon I) in the anomaly type list reveals three group of hosts highlighted in red in Figure 4.18. They are all workstations having enormous IRC connections to a portion of 12 websites (10.32.5.*), with the potential to be compromised botnet clients. Further selecting two typical workstations (172.23.123.105, 172.23.231.174) and websites (10.32.5.50, 10.32.5.52), their temporal anomaly distributions are plotted in the right panel of Figure 4.18.

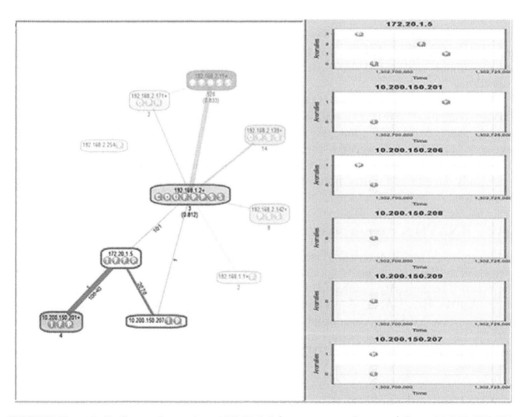

FIGURE 4.17: DoS attacks against 172.20.1.5 (corporate web server) from 10.200.150.201, 206–209. (From Lei Shi et al., *IEEE International Conference on Big Data*, pages 606–612, 2013 [66].)

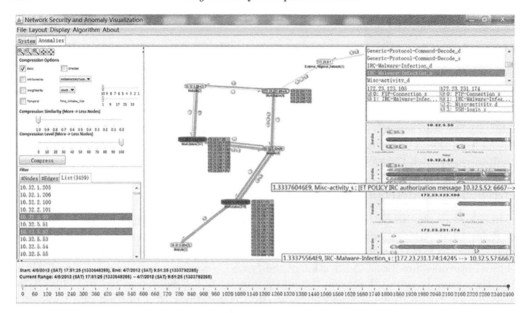

FIGURE 4.18: Three groups of machines in heavy IRC traffic with the websites through port 6667 suggesting suspicious botnet infections. (From Lei Shi et al., *IEEE International Conference on Big Data*, pages 606–612, 2013 [66].)

It is shown that the IRC traffic exchanged with the websites overwhelm in the whole inspected period. Note that nine of the websites (10.32.5.51-59) reply with the IRC authorization message (icon *M*), indicating the establishment of potential botnet server-client connections.

The host 172.23.231.174 (all-time IRC client) shows fine-grained patterns: the connections are composed of two temporal stages, indicated by a small gap in the middle of the anomaly panel. A drill-down analysis at this gap shows that the first stage ends up with a large port (43325) and the second stage starts with a small port (1185). After checking the anomaly file, we conclude that the IRC traffic from the workstations are programmed, with sequentially enumerated source ports. It verifies the hypothesis that these hosts have been compromised as botnet clients.

4.4.2   ENAVis for enterprise network security management

Enterprise Network Activities Visualization (ENAVis) [73] is a tool for visualizing the network activities among hosts (domains), users and applications (HUA) graph model. The network connectivity graphs can be built based on the various combinations of states (H, U, A, HU, HA, UA, HUA). The hierarchical, heterogeneous graph representation of HUA data, enabled by the local context information collected from enterprise networks, captures the dynamic relationship and interaction between machines and user applications. Figure 4.19 illustrates such HUA graph visualization in enterprise network security management.

4.4.2.1   User and application-level policy compliance

In this example, suppose an enterprise management team needs to know whether their employees have complied with the company's network usage policy and whether the current mechanisms are adequate for enforcing the policy. A financial intranet server's access policy

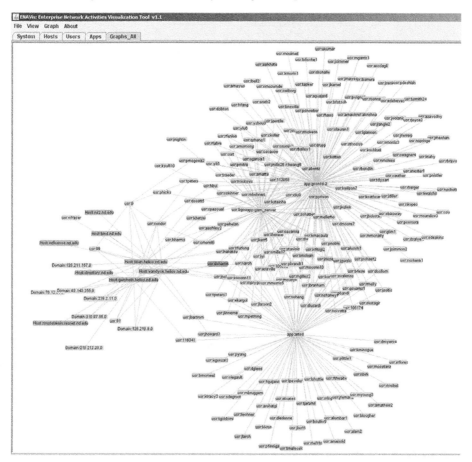

FIGURE 4.19: Hierarchical host-user-application (HUA) graph model in the *ENAVis* tool. (From Qi Liao et al., *Proceedings of the USENIX 22nd Large Installation System Administration Conference (LISA '08)*, pages 59-74, San Diego, CA, November 9–14 2008 [73].)

is defined such that only authorized users can access or even see the financial system. To that end, a set of host-based firewall rules are put in place on the finance server (host name *finance*) with restrictions to the hosts (host name *concert*) of authorized finance personnel on the company campus.

Since host to user mapping is dynamic, the notion of host as identity quickly breaks down (see Figure 4.20). Suppose in the same environment that *ssh* connectivity was enabled on the network. In the scenario of Figure 4.20, an unauthorized user *qliao* connects from *IrishFB.nd.edu* with X11 forwarding to *concert.cse.nd.edu* and launches an instance of *firefox* to access the finance web server. In a multiuser environment, where multiple users are logged on to the same machine and make network connections, other tools have no way of differentiating those connections because the connections all have the same source IP. Similarly, the legitimate user *striegel* may carelessly connect from a *Starbucks* shop to his office desktop *concert* in order to access a financial account just for convenience. Neither of these two cases is desirable and is a violation of the policy because the original intent of the policy was that anyone not part of the finance department should not be able to access the finance host.

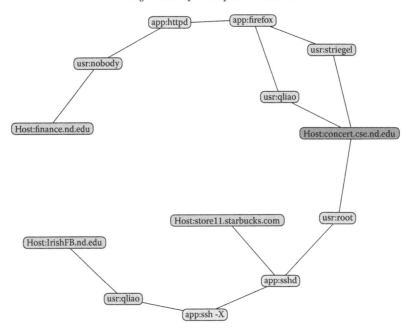

FIGURE 4.20: HUA graph captures two possible host IP ACL policy violations caused by the unintentional configuration on host *concert* without `ssh` restriction. (From Qi Liao et al., *Proceedings of the USENIX 22nd Large Installation System Administration Conference (LISA '08)*, pages 59–74, San Diego, CA, November 9–14 2008 [73].)

4.4.2.2   Machine compromise and attack analysis

In this example, a phishing e-mail pretending to be from the IT department in an enterprise network claims the system has been upgraded and requires all users to send in their passwords, or their accounts will be suspended. Some users are suspected to be victims of the scam and may have their system compromised.

An administrator generates HUA network graphs and highlights the problem user node (*jdoe*) (as in Figure 4.21). It is straightforward to see which hosts the user has touched during the time frame and what applications the user used. The file access information logged by `lsof` is kept in the master database and a single query would reveal all files that user ID has touched among all the hosts. In this case, a visual HUA graph is helpful to see what hosts and users that a compromised user account has contacted and which applications it has attempted to launch. This helps expedite significantly such an investigation should it occur.

4.4.3   OnionGraph for heterogeneous networks

OnionGraph [71] allows a hierarchical grouping of nodes based on not only topological structure, but node attributes as well. The graph model is useful in a network management setting. Similar to the earlier mentioned HUA graph model, OnionGraphs can be applied to visually analyze HUA network graphs. Suppose a network administrator needs to analyze log data collected from his or her managed network. Since there must be a starting point, he or she starts with the most concise HUA network visualization in Figure 4.22. Not overwhelmed by the number of nodes and connections, from this top-level graph, the administrator easily finds that there were 128 users logged on 601 internal hosts running 298 unique applications, which connected to either internal hosts or 2802 external domains.

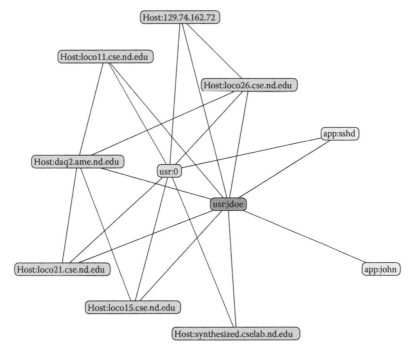

FIGURE 4.21: The HUA network graph reveals a highlighted user (jdoe) has logged on seven machines via `ssh` and has used the application John the Ripper to crack password files on those machines. (From Qi Liao et al., *Proceedings of the USENIX 22nd Large Installation System Administration Conference (LISA '08)*, pages 59–74, San Diego, CA, November 9–14 2008 [73].)

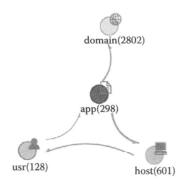

FIGURE 4.22: A top-level HUA OnionGraph gives the most concise and scalable visualization on large enterprise activity networks. (From Lei Shi et al. *IEEE Pacific Visualization Symposium*, pages 89–96, 2014 [71].)

The administrator makes a few interesting observations when moving from the initial "heterogeneous groups" to "RRE groups," on each node type. First, the *app* nodes are split into five subgroups, as shown in Figure 4.23, displayed by neighborhood charts: (1) the majority of applications [apps (217) node] are connected to only internal hosts by users (focused node in the graph); (2) six applications [apps (6) node] are connected to only external domains by users; (3) 69 applications [apps (69) node] are connected to both internal hosts and external domains by users; (4) five applications [apps (5) node] do not

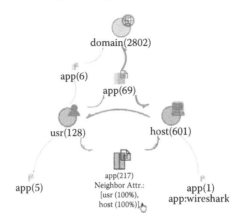

FIGURE 4.23: Expanded view on application nodes in OnionGraphs reveals a suspicious application (wireshark) possibly sniffing packets on the network by a malicious user. (From Lei Shi et al. *IEEE Pacific Visualization Symposium*, pages 89–96, 2014 [71].)

make network connections; and (5) one application [app (1) app: wireshark] run by an unknown user talked to a few internal hosts. Type-1 apps contain predominantly scientific computing programs while Type-2 and Type-3 have significantly more generic network applications such as ssh, firefox, ftp, and so on. In particular, the Type-5 node containing only one app (wireshark) is clearly suspicious, possibly leveraged by a malicious user to sniff packets on the network.

In addition to expansion of application nodes, the administrator may further divide the user nodes into two groups (Figure 4.24): 127 users that had run applications to connect to other computers; and the only user who does not run applications to make network connections. The Type-1 users are primarily enterprise users who are allowed to run scientific

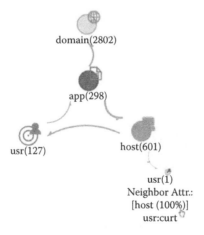

FIGURE 4.24: Expanded view on user nodes in OnionGraphs separates enterprise users with privileged users in network activities. (From Lei Shi et al. *IEEE Pacific Visualization Symposium*, pages 89–96, 2014 [71].)

programs. The Type-2 user is the system administrator. As policy compliance auditing, it is clear that normal users and privileged users have distinguished activity patterns.

4.4.4   OntoVis for terrorism networks

OntoVis [74] prunes a large, heterogeneous network according to both structural abstraction and semantic abstraction, thus making large networks manageable and reducing visual analytic complexity. Graph visualization has applications in social networks as well as for national security, such as terrorism domains. The original graph of a terrorism network (Figure 4.25a) is too large to analyze effectively while the ontology graph of the same network (Figure 4.25b) reduces such a network into only nine node types including terrorist organization, classification, terrorist, legal case, country/area, attack, attack target, and weapon and tactic.

Al-Qaeda is one of the most dominant organizations in the network and is related to most terrorist organizations, which can be further clustered. Terrorist organizations are consistent with their locations. For example, Al-Qaeda connects to the cluster at Gaza Strip, West Bank, and another cluster at Kashmir, India, and Pakistan. By investigating the attributes of related organizations, the researchers find that Al-Qaeda and most organizations related to it are classified as religious and nationalist/separatist organizations. In Figure 4.26a, organization clusters with the same classifications tend to connect to each other. Organizations that are located in the same areas tend to be of the same type.

In order to characterize the terrorist attacks, the researchers use attacking targets, tactics, and weapons to describe attacks. In Figure 4.26(b), the green nodes are attacks, the light blue ones are targets, the purple ones are weapons, and the pink ones are tactics. Most attacks by Al-Qaeda are bombing attacks using explosive weapons. The attacking targets vary from business, airport, and government to diplomatic objects. The 9/11 attacks can also be classified as unconventional attacks.

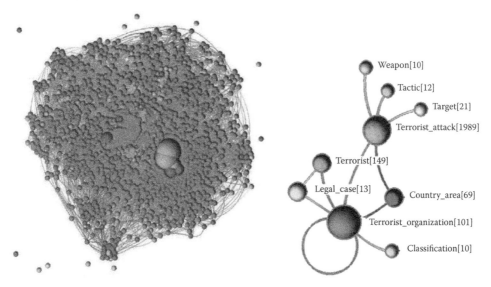

FIGURE 4.25: Ontology graph of the terrorism network. (a) Graphs of over 2000 nodes and 8000 edges creates an overwhelming visual complexity; (b) Onto-graph of only nine nodes. (From Zeqian Shen et al., *IEEE Transactions on Visualization and Computer Graphics*, 12(6):1427–1439, 2006. [74].)

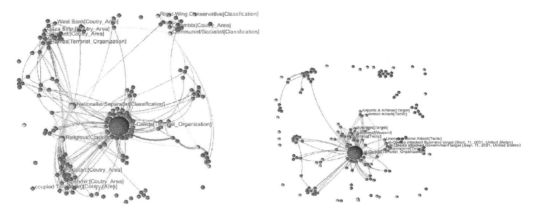

FIGURE 4.26: Visualization of various components of terrorist networks and their relationships. (a) Terrorist organizations, their locations, and classifications; (b) Terrorist attacks, their targets, weapons used, and attack tactics. (From Zeqian Shen et al., *IEEE Transactions on Visualization and Computer Graphics*, 12(6):1427–1439, 2006. [74].)

4.5   Summary and Future Directions

In this chapter, we have discussed the state-of-the-art visualization methods to display large complex networks. The main challenge lies in the unique characteristics of these networks collected in the Big Data era: the huge volume in both network size and complex connections; the increasing variety on the data source, network topology, and the node/edge attributes and the changing dynamics of the network structure. We focus on the first two challenges and describe two kinds of visualization approaches: the large graph drawing method, and the visual abstraction method on large multivariate networks. The graph drawing methods propose new algorithms to support the layout of large graphs with up to millions of nodes. The visual abstraction methods study the algorithms to cluster, compress, and aggregate large graphs into smaller abstractions. These abstractions are visualized by new metaphor designs and manipulated by users through novel interaction models. We also introduce one application domain in security where the large complex network visualization methods are successfully applied to support the visual analysis from end users.

Despite the many existing works in this area, we are far from perfection in connecting the dots between large complex networks and domain users through visualizations. In the future, we foresee some important topic threads where many questions remain unanswered, though the area is largely open and new efforts should not be limited to the following list.

Theory. While the Big Data surge keeps on going, how people visually understand the result on large complex networks remains open. In particular, given that the graph drawing algorithms can not scale to an ultra-large network size and visual abstraction is inevitable, what is the best method and visual design for the abstraction to meet the goal of effectively analyzing big complex networks?

Algorithm. Rather than focusing on the layout algorithm to place nodes for the large network, new transformation and mining algorithms on graphs and networks can be more important in the Big Data context. Compared to the general-purpose graph mining algorithms, the focus on visualization calls for novel mining algorithms to detect interpretable patterns from the large complex network. In this perspective, the joint optimization of both the data mining objective and the visualization objective can be an interesting direction.

Framework and system. In this area of the application research, the design and implementation of software frameworks and systems to visualize the incoming large complex newtork, presumptively in a web context, are in most cases more visible than the theoretical and algorithmic studies. We need online systems that allow users to upload their large networks for visualization and/or to visually navigate the existing large network collections for valuable insights.

References

1. Peter Eades. A heuristic for graph drawing. In *Congressus Numerantium*, volume 42, pages 149–160, 1984.

2. Thomas M. J. Fruchterman and Edward M. Reingold. Graph drawing by force-directed placement. *Software, Practice & Experience*, 21(11):1129–1164, 1991.

3. Tomihisa Kamada and Satoru Kawai. An algorithm for drawing general undirected graphs. *Information Processing Letters*, 31(1):7–15, 1989.

4. Emden R. Gansner, Yehuda Koren, and Stephen North. Graph drawing by stress majorization. In *Proceedings of International Symposium on Graph Drawing*, pages 239–250, 2004.

5. Colin Ware. *Information visualization: Perception for design*. Elsevier, Amsterdam, the Netherlands, 2012.

6. Mohammad Ghoniem, Jean-Daniel Fekete, and Philippe Castagliola. A comparison of the readability of graphs using node-link and matrix-based representations. In *IEEE Symposium on Information Visualization*, pages 17–24, 2004.

7. Hyunmo Kang, Catherine Plaisant, Bongshin Lee, and Benjamin B. Bederson. Netlens: Iterative exploration of content-actor network data. *Information Visualization*, 6(1):18–31, 2007.

8. Giuseppe Di Battista, Peter Eades, Roberto Tamassia, and Ioannis G. Tollis. *Graph Drawing: Algorithms for the Visualization of Graphs*. Prentice Hall PTR, Upper Saddle River, NJ, 1998.

9. Emden R. Gansner and Stephen North. An open graph visualization system and its applications to software engineering. *Software, Practice & Experience*, 30:1203–1233, 2000.

10. Mathieu Bastian, Sebastien Heymann, Mathieu Jacomy, et al. Gephi: an open source software for exploring and manipulating networks. *ICWSM*, 8:361–362, 2009.

11. OGDF. Open graph drawing framework, http://www.ogdf.net/.

12. David Auber. Tulip—a huge graph visualization framework. In *Graph Drawing Software*, pages 105–126, 2004.

13. Fabian Beck, Michael Burch, Stephan Diehl, and Daniel Weiskopf. The state of the art in visualizing dynamic graphs. In *Euro-Vis'14 State-of-the-Art Reports*, pages 83–103, 2014.

14. A. Noack. An energy model for visual graph clustering. In *Proceedings of the 11th International Symposium on Graph Drawing (GD 2003)*, volume 2912 of *LNCS*, pages 425–436. Springer-Verlag, Berlin, Germany, 2004.

15. I. Bruss and A. Frick. Fast interactive 3-D graph visualization. *LNCS*, 1027:99–11, 1995.

16. Yifan Hu. Efficient and high quality force-directed graph drawing. *Mathematica Journal*, 10:37–71, 2005.

17. J. Barnes and P. Hut. A hierarchical O(NlogN) force-calculation algorithm. *Nature*, 324:446–449, 1986.

18. D. Tunkelang. *A numerical optimization approach to general graph drawing.* PhD thesis, Carnegie Mellon University, 1999.

19. A. Quigley. *Large scale relational information visualization, clustering, and abstraction.* PhD thesis, Department of Computer Science and Software Engineering, University of Newcastle, Australia, 2001.

20. S. Pfalzner and P. Gibbon. *Many-body tree methods in physics.* Cambridge University Press, Cambridge, 1996.

21. A. Burton, A. J. Field, and H. W. To. A cell-cell Barnes Hut algorithm for fast particle simulation. *Australian Computer Science Communications*, 20:267–278, 1998.

22. L. F. Greengard. *The rapid evaluation of potential fields in particle systems.* The MIT Press, Cambridge, Massachusetts, 1988.

23. S. Hachul and M. Jünger. Drawing large graphs with a potential field based multilevel algorithm. In *Proc. 12th Intl. Symp. Graph Drawing (GD '04)*, volume 3383 of *LNCS*, pages 285–295. Springer-Verlag, Berlin, Germany, 2004.

24. A. Gupta, G. Karypis, and V. Kumar. Highly scalable parallel algorithms for sparse matrix factorization. *IEEE Transactions on Parallel and Distributed Systems*, 5:502–520, 1997.

25. Bruce Hendrickson and Robert Leland. A multilevel algorithm for partitioning graphs. In *Proceedings of the 1995 ACM/IEEE Conference on Supercomputing*, New York, NY, USA, 1995. ACM.

26. C. Walshaw, M. Cross, and M. G. Everett. Parallel dynamic graph partitioning for adaptive unstructured meshes. *Journal of Parallel and Distributed Computing*, 47:102–108, 1997.

27. Y. Hu and J. A. Scott. A multilevel algorithm for wavefront reduction. *SIAM Journal on Scientific Computing*, 23:1352–1375, 2001.

28. I. Safro, D. Ron, and A. Brandt. Multilevel algorithms for linear ordering problems. *J. Exp. Algorithmics*, 13:1.4–1.20, 2009.

29. C. Walshaw. A multilevel approach to the travelling salesman problem. *Oper. Res.*, 50:862–877, 2002.

30. C. Walshaw. Multilevel refinement for combinatorial optimisation problems. *Annals of Operations Research*, pages 325–372, 2004.

31. P. Gajer, M. T. Goodrich, and S. G. Kobourov. A fast multi-dimensional algorithm for drawing large graphs. *LNCS*, 1984:211–221, 2000.

32. R. Hadany and D. Harel. A multi-scale algorithm for drawing graphs nicely. *Discrete Applied Mathematics*, 113:3–21, 2001.

33. D. Harel and Y. Koren. A fast multi-scale method for drawing large graphs. *Journal of graph algorithms and applications*, 6:179–202, 2002.

34. C. Walshaw. A multilevel algorithm for force-directed graph drawing. *J. Graph Algorithms and Applications*, 7:253–285, 2003.

35. Y. Hu. A gallery of large graphs. http://research.att.com/yifanhu/GALLERY/GRAPHS/index.html.

36. T. A. Davis and Y. Hu. University of Florida Sparse Matrix Collection. *ACM Transaction on Mathematical Software*, 38:1–18, 2011.

37. J. B. Kruskal. Multidimensioal scaling by optimizing goodness of fit to a nonmetric hypothesis. *Psychometrika*, 29:1–27, 1964.

38. J. B. Kruskal and J. B. Seery. Designing network diagrams. In *Proceedings of the First General Conference on Social Graphics*, pages 22–50, Washington, D.C., July 1980. U. S. Department of the Census. Bell Laboratories Technical Report No. 49.

39. U. Brandes and C. Pich. An experimental study on distance based graph drawing. In *Proc. 16th Intl. Symp. Graph Drawing (GD '08)*, volume 5417 of *LNCS*, pages 218–229. Springer-Verlag, Berlin, Germany, 2009.

40. W. S. Torgerson. Multidimensional scaling: I. theory and method. *Psychometrika*, 17:401–419, 1952.

41. G. Young and A. S. Householder. Discussion of a set of points in terms of their mutual distances. *Psychometrica*, 3:19–22, 1938.

42. V. de Silva and J. B. Tenenbaum. Global versus local methods in nonlinear dimensionality reduction. In *Advances in Neural Information Processing Systems 15*, pages 721–728. MIT Press, Cambridge, MA, 2003.

43. Ulrik Brandes and Christian Pich. Eigensolver methods for progressive multidimensional scaling of large data. In *Proc. 14th Intl. Symp. Graph Drawing (GD '06)*, volume 4372 of *LNCS*, pages 42–53, 2007.

44. Emden R. Gansner, Yifan Hu, and Stephen C. North. A maxent-stress model for graph layout. *IEEE Trans. Vis. Comput. Graph.*, 19(6):927–940, 2013.

45. Marc Khoury, Yifan Hu, Shankar Krishnan, and Carlos Eduardo Scheidegger. Drawing large graphs by low-rank stress majorization. *Comput. Graph. Forum*, 31(3):975–984, 2012.

46. Emden R. Gansner, Yifan Hu, and Shankar Krishnan. Coast: A convex optimization approach to stress-based embedding. In Stephen Wismath and Alexander Wolff, editors, *Graph Drawing*, volume 8242 of *Lecture Notes in Computer Science*, pages 268–279. Springer-Verlag, Berlin, Germany, 2013.

47. Y. Koren, L. Carmel, and D. Harel. Ace: A fast multiscale eigenvectors computation for drawing huge graphs. In *INFOVIS '02: Proceedings of the IEEE Symposium on Information Visualization (InfoVis'02)*, pages 137–144, Washington, DC, USA, 2002. IEEE Computer Society.

48. D. Harel and Y. Koren. Graph drawing by high-dimensional embedding. *LNCS*, pages 207–219, 2002.

49. Cytoscape. An open source platform for complex network analysis and visualization. //http://www.cytoscape.org/.

50. Shailesh Tripathi, Matthias Dehmer, and Frank Emmert-Streib. NetBioV: an R package for visualizing large network data in biology and medicine. *Bioinformatics (Oxford, England)*, 30(19):2834–2836, October 2014.

51. D3.js. Data-driven documents. http://d3js.org.

52. Tim Dwyer, Kim Marriott, and Michael Wybrow. Dunnart: A constraint-based network diagram authoring tool. In IoannisG. Tollis and Maurizio Patrignani, editors, *Graph Drawing*, volume 5417 of *Lecture Notes in Computer Science*, pages 420–431. Springer, Berlin, Heidelberg, 2009.

53. SNAP. Stanford large network dataset collection. http://snap.stanford.edu/data.

54. Koblenz. The Koblenz network collection. http://konect.uni-koblenz.de.

55. GraphArchive. Exchange and archive system for graphs. http://www.graph-archive.org/.

56. House of Graphs. A database of interesting graphs. https://hog.grinvin.org/.

57. Ivan Herman, Guy Melancon, and M. Scott Marshall. Graph visualization and navigation in information visualization: A survey. *IEEE Transactions on Visualization and Computer Graphics*, 6(1):24–43, 2000.

58. T. von Landesberger, A. Kuijper, T. Schreck, J. Kohlhammer, J. J. van Wijk, J.-D. Fekete, and D. W. Fellner. Visual analysis of large graphs. *EuroGraphics - State of the Art Report*, pages 37–60, 2010.

59. Andreas Kerren, Helen Purchase, and Matthew O. Ward. *Multivariate Network Visualization (Proc. Dagstuhl Seminar 13201)*. Springer-Verlag, Berlin, Germany, 2014.

60. Douglas Brent West. *Introduction to graph theory*, volume 2. Prentice Hall, Upper Saddle River, NJ 2001.

61. Glen Jeh and Jennifer Widom. Simrank: a measure of structural-context similarity. In *Proceedings of the Eighth ACM SIGKDD International Conference on Knowledge Discovery and Data Mining*, pages 538–543, 2002.

62. Jiawei Han, Micheline Kamber, and Jian Pei. *Data mining: concepts and techniques*. Elsevier, Amsterdam, the Netherlands, 2011.

63. Lei Shi, Nan Cao, Shixia Liu, Weihong Qian, Li Tan, Guodong Wang, Jimeng Sun, and Ching-Yung Lin. HiMap: Adaptive visualization of large-scale online social networks. In *Proceedings of the IEEE Pacific Visualization Symposium*, pages 41–48, 2009.

64. M. E. J. Newman and M. Girvan. Finding and evaluating community structure in networks. *Physical Review E*, 69:026113, 2004.

65. Weiwei Cui, Hong Zhou, Huamin Qu, Pak Chung Wong, and Xiaoming Li. Geometry-based edge clustering for graph visualization. *IEEE Transactions on Visualization and Computer Graphics*, 14(6):1277–1284, 2008.

66. Lei Shi, Qi Liao, Xiaohua Sun, Yarui Chen, and Chuang Lin. Scalable network traffic visualization using compressed graphs. In *IEEE International Conference on Big Data*, pages 606–612, 2013.

67. Francois Lorrain and Harrison C. White. Structural equivalence of individuals in social networks. *The Journal of Mathematical Sociology*, 1(1):49–80, 1971.

68. Charis Papadopoulos and Costas Voglis. Drawing graphs using modular decomposition. *Journal of Graph Algorithms and Applications*, 11(2):481–511, 2007.

69. Cody Dunne and Ben Shneiderman. Motif simplification: Improving network visualization readability with fan and parallel glyphs. In *Proceedings of the International Conference on Human Factors in Computing Systems (CHI'13)*, pages 3247–3256, 2013.

70. Tim Dwyer, Nathalie Henry Riche, Kim Marriott, and Christopher Mears. Edge compression techniques for visualization of dense directed graphs. *IEEE Transactions on Visualization and Computer Graphics*, 19(12):2596–2605, 2013.

71. Lei Shi, Qi Liao, Hanghang Tong, Yifan Hu, Yue Zhao, and Chuang Lin. Hierarchical focus+context heterogeneous network visualization. In *IEEE Pacific Visualization Symposium*, pages 89–96, 2014.

72. Douglas R. White and Karl P. Reitz. Graph and semigroup homomorphisms on networks of relations. *Social Networks*, 5(2):193–234, 1983.

73. Qi Liao, Andrew Blaich, Aaron Striegel, and Douglas Thain. ENAVis: Enterprise network activities visualization. In *Proceedings of the USENIX 22nd Large Installation System Administration Conference (LISA '08)*, pages 59–74, San Diego, CA, November 9-14 2008.

74. Zeqian Shen, Kwan-Liu Ma, and Tina Eliassi-Rad. Visual analysis of large heterogeneous social networks by semantic and structure. *IEEE Transactions on Visualization and Computer Graphics*, 12(6):1427–1439, 2006.

Chapter 5

Finding Small Dominating Sets in Large-Scale Networks

Liang Zhao

5.1 Introduction

We are connected by various kinds of networks. Some are explicit, such as road networks, power grids, and communication networks. Others are rather implicit, such as social networks, citations, collaborations, product purchasing, and chemical reactions. However, all of these networks share a common feature: *Big Data*, that is, they involve high volume, high velocity, and a high variety of information, which requires new types of processing methods to unearth the value hidden within.

Traditionally, networks are considered by *graph theory* and *combinatorial optimization*. Usually researchers in graph theory are interested in finding beautiful structures and properties for various types of graphs, whereas those researching combinatorial optimization aim to find *efficient*, that is, *polynomial-time* algorithms for specific problems. There is no doubt that these studies are important. However, the fact is that we remain in need of *practical* tools capable of solving real problems in *reasonable* time. An algorithm of running time $O(n^2)$ may be significantly efficient in theory, but this may not be the case in practice. Motivated by recent Big Data studies, researchers began to focus on practical (i.e., linear-time or near linear-time) algorithms to treat large-scale instances.

This chapter considers a fundamental issue in network theory, namely the "dominating set (DS)." The main contribution includes some novel applications and simple algorithms based on previous works conducted by the author and others for large-scale (larger than one billion $= 10^9$ in size) networks. Let us start with some applications.

Firstly, power grids and computer networks are two of the most important kinds of infrastructure networks around us. In networks such as these, the failure of a single node (which may be a power station or a router) degrades the performance of the network, and sometimes its failure has fatal consequences for many people. Therefore, it is important to monitor the health of such large networks.

There are several approaches to monitoring a network. Here, we are interested in a *distributed* approach as it can be applied to large-scale (and autonomous) networks. In this approach, we suppose that every node can be configured to monitor its own status and that of its neighbors, and our task is to determine a set of nodes that can be configured to monitor the whole network.

Of course, our aim is to use a small set of nodes to simplify the task. Thus, we want to know the minimum number of nodes that need to be configured to monitor the whole network. The smaller this number, the easier to monitor the network. This is an *optimization problem*.

Let us look at another application. Consider a social network, for example, Facebook or Twitter. When someone says (posts) something, his or her connections (followers) are notified. This means that if a group of people discuss a common topic about, for example, a recent social problem or a new product, then all of their connections would receive the information. As a result, they may join the discussion too. This expands the scale of the group.

Let us say a node in a social network *dominates* its neighbors (and itself, of course). Similar to the first example, we would like to know the minimum number of members in a *DS*, which would provide an index to estimate how easy it would be to control (or to spread a virus, disease, advertisement, etc., in) the network. This is the *same type* of optimization problem we encountered before.

This problem is known as the (minimum) *DS* problem in the literature. The minimum number of a DS, called the *domination number*, is used as an index to estimate the ease with which the network can be dominated. In general, the object is to find a minimum DS for a given graph, rather than just the domination number. This problem has gathered much attention during the past half century (see the reviews [1–4]). Yet recently, it attracted renewed interest in terms of *controlling complex networks* and other types of networks in Big Data research (e.g., [5–9]). Here, let us illustrate the concepts by considering the example of a drug-target protein network.

Inspired by several pioneering research efforts, Nacher and Akutsu [8] presented a DS-based approach, which "identifies the topologies that are relatively easy to control with a minimum number of driver nodes" ([8], p. 1). They used Figure 5.1 to motivate the problem: "Although 11 drugs interact with these proteins, only three drugs (red hexagons) are required to control the system simultaneously. Drugs belonging to the DS are indicated in red. These three drugs are called the cover set (or DS) of the network. Interactions between the drugs from the DS and the disease-gene products are represented as wavy red arrows." They even used a rather large network ([8], Figure 4) to demonstrate their research results.

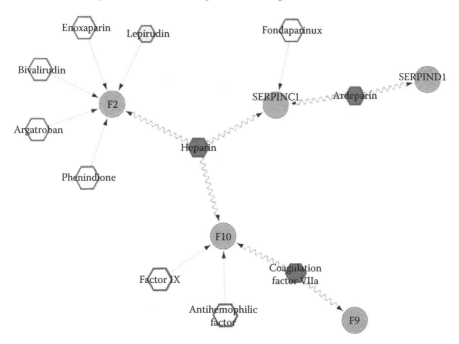

FIGURE 5.1: An application of dominating set research in a drug-target protein network: "A small component of the drug-target protein network consisting of 11 drugs (hexagons) interacting with five disease-gene products corresponding to cardiovascular disorder class" (From Jose C. Nacher and Tatsuya Akutsu. Structural controllability of unidirectional bipartite networks. *Scientic Reports*, 3(1647), 2013; Figure 1).

On the other hand, in real applications we often require a DS to satisfy some additional constraint(s). For example, in applications involving the design of a backbone for wireless networks, we require the DS to be *connected*, that is, it should induce a connected subgraph [10]. Moreover, the application of a wireless network for energy-efficient computing requires the DS to be *independent*, that is, the distance between any two of its elements should be at least two [11]. On the other hand, in applications of communication networks design and facility location, we extend the term "domination" from only covering the neighborhood to covering nodes within a given *distance* (Henning, Distance Domination in Graphs, Chapter 12 in [2]). We note that, by generalizing the works in [7–9], which only considered DSs, distance dominating can also play an interesting role in controlling complex networks. Furthermore, sometimes we may only require a given number of nodes (e.g., 50% of the network) to be dominated. For more variants, we refer the readers to the books considered to be the Bible of reference works [2,3], which cited more than 75 variations by reviewing more than 1200 papers on domination in graphs published before 1998.

Before going into more details, let us use an example to illustrate the definitions of domination. In Figure 5.2, node set $\{3, 6, 8, 11\}$ is a DS but not connected; $\{4, 6, 8, 11\}$ is a connected DS; and $\{3, 6, 9, 11\}$ is an independent DS. On the other hand, $\{6, 8\}$ is a distance-2 DS, that is, all nodes are within distance 2 from nodes 6 or 8, whereas $\{6, 11\}$ is an independent distance-2 DS.

Now let us provide a general formulation. Let $G = (V, E)$ denote a graph with a set V of nodes and a set E of edges, and let $\text{dist}(u, v)$ denote the distance from a node u to a node v in G, that is, the minimum number of edges required to reach v from u. In addition to a distance and independence generalization, we consider it natural to require a

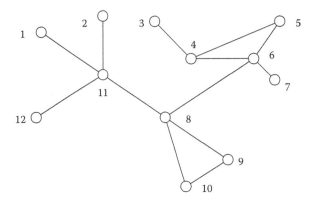

FIGURE 5.2: Graph to illustrate various dominations.

designated *target* set $T \subseteq V$ to be dominated using only nodes from a given *source* node set $S \subseteq V$. For example, we often want to find the most "important" packages for an intensive security check in a large-scale software distribution, for example, a Linux distribution. In this case we can let G be the package dependence graph among all packages, and let S be the "important" packages. See also [8] for another example.

We formulate the following *general dominating set problem* (GDS):

> **Problem GDS**: Given a graph $G = (V, E)$, two sets S and $T \subseteq V$, and two integers $R \geq 1$ and $z \geq 1$, find a minimum set $D \subseteq S$ such that it satisfies the following two requirements.
>
> 1. (Distance) $\forall t \in T$, $\exists s \in S$, $\text{dist}(s, t) \leq R$.
> 2. (Independence) $\forall v_1, v_2 \in D, v_1 \neq v_2$ implies $\text{dist}(v_1, v_2) \geq z$.

In particular, problem DS is a special case of $S = T = V$ and $R = z = 1$. In fact, it is easy to see that GDS contains the following special cases:

- *Set cover* (SC): This is one of the most well-known optimization problems (see [12]). This problem enables us to use a bipartite graph G with $V = S \cup T$, $S \cap T = \emptyset$, $E \subseteq S \times T$, and $R = z = 1$ to reduce the SC problem to GDS.

- *Vertex cover* (VC): This is another well-known (SC-type) NP-hard problem, in which an edge is covered by an incident node. This problem is able to model the problem of monitoring links in a computer network (see [13]), and so on.

- *Edge cover* (EC): This is a polynomial-time solvable (SC-type) graph problem in which an edge is covered by an incident edge.

- Nine *Fundamental Dominations* defined in [14] including VC and EC.

- *Distance dominating* ($S = T = V$, $R \geq 2$, and $z = 1$, see [2,13]).

- *Independent dominating* ($S = T = V$, $R = 1$, and $z \geq 2$, see [1,15]).

All of these dominating problems, except EC, are known to be NP-hard, which means it is generally difficult (if not impossible) to develop efficient exact algorithms for them. Therefore, it is natural to design efficient *approximate* or *heuristic* algorithms for large-scale problems.

For instance, GDS with $z = 1$ (i.e., without any independence requirement) can be reduced to an SC problem (in which a node covers all nodes within distance R); hence, any algorithm designed to solve an SC problem can be employed to solve a GDS problem. The well-known and standard greedy algorithm (see [12]), for example, can always find a solution with size no more than $1 + \log |S|$ times of the optimal.

The problem is, this algorithm requires $O(|V||E|)$ time if $R \geq 2$ (for $R = 1$, however, it is linear time). Therefore, it cannot be applied to large-scale networks. Nevertheless, it is possible to design more efficient (more precise or faster) algorithms by exploiting the graph structure (see an example in [5]), which is also true for $z \geq 2$. In the remainder of this chapter, we first review some complexity results and approaches for finding small solutions efficiently by considering the graph structure (Sections 5.2 through 5.7), including some possible variations. Extensive empirical results are presented in Section 5.8. Finally, we summarize and conclude in Section 5.9.

5.2   Notations and Preliminaries

Unless otherwise stated, graphs in this chapter are undirected. We assume without loss of generality that graphs are simple, that is, they contain neither self-loops nor multiedges, otherwise a single scan of the graph (the adjacent lists) would be able to remove them in linear time. Given an instance $I = (G = (V, E), S, T, R, z)$ of GDS, let $n = |V|$ and $m = |E|$ denote the number of nodes and the number of edges, respectively. For ease of notation, we do not distinguish a set $\{v\}$ from its unique element v.

A *path* $P = v_1 v_2 \cdots v_k$ is a sequence of nodes such that $(v_i, v_{i+1}) \in E$, $i = 1, \ldots, k - 1$. The *length* of P is defined by $k - 1$, that is, the number of edges. P is an s, t-*path* if it starts from s and ends at t. The *distance* $\mathrm{dist}(s, t)$ from a node s to a node t is defined as the minimum length of an s, t-path.

For two sets $U, W \subseteq V$, let

$$\mathrm{dist}(U, W) = \min\{\mathrm{dist}(u, w) \mid u \in U, w \in W\}$$

denote the (minimum) *distance* from U to W in G. In particular, we have $\mathrm{dist}(U, W) = 0$ if $U \cap W \neq \emptyset$. Moreover, let

$$\Gamma_r(U) = \{v \in V \mid \mathrm{dist}(U, v) \leq r\}$$

denote the set of nodes that can be reached from U via at most r edges. Thus, $\Gamma_0(v) = v$, and $\Gamma_1(v) \setminus v$ is the set of neighbors of v, and so on. The *kth power* G^k of G is a graph defined by

$$G^k = (V, E^k = \{(u, v) \mid u \in V, v \in V, u \neq v, \mathrm{dist}(u, v) \leq k\})$$

In other words, there is an edge (u, v) in G^k if and only if $u \neq v$ and v is within distance k from u in graph G. Clearly, a distance-R DS in G is a DS in G^R and vice versa. Similarly, a distance-z ($z \geq 2$) independent set in G is an independent set in G^{z-1}, and vice versa.

Let $\deg(v) = |\Gamma_1(v) \setminus v|$ denote the degree of a node v. Then, we have

$$0 \leq \deg(v) \leq n - 1, \ \forall v \in V, \quad \text{and} \quad \sum_{v \in V} \deg(v) = 2m$$

For many years researchers studied the DS problem and its variants in numerous fields, mostly from a theoretical point of view (see, e.g., [1–4,11,16,17]). It is known that even the simplest $(G = (V, E), V, V, 1, 1)$ GDS problem (i.e., the DS problem) is quite hard to solve. In fact, it has been shown that no algorithm can achieve an approximation factor more accurate than $c \log n$ for some constant $c > 0$ unless P = NP [18]. On the other hand, $(G = (V, E), V, V, 1, 2)$ GDS cannot be approximated within a factor of $n^{1-\epsilon}$ for any $\epsilon > 0$ unless P = NP [19]. However, for graphs obeying the power law, there is a positive result showing that the $(G = (V, E), V, V, 1, 1)$ problem can be approximated within a constant factor, although it is still APX-hard [20], that is, there exists some constant $c > 1$ such that it is not possible to approximate the problem within factor c unless P = NP.

As stated before, GDS with $z = 1$ can be reduced to an SC problem. Thus, the well-known greedy algorithm [21] is a $1 + \log n$ approximation, and almost matches the best bound $c \log n$. However, this approach requires $O(mn)$ running time which generally cannot be improved, because the number of edges in G^R can be $\Omega(mn)$ even for $R = 2$ (for $R = 1$, however, the running time is $O(m + n)$); thus, the approach is not suitable for large-scale instances except for $R = z = 1$. Some empirical studies have also been carried out [5,6,22,23], but unfortunately they are limited to small-size instances (fewer than one million nodes).

For the case of $z = 2$, there exists a 5-approximation algorithm for unit disk graphs [17], and an empirical study [22] considered small instances with fewer than 1500 nodes. For $z \geq 3$, as far as we know, an empirical study has not been conducted.

In an attempt to study networks with as many as a billion nodes, Sasaki [24] proposed the heuristic *Sieve* algorithm for distance VC when we were investigating the problem of monitoring links for large-scale computer networks. An interesting characteristic of Sieve is that it is fast. In fact, the running time is $O(R(m + n))$, where R is small in practice. In addition, even though there is no approximation guarantee, in practice its accuracy is acceptable, and the solutions are usually within two times of the lower bounds. Related works were published in [13,24].

Subsequently, in 2009, Gim and Wang generated Sieve to provide a robust covering (distance links) [25]. Then, more recently, Kadowaki suggested a novel approach based on what has become known as the *Quasi-Greedy* algorithm [15], using the same labelling method used in Sieve to dominate distance *nodes*. The idea is to repeatedly select a node v with the largest *residual degree*, that is, the number of neighbors of v that have not been dominated up to that point. Notice that for $R = 1$, this is nothing but the greedy algorithm. For $R \geq 2$, however, these two algorithms are different (which is why we call the new algorithm as Quasi-Greedy). An interesting observation is that, if $S = T = V$, then any two nodes selected in this way must be of distance R or more; hence, it is an R-distance and z-independent algorithm for all $z \leq R$. We published this work in 2013 [15].

In the following sections, based on the work [13,15,24,25], we provide a generalized and detailed description of the Sieve and Quasi-Greedy algorithms together with more thorough experimental studies. In particular, new results obtained for networks with as many as a billion nodes are first shown in this chapter. Finally, we discuss variants of these algorithms for other dominating problems.

5.3   A General Algorithm Framework for GDS with $z = 1$

Let us first show a general algorithm framework for GDS for the case $z = 1$ (i.e., there is no independence requirement). We discuss GDS with $z \geq 2$ later in Section 5.7.

The idea is simple and is based on simply repeatedly finding a source node that dominates some undominated node(s). If we failed in finding such a node even though an undominated target exists, then it would enable us to conclude that there is no feasible solution (see Table 5.1).

Clearly, the most important question relating to this framework is to determine how to select an appropriate node s. Three strategies are possible:

1. *Maximizing* $|\Gamma_R(s) \cap T - \Gamma_R(D)|$. This is the well-known greedy algorithm (greedy in the following). As we have stated before, in general it is a $1 + \log n$ approximation but requires quadratic time and even quadratic space for all $R \geq 2$, because the number of edges in G^R is $\Omega(mn)$ in general. However, for $R = 1$ (the DS problem), the problem can be solved in linear time, because $\sum_{v \in V} |\Gamma_1(v)| = n + 2m$. The implementation of the greedy algorithm is straightforward; hence, it is not shown in this chapter.

2. *Choosing an arbitrary s such that it satisfies* $\Gamma_R(s) \cap T - \Gamma_R(D) \neq \emptyset$.

    This is the principle underlying algorithm Sieve [13,25], which, in search of a good solution, checks nodes in the fixed *degree-descending* order which it determines in the beginning by using a bucket-sort in $O(m + n)$ time. We remark that this is a constant approximation algorithm for the DS problem in power-law graphs (see [20], p274, Corollary 10.4), but in general we have no such a guarantee. Sieve also has a labelling method to complete other tasks in a total of $O(R(m + n))$ time (see Section 5.4). We note that R is usually small in practice. An interesting observation is that the running time of Sieve actually does not increase that much when R increases, see Section 5.8. However, theoretically the worst running time is $\Omega(R(m + n))$. Algorithm Sieve is explained in Section 5.4.

3. *Maximizing the residual degree* $|\Gamma_1(s) \cap T - \Gamma_R(D)|$ and falling back to Sieve if $|\Gamma_1(s) \cap T - \Gamma_R(D)| = 0$ for all remaining sources $s \in S$. This is the principle on which the Quasi-Greedy heuristic algorithm [15] (QGreedy in the following) is based, as it is greedy to some extent, but not exactly (however, when $R = 1$, it is greedy). Unlike Greedy, QGreedy can obtain solutions in $O(R(m + n))$ time by combining a (bucket-based) priority queue with the labelling method in [13,25]. An interesting fact is that, when $S \subseteq T$, it always finds an *independent* DS. We explain the detail later in Section 5.5.

Now, let us introduce a *reverse-delete step*. This idea, first proposed by Sasaki et al. [13,24], is to remove a selected node s if all nodes within distance R from s can be dominated by other selected nodes. Applying the reverse-delete step enables us to obtain a minimal solution D, that is, the removal of any node in D will break the feasibility of D (see Table 5.2).

Here, the *reverse* order is used, because each node was selected to dominate some targets that were not dominated by *previously selected nodes*. However, these targets could possibly

TABLE 5.1: A general algorithm framework for GDS with $z = 1$

Input	An Instance $(G = (V, E), S, T, R, z = 1)$
Output	A Solution $D \subseteq S$ or "No Solution"
1	$D = \emptyset$;
2	while $\Gamma_R(D) \cap T \neq T$ {
3	select a node $s \in S$ such that $\Gamma_R(s) \cap T - \Gamma_R(D) \neq \emptyset$;
4	if such an s does not exist, output "no solution" and **halt**;
5	else $D = D \cup \{s\}$
6	}

TABLE 5.2:   Reverse-delete step for the framework in
Table 5.1

7   **for** all $s \in D$ in the reverse order they were selected {
8   **if** $(\Gamma_R(s) \cap T \subseteq \Gamma_R(D \setminus s))$ $D = D \setminus s$
9   }

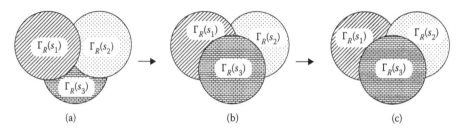

    (a)   (b)   (c)

FIGURE 5.3: Illustration of the reverse-delete step: (a) shows that s_1, s_2, and s_3 were
selected before the reverse-delete step; we check if s_2 is redundant by expanding the domination area of s_3 as illustrated in (b), then check if s_1 is redundant as illustrated by (c).

be dominated by *nodes that are subsequently selected*. Therefore, we need the reverse order
to remove redundant nodes from D. This is illustrated in Figure 5.3. We remark that, again,
this step can be completed in $O(R(m + n))$ time.

In the following, we assume that readers are familiar with fundamental data structures
and algorithms such as *bucket sort*, *adjacency lists* for storing a graph, *queue* and *priority
queue*, *breadth-first search* (BFS) for graphs, *linear programming* (LP), and *weakly duality
theorem* for LP. Otherwise, we refer readers to the textbook by Coremen et al. [26].

5.4   Algorithm Sieve for GDS with $z = 1$

We store the graph by using adjacency lists and use BFS with radius R to search the
targets. Each node $v \in V$ is assigned a label $\ell(v)$, $0 \leq \ell(v) \leq R + 1$, to keep track of
the distance measured from v that has been searched up to that point. Speaking precisely,
$\ell(v) = r$ means that nodes in $\Gamma_{r-1}(v)$ have been searched, and $\ell(v) \geq 1$ if and only if v is
dominated by some selected s. For each root node $s \in S$ of a BFS, label $n(s)$ stores the
number of newly found targets during this BFS search. For nonroot nodes $s \in S$, $n(s) = 0$.
Finally, we use $p = \sum_{v \in S} n(s)$ to store the total number of dominated targets. Following
the framework in Table 5.1, Algorithm Sieve is described in Table 5.3.

Theorem 1 *Algorithm Sieve is either able to correctly find a feasible GDS or determine
the nonexistence of such a solution. The space and time complexities are $O(m + n)$ and
$O(R(m + n))$, respectively.*

Proof 1 *Notice that, for any $v \in V$, $\ell(v) = r$ always means that the nodes in $\Gamma_{r-1}(v)$ have
been searched, whereupon the correctness becomes clear. The space complexity is trivial. Let
us consider the running time.*

TABLE 5.3: Algorithm Sieve for GDS with $z = 1$ (without the reverse-delete step). For source nodes in the degree-descending order, a BFS with radius R is performed to discover undominated target(s). Label $\ell(v)$, $0 \leq \ell(v) \leq R + 1$, keeps track of the distance from v that has already been searched (precisely, $\Gamma_{\ell(v)-1}(v)$ has been searched). On the other hand, $n(s_i)$ contains the number of targets newly dominated by s_i, and is calculated by the BFS starting from s_i. Label $s(w)$, which is determined by the BFS starting from s_i, contains the source node s_i responsible for dominating w. Algorithm Sieve either terminates when all targets have been dominated ($p == q$) or when all source nodes have been used ($i > k$).

Input	An Instance $(G = (V, E), S, T, R, z = 1)$. $\lvert S \rvert = k, \lvert T \rvert = q$
Output	A Solution $D \subseteq S$ or "No Solution"
1	Sort $s_i \in S$ such that $\deg(s_1) \geq \deg(s_2) \geq \cdots \geq \deg(s_k)$;
2	Let $\ell(v) = 0$ for all $v \in V$;
3	$p = 0$, $D = \emptyset$, $i = 1$;
4	**while** ($i \leq k$ and $p < q$) {
5	$\quad n(s_i) = 0$;
6	$\quad$ **if** ($\ell(s_i) == 0$ and $s_i \in T$) {
7	$\quad\quad n(s_i) = 1$, $s(s_i) = s_i$;
8	$\quad$ }
9	$\quad Q =$ an empty queue;
10	$\quad \ell(s_i) = R + 1$, enqueue(Q, s_i);
11	$\quad$ **while** Q is not empty {
12	$\quad\quad v =$ dequeue(Q);
13	$\quad\quad$ **for** all neighbors w of v **do**
14	$\quad\quad\quad$ **if** ($\ell(w) == 0$ and $w \in T$) {
15	$\quad\quad\quad\quad n(s_i) = n(s_i) + 1$, $s(w) = s_i$;
16	$\quad\quad\quad$ }
17	$\quad\quad\quad$ **if** ($\ell(w) < \ell(v) - 1$) {
18	$\quad\quad\quad\quad \ell(w) = \ell(v) - 1$;
19	$\quad\quad\quad\quad$ **if** ($\ell(w) \geq 2$) enqueue(Q, w);
20	$\quad\quad\quad$ }
21	$\quad\quad$ }
22	$\quad$ }
23	$\quad$ **if** ($n(s_i) > 0$) $D = D \cup \{s_i\}$;
24	$\quad p = p + n(s_i)$, $i = i + 1$;
25	}
26	**if** ($p < q$) output "no solution";

Sorting nodes in S by their degrees requires $O(m + n)$ time with a bucket sort (notice $0 \leq \deg(v) \leq n - 1$). A node v can be searched more than once by the algorithm during the search procedure. However, because $0 \leq \ell(v) \leq R + 1$ and $\ell(v)$ increases by at least one each time v is searched (i.e., by enqueue(Q, v)), it is obvious that node v can be searched by at most $R + 1$ times; hence, an edge (v, w) is checked by at most $2(R + 1)$ times. This dominates other operations. Therefore, the total running time is $O(R(m + n))$. ∎

Remark 1 *Sieve is a constant approximation algorithm for $R = 1$ and power-law graphs. See [20] (p. 274 Corollary 10.4).*

It is not difficult to implement a reverse-delete step. See Table 5.4.

TABLE 5.4:   Reverse-delete step for Algorithm Sieve

27	**for** all $s_i \in D$ in the reverse order they were selected {
28	// For ease of notation, suppose $s_i < s_j \Leftrightarrow s_i$ was selected before s_j.
29	**if** $(n(s_i) == 0)$ **continue**; // i.e., $\Gamma_R(s) \cap T \subseteq \Gamma_R(D \setminus s)$
30	// Otherwise $n(s_i) \geq 1$.
31	**if** $(s_i \in T$ and $s(s_i) < s_i)$ { // s_i was dominated by a prior source
32	$n(s(s_i)) = n(s(s_i)) - 1$; // change the dominator for s_i to s_i
33	$s(s_i) = s_i$;
34	$n(s_i) = n(s_i) + 1$;
35	}
36	Q = an empty queue;
37	$\ell(s_i) = 2R + 2$; enqueue(Q, s);
38	**while** Q is not empty {
39	$v = $ dequeue(Q);
40	**for** all neighbors w of v **do**
41	**if** $(\ell(w) < \ell(v) - 1)$ {
42	$\ell(w) = \ell(v) - 1$;
43	**if** $(w \in T$ and $s(w) < s_i)$ {
44	$n(s(w)) = n(s(w)) - 1$;
45	$s(w) = s_i$;
46	$n(s_i) = n(s_i) + 1$;
47	}
48	**if** $(\ell(w) \geq R + 2)$ enqueue(Q, w);
49	}
50	}
51	}
52	}
53	Remove all $s_i \in D$ with $n(s_i) == 0$ from D.

5.5   Algorithm QGreedy (Quasi-Greedy)

Following the framework in Table 5.1, we explain how to find a source $s \in S$ that maximizes the residual degree $|\Gamma_1(s) \cap T - \Gamma_R(D)|$.

Again, we use adjacent lists to store the graph. Then, different from Sieve, we employ a priority queue to store the candidate sources $s \in S$ with keys $|\Gamma_1(s) \cap T - \Gamma_R(D)|$. A standard bucket-based priority queue can do this efficiently. At the beginning, $key(s) = |\Gamma_1(s) \cap T|$ can be calculated in a total of $O(n + m)$ time (notice that $\sum_{s \in S} |\Gamma_1(s)| \leq n + 2m$). As the value of a key does not exceed the maximum degree plus 1 and never increases in the algorithm, the combined operations of the priority queue (**insert**, **deletemax**, and **decreasekey**) require only $O(n + m)$ time (each operation requires $O(1)$ time and there are at most $n + m$ operations).

Updating $\Gamma_R(D)$ and $|\Gamma_1(s) \cap T - \Gamma_R(D)|$ for all sources s is not trivial. Fortunately, we can use the labelling method of Algorithm Sieve. Therefore, we share the same running time as Sieve, that is, $O(R(m + n))$. In terms of the performance ratio, when $R = 1$, (the case for which QGreedy is the same as Greedy) the solution is a $1 + \log n$ approximation. Otherwise, we cannot prove a nontrivial guarantee for $R \geq 2$.

Notice that, at some stage while the algorithm is running, $|\Gamma_1(s) \cap T - \Gamma_R(D)|$ can be zero for all candidate sources, even if an undominated target exists. In that case, we may have several fallbacks.

- Fallback 1: Maximizing $|\Gamma_1(s) - \Gamma_R(D)|$ instead. This can be done in the same fashion as QGreedy (in fact, this is exactly the approach followed in [15], because in their formulation $S = T = V$). If $|\Gamma_1(s) - \Gamma_R(D)| = 0$ for all candidate sources s, we can switch to the next Fallback 2.

- Fallback 2: Maximizing $|\Gamma_1(s)| = \deg(s) + 1$ instead. This is the approach followed by Algorithm Sieve. This fallback can also be combined with Fallback 1, that is, when $|\Gamma_1(s) - \Gamma_R(D)| = 0$ for all candidate sources s.

Each of these fallbacks, or their combination, has a running time of $O(R(m + n))$. The pseudo-code for QGreedy is omitted here, because it is rather long and it shares much code with Sieve. We believe readers should be able to implement it easily based on the explanations of Sieve and those in this section.

Now let us consider an interesting property of QGreedy.

Theorem 2 *If $\Gamma_1(s) - \Gamma_R(D) \neq \emptyset$, then* $\text{dist}(D, s) \geq R$.

Proof 2 $\Gamma_1(s) - \Gamma_R(D) \neq \emptyset$ *implies either s or at least one neighbor of s is not dominated by D. This implies $\ell(s) \leq 1$; hence, $\text{dist}(D, s) \geq R$.* ∎

Theorem 3 *If Algorithm QGreedy finds a set D without using Fallback 2, then $\text{dist}(s, s') \geq R$ for any pair $s \neq s', s, s' \in D$. In other words, the set D is a feasible solution for problem instance (G, S, T, R, z) for all $z \leq R$ if Fallback 2 was not used.*

Proof 3 *Assume QGreedy finds $s_1, s_2, \ldots, s_k$ in that sequence. Let $D_0 = \emptyset$ and $D_i = \{s_1, \ldots, s_i\}$, $i = 1, 2, \ldots, k$. Because QGreedy finds $D = D_k$ without using Fallback 2, we see that for any $i = 1, 2, \ldots, k$, $\Gamma_1(s_i) - \Gamma_R(D_{i-1}) \neq \emptyset$. Hence, $\text{dist}(s_i, s_j) \geq R$ for all $j < i$ by the previous theorem.* ∎

We remark that there are many GDS instances that do not require Fallback 2. One example is when $T \subseteq S$, because if there exists a target $t \in T$ that has not been dominated before, then $\Gamma_1(t) - \Gamma_R(D) \supseteq \{t\} \neq \emptyset$; hence, in this case even Fallback 1 is not used. Another example is when the graph is bipartite with two sides, that is, source side S and target side T, of nodes (recall that problem SC is such a special case). In this case, assuming the existence of a feasible solution, if there exists a target $t \in T$ that has not been dominated, then, for each neighbor $s \in S$ of t, $\Gamma_1(s) - \Gamma_R(D) \supseteq \{t\} \neq \emptyset$; hence, neither Fallback 1 nor Fallback 2 is used.

5.6 Finding Lower Bounds for GDS with $z = 1$

Next, we discuss how to find lower bounds, which is used in the empirical study to estimate the quality of solutions. Let us first consider GDS with $z = 1$, which can be formulated as the following integer programming (IP) problem:

$$
\begin{aligned}
\text{(IP)} \quad & \min \sum_{s \in S} x_s \\
\text{s.t.} \quad & \sum_{t \in \Gamma_R(s)} x_s \geq 1, \quad \forall t \in T \\
& x_s \in \{0, 1\}, \quad \forall s \in S
\end{aligned}
$$

A dual problem (D) can be written as the following:

$$
\begin{aligned}
(\text{D}) \qquad & \max \sum_{t \in T} y_t \\
s.t. \quad & \sum_{t \in \Gamma_R(s)} y_t \leq 1, \quad \forall s \in S \\
& y_t \in \{0, 1\}, \quad \forall t \in T
\end{aligned}
$$

Let x be a feasible solution of (IP), and let y be a feasible solution of (D).

Theorem 4 (Weakly duality theorem) $\sum_{t \in T} y_t \leq \sum_{s \in S} x_s$.

Proof 4

$$
\begin{aligned}
\sum_{t \in T} y_t &= \sum_{t \in T} y_t \cdot 1 \leq \sum_{t \in T} y_t \sum_{t \in \Gamma_R(s)} x_s \\
&= \sum_{s \in S} x_s \sum_{t \in \Gamma_R(s)} y_t \leq \sum_{s \in S} x_s \cdot 1 = \sum_{s \in S} x_s
\end{aligned}
$$

In other words, any feasible solution of (D) is a lower bound of (P). ∎

Notice that Problem (D) is a distance independent set problem, in which the distance between any two selected nodes is required to be at least $2R + 1$. It is known that, even for the special case $S = T = V$, finding such a maximum set is NP-hard even for chordal graphs or bipartite planar graphs of maximum degree three [27]. Moreover, with $\forall \epsilon > 0$, approximating the problem within factor $n^{1/2-\epsilon}$ is also hard (i.e., impossible unless P = NP) even for bipartite graphs [27]. Nevertheless, it is possible to use a greedy method to calculate a maximal solution. A simple implementation using BFS with distance $2R + 1$ as subroutine takes $O(m + n)$ time, as first shown in [13,25]. An interesting observation in our experiments is that the gap between (D) and (IP) is usually not as large in practice.

5.7   Extension to GDS with $z \geq 2$ and Other Variations

As stated before, if QGreedy succeeds in finding a set D without using Fallback 2, then $dist(s, s') \geq R$ for all $s \neq s' \in D$. This implies that the distance between any pair of selected nodes is at least R; hence, QGreedy finds a solution for all $1 \leq z \leq R$. We note that neither Greedy nor Sieve has this feature, because they are unable to construct a distance DS satisfying the independence requirement (without modification).

For the case $z > R$, for which we are presently not aware of any application, we can use another label $s(v) = \max\{0, z + 1 - \min_{d \in D} dist(d, v)\}$ and only adopt node d with $s(d) \leq 1$ that maximizes its residual degree. Label $s(v)$ can be updated in a way similar to $\ell(v)$ in a total running time of $O(z(m+n))$. Therefore, the total running time is $O((R+z)(m+n))$, an extension which may, however, fail to find a feasible solution. See an example in Figure 5.4. Therefore further study is required for this case.

We remark that it is not difficult to extend the heuristic algorithms to directed graphs, robust dominating, edge lengths, connected dominating, and so on. Let us discuss in the following.

- It is trivial to treat partial dominating (i.e., we only need to dominate part of the target sets)—simply stop as soon as the required number of targets have been dominated.

FIGURE 5.4: Illustration of the possible failure of the extended QGreedy algorithm for $z > R$. Here, $S = T = V = \{a, b, c, d\}$, $E = \{(a,b), (b,c), (c,d)\}$, $R = 1$, and $z = 3$. The only feasible solution is $\{a, d\}$, whereas the extended QGreedy (and of course Greedy and Sieve too) first finds b or c and then terminates without finding a feasible solution.

- For directed graphs, the generalization is straightforward as we did not require the graph to be undirected in the earlier-mentioned algorithms. Minor modification of the pseudocode presented in this chapter might be necessary.

- For robust dominating, that is, where either some or all of the target nodes are required to be dominated more than once, we can employ a counter for each target node to remember the number of times it has been dominated and update it each time the node is dominated by a new source. In this case the running time would be $O(dR(m+n))$, where d denotes the maximum requirement of multiple domination. We refer those readers who are interested to [25] for more detail. Although this source only discusses the distance VC problem, it is not difficult to extend the idea to distance DS.

- For heterogeneous edge lengths, we could extend the algorithms to process nonconstant edge lengths, see [25] for details. The running time, however, depends on the edge lengths; thus, it is weakly polynomial rather than strongly polynomial. How to design a strongly polynomial time heuristic is a future challenge.

- For connected dominating, we can choose a source only if it is connected to the previously selected nodes. The problem is that solutions found by this simple heuristic could be infeasible even if a feasible solution existed. This is one of our future projects.

5.8 Empirical Study

In this section, we present a comparative study of the performance of an exact algorithm, named Exact, with Greedy ([21]), Sieve, and QGreedy. Algorithm Exact simply uses a free and widely available Integer Programming solver GLPK (see http://www.gnu.org/software/glpk/). Greedy, Sieve, QGreedy, and the lower-bound algorithm shown in Section 5.6 are implemented in C language and compiled by gcc (http://gcc.gnu.org/). The running time is determined by omitting the time required to read the data file from the hard drive into the main memory, which usually involves more than a single computation.

First of all, we illustrate the five algorithms with a small network, which is a circle (#3980 circle) in the ego-Facebook network of the SNAP data set (see http://snap.stanford.edu/data/egonets-Facebook.html). For easy understanding, the three tiny connected components $\{40, 41\}$, $\{42, 3, 45, 48\}$, and $\{50, 51\}$ were removed. See Figure 5.5 for the comparison.

Next, we test the algorithms with random graphs to assess their scalability. The graphs were generated by GTgraph (http://www.cse.psu.edu/~madduri/software/GTgraph/) with the parameter $m = 10n$. These graphs are supposed to represent small-world complex networks. Because graphs generated by GTgraph are directed and may have multiedges, we added reverse edges to obtain undirected graphs, from which we then removed the multiedges. The final number of edges is roughly $20n$ for all graphs. Details of these graphs are listed in Table 5.5.

The results of the scalability experiment can be found in our paper [15]. All tests were done on a laptop, a Fujitsu Lifebook SH76/HN with an Intel core i5-3210M CPU and 8 GB

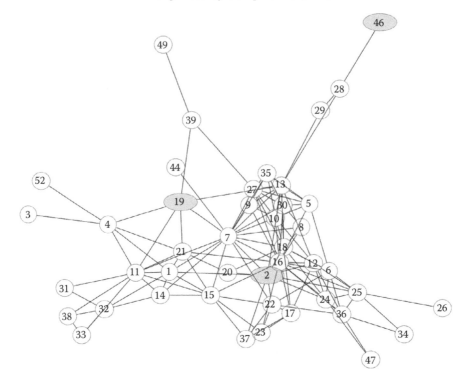

FIGURE 5.5: The #3980 circle in the ego-Facebook network of the SNAP data set. Filled nodes show the optimal solution for $R = 2$ (where $S = T = V$ and $z = 1$). On the other hand, Greedy, Sieve, QGreedy and Lowerbound found $\{7, 19, 16, 13\}$, $\{16, 13, 27, 11\}$, $\{16, 11, 13, 39\}$, and $\{52, 46, 26\}$, respectively.

TABLE 5.5: GTgraph-generated random graphs used in the experiment. As we use adjacent lists to store the graph, m denotes the number of *directed* edges stored in memory, which is twice the number of *undirected* edges

n	100	110	120	130	140	150	160	170	180
m	1,812	1,992	2,180	2,408	2,628	2,824	2,974	3,202	3,402
n	190	200	1,000	10,000		100,000		1,000,000	
m	3,600	3,786	19,824	199,798		1,999,792		19,999,798	

RAM running Linux OS. We cite the results obtained for small graphs in Figure 5.6. For larger instances, the calculation results are shown in Table 5.6. The results indicate that both Sieve and QGreedy usually deliver a reasonable performance. Despite its simplicity, the algorithm used to find the lower bounds performs quite well, with the gap between the lower bound and the solution not being large. We remark that the running time of QGreedy is usually twice as fast as that of Sieve, because there is no need to run a reverse-delete step, and it is usually only several times slower than that of a pure breadth-first-search (BFS).

We also tested many other kinds of real networks and prepared more large-scale instances specifically for this book by employing a small server machine with an Intel(R) Xeon(R) CPU E5-2687W v2 @ 3.40 GHz and 128 GB RAM, running Linux OS (Ubuntu Server 14.04.3 LTS). The data sources include:

- 9th DIMACS: Road networks (to enable a comparison with complex networks). See `http://www.dis.uniroma1.it/challenge9/download.shtml`.

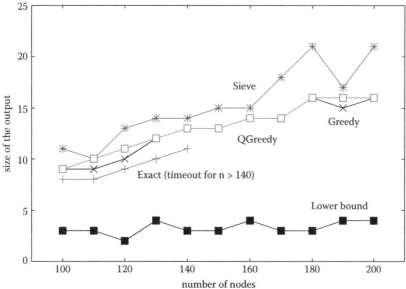

FIGURE 5.6: Results for the small random graphs with a time limit of 180s, where we set $S = T = V$ and $R = z = 1$. The optimal values (Exact) show that the lower bounds are not tight, and it can be seen that the performance of both QGreedy and Greedy is superior to that of Sieve. The running time of all the algorithms, except Exact, was 0.00s. For $R \geq 2$, as the diameters of these instances are at most 2, all algorithms find optimal solutions in almost 0s; hence, they are not shown in this figure. Note that Exact was only able to proceed for $n \leq 140$; thus, it is not considered in subsequent experiments.

TABLE 5.6: Results for large random graphs, where $S = T = V$ and $z = 1$. Each record contains two values "size/time," where size is the size of the output and time is the running time in seconds. As for small GTgraph-generated random graphs, the lower bounds calculated by our algorithm are quite weak. We remark that QGreedy finds distance-R independent sets

n	$1,000$	$10,000$	$100,000$	$1,000,000$
Greedy, $R = 1$	85/0.00	844/1.22	memory out	
Greedy, $R = 2$	7/0.04	72/1.38	memory out	
Greedy, $R = 3$	1/0.10	6/5.52	memory out	
Sieve, $R = 1$	109/0.00	1,013/0.00	10,243/0.06	103,037/0.70
Sieve, $R = 2$	8/0.00	89/0.00	906/0.06	9,181/0.88
Sieve, $R = 3$	1/0.00	7/0.00	70/0.06	679/0.92
QGreedy, $R = 1$	84/0.00	843/0.02	8,469/0.12	84,691/2.38
QGreedy, $R = 2$	13/0.00	122/0.00	1,149/0.14	11,638/2.48
QGreedy, $R = 3$	1/0.00	12/0.00	113/0.14	1,104/2.66
Lowerbound, $R = 1$	20/0.00	191/0.02	1,891/0.10	18,965/1.66
Lowerbound, $R = 2$	1/0.00	2/0.00	13/0.10	147/1.82
Lowerbound, $R = 3$	1/0.00	1/0.00	1/0.04	2/0.96

- SNAP ([28]): Various types of complex networks. See http://snap.stanford.edu/data/.

- GTgraph: GTgraph was used to generate random graphs as before.

We list the networks in Table 5.7, sorted according to size.

TABLE 5.7:   Large-scale networks used in experiments (sorted by size). Again, as we store each graph by using adjacent lists, here m denotes the number of *directed* edges stored in the memory, which matches the number of edges for DIMACS graphs, but contains twice the number of (undirected) edges for SNAP graphs. We remark that for the instance ca-AstoPh, the values of n and m differ from the description provided on SNAP, which we consider a mistake of the SNAP data set.

Name	Source	n	m	File Size
ego-Facebook	SNAP	4,039	176,468	2.0M
ca-AstroPh	SNAP	37,544	792,320	9.9M
loc-Gowalla	SNAP	196,591	1,900,654	25M
com-Amazon	SNAP	334,863	1,851,744	27M
flickr	SNAP	105,938	4,633,896	57M
com-Youtube	SNAP	1,134,890	5,975,248	85M
as-Skitter	SNAP	1,696,415	22,190,596	327M
com-LiveJournal	SNAP	3,997,962	69,362,378	1.1G
USA-road	9th DIMACS	23,947,347	58,333,344	1.3G
com-Orkut	SNAP	3,072,441	234,370,166	3.8G
GTgraph-RMAT	GTgraph	60,000,000	1,199,951,796	22G
com-Friendster	SNAP	65,608,366	3,612,134,270	65G

Name	Description
ego-Facebook	Social circles from Facebook (anonymized)
ca-AstroPh	Collaboration network of Arxiv Astro Physics
loc-Gowalla	Gowalla location-based online social network
com-Amazon	Amazon product network
flickr	Images sharing common metadata on Flickr
com-Youtube	Youtube online social network
as-Skitter	Internet topology graph (2005)
com-LiveJournal	LiveJournal online social network
USA-road	Road network of full USA
com-Orkut	Orkut online social network
GTgraph-RMAT	Random graph generated by GTgraph-RMAT
com-Friendster	Friendster online social network

Experimental results are shown in Figures 5.7 through 5.18. In these tests, we set $S = T = V$ and $z = 1$. However, as mentioned before, we remark that QGreedy finds solutions for $z = R$.

These results indicate that both Sieve and QGreedy usually perform nicely, because the difference between the results and the lower bounds are usually not large. Sieve usually delivers a solution of which the quality exceeds that of QGreedy (we remark that QGreedy finds distance-R independent sets as well which is not available for Sieve.) Depending on the type of network, the running time of Sieve is not as stable as that of QGreedy; in fact, QGreedy is usually faster than Sieve, because it checks far fewer redundant nodes than Sieve. We also compared the performance of our algorithms against that of a pure BFS algorithm for the same networks and found our algorithms to usually require no more than several times the running time of the pure BFS algorithm.

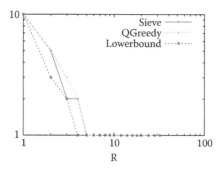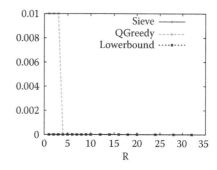

FIGURE 5.7: Experimental result for network ego-Facebook (diameter 8). Left: size of the output (smaller is better); right: running time in seconds (cf. the running time of pure BFS algorithm: 0.00s). Here, only QGreedy took 0.01s for $R = 1, 2, 3$. Other records were all 0.00s. Notice that 10 (i.e., $10/4039 \approx 0.25\%$) nodes is enough to dominate (with distance $R = 1$) the whole network. Further experiment shows that, if we are satisfied with 50% domination of the nodes, then only 3, that is, $3/4039 \approx 0.07\%$, nodes is enough. To some extent, this matches a survey conducted by Yahoo! research [29] saying that 50% of tweets come from only 0.05% of twitter uses.

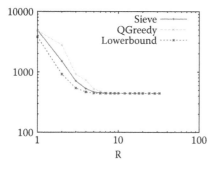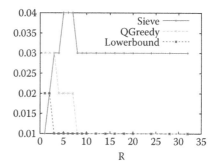

FIGURE 5.8: Experimental result for network ca-AstroPh (diameter 14). Left: size of the output (smaller is better); Right: running time in seconds (cf. the running time of a pure BFS algorithm: 0.01s). Notice that the size of the output converges on the number of the connected components, which is 442. We remark that Sieve is not as stable as QGreedy when R increases. This can also be observed in the following results.

From these results, it can be said that the size of a minimum distance-R DS decreases exponentially as R increases (notice that the lower bounds are also calculated). This is reasonable as the dominating size of a node increases exponentially as the size of R increases.

Another observation concerns the relationship between R and the diameter d^* of the network. Let C denote the number of connected components (1 if the network is connected). Obviously, the size of an *optimal* solution for $R = \lceil d^*/2 \rceil$ is required to equal C, and the size of any *feasible* solution for $R \geq d^*$ would have to equal C as well. This can be observed in the figures in which the experimental results are shown.

Finally, we note again that a solution can be viewed as representing the set of "important" nodes in the network, because it dominates (with respect to distance R) the network. However, there can exist many such sets of nodes. How to find a stable or unique DS is left as a future work.

To finish this section, we show some visualizations of the experimental results in Figures 5.19 and 5.20.

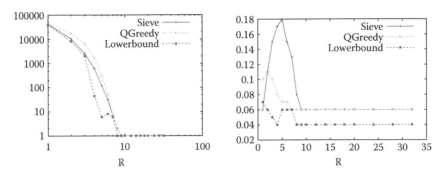

FIGURE 5.9: Experimental result for network loc-Gowalla (diameter 14). Left: size of the output (smaller is better); right: running time in seconds (cf. the running time of a pure BFS algorithm: 0.03s).

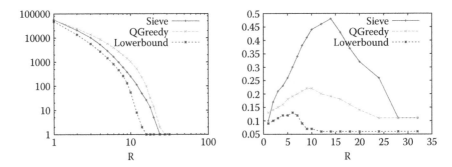

FIGURE 5.10: Experimental result for network com-Amazon (diameter 44). Left: size of the output (smaller is better); right: running time in seconds (cf. the running time of a pure BFS algorithm: 0.05s).

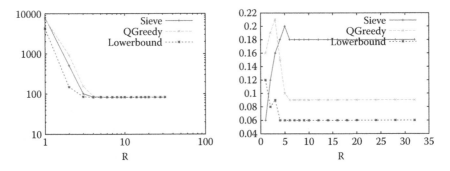

FIGURE 5.11: Experimental result for network flickr (diameter 9). Left: size of the output (smaller is better); right: running time in seconds (cf. the running time of a pure BFS algorithm: 0.04s).

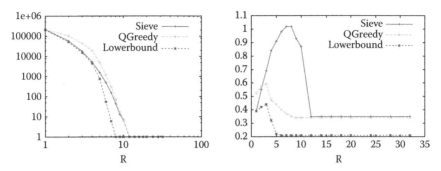

FIGURE 5.12: Experimental result for network com-Youtube (diameter 20). Left: size of the output (smaller is better); right: running time in seconds (cf. the running time of a pure BFS algorithm: 0.16s).

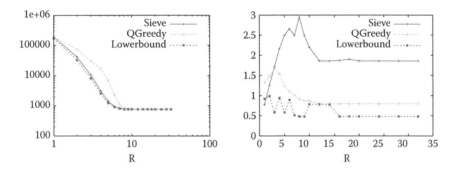

FIGURE 5.13: Experimental result for network as-Skitter (diameter 25). Left: size of the output (smaller is better); right: running time in seconds (cf. the running time of a pure BFS algorithm: 0.24s).

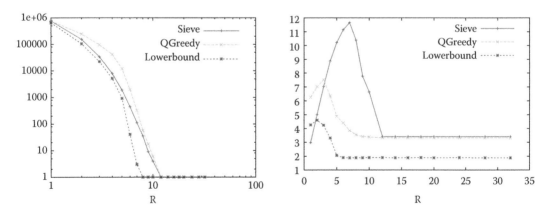

FIGURE 5.14: Experimental result for network com-LiveJournal (diameter 17). Left: size of the output (smaller is better); right: running time in seconds (cf. the running time of a pure BFS algorithm: 1.41s).

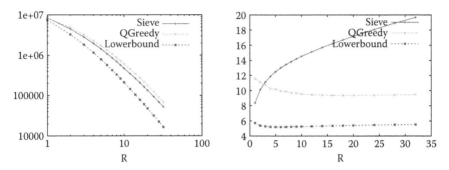

FIGURE 5.15: Experimental result for network USA-road. Left: size of the output (smaller is better); right: running time in seconds (cf. the running time of a pure BFS algorithm: 2.73s).

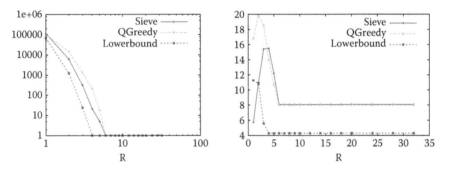

FIGURE 5.16: Experimental result for network com-Orkut (diameter 9). Left: size of the output (smaller is better); right: running time in seconds (cf. the running time of a pure BFS algorithm: 2.81s).

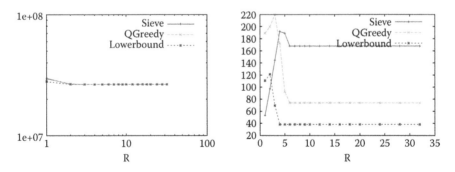

FIGURE 5.17: Experimental result for network GTgraph-RMAT. Left: size of the output (smaller is better); right: running time in seconds (cf. the running time of a pure BFS algorithm: 27.93s).

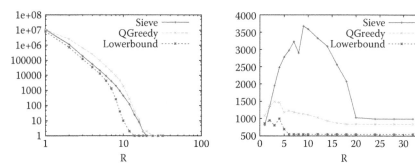

FIGURE 5.18: Experimental result for network com-Friendster (diameter 32). Left: size of the output (smaller is better); right: running time in seconds (cf. the running time of a pure BFS algorithm: 338.89s). Notice that Sieve is not as stable as QGreedy when R increases.

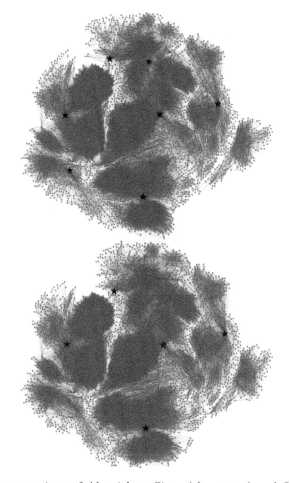

FIGURE 5.19: A comparison of Algorithms Sieve (the upper) and QGreedy (the lower). The network is ego-Facebook with parameters $S = T = V$ and $R = z = 1$ and was visualized by GraphViz. We stopped both algorithms when 90% of nodes had been dominated. Star nodes (seven for Sieve and six for QGreedy) show the solutions found.

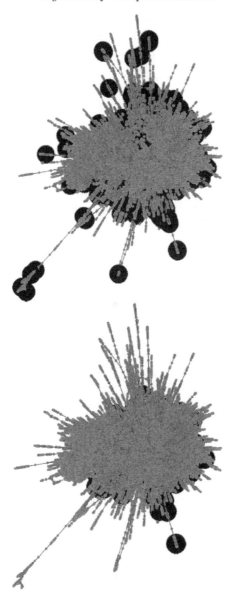

FIGURE 5.20: A comparison of domination distances $R = 1$ (the upper) and $R = 2$ (the lower). The network is com-Amazon with $S = T = V$ and $z = 1$ and was visualized by GraphViz. We stopped the calculation when 50% of nodes had been dominated. Big nodes $(10,956$ for $R = 1$ and $2,940$ for $R = 2)$ show the solutions found.

5.9 Summary and Conclusion

In this chapter, we defined a GDS problem and discussed its applications in Big Data analysis. We demonstrated two very efficient $(O(R(m + n))$ time) algorithms, Sieve and Quasi-Greedy, for finding small (independent) DSs. The experimental results showed that the running time is usually less than several times that of a pure BFS algorithm, and the size of the solution is usually found to be within several times of the lower bound. Although Sieve usually succeeds in finding smaller distance DSs, QGreedy has the advantages of speed and independence, which are considered important in many applications (e.g., wireless sensor networks [11,22] and parallel computing [16]). In conclusion, these two tools are suitable for estimating the ease with which (independent) dominating can be achieved for a given large-scale network.

Our future plans include achieving some theoretical guarantee, and extending the algorithms to other types of dominating problems. Of course, applying the tools described in this chapter to analyze other Big Data problems would be another exciting task.

Acknowledgments

This work was partially supported by JSPS KAKENHI Grant Numbers 23700018 and 25330026. I would like to thank all the coauthors of our papers [13,15,25] for their contributions to this work. They are (in alphabetical order of family name): Jaeseong Gim, Hiroshi Kadowaki, Hiroshi Nagamochi, Masahiro Sasaki, Dorothea Wagner, and Jiexun Wang. I am also deeply grateful to Prof. Matthias Dehmer for his proposal of this book, and his patience and comments.

References

1. Wayne Goddard and Michael A. Henning. Independent domination in graphs: a survey and recent results. *Discrete Math.*, 313(7):839–854, 2013.

2. T.W. Haynes, S. Hedetniemi, and P. Slater. *Domination in Graphs: Volume 2: Advanced Topics*. Chapman & Hall/CRC Pure and Applied Mathematics. Marcel Dekker Inc., New York, 1998.

3. T.W. Haynes, S. Hedetniemi, and P. Slater. *Fundamentals of Domination in Graphs*. Chapman & Hall/CRC Pure and Applied Mathematics. Marcel Dekker Inc., New York, 1998.

4. Anush Poghosyan. *The Probabilistic Method for Upper Bounds in Domination Theory*. PhD thesis, University of the West of England, Bristol, UK, 2010.

5. Matthew Beckerich Alina Campan, Traian Marius Truta. Fast dominating set algorithms for social networks. In *Proceedings of the 26th Modern AI and Cognitive Science Conference 2015*, pages 55–62, 2015.

6. F. Molnár, S. Sreenivasan, B. K. Szymanski, and G. Korniss. Minimum dominating sets in scale-free network ensembles. *Scientific Reports*, 3:1736, April 2013.

7. Jose C. Nacher and Tatsuya Akutsu. Analysis on controlling complex networks based on dominating sets. In *Journal of Physics: Conference Series 410 (2013) 012104*, 2013.

8. Jose C. Nacher and Tatsuya Akutsu. Structural controllability of unidirectional bipartite networks. *Scientific Reports*, 3(1647), 2013.

9. Xiao-Fei Zhang, Le Ou-Yang, Yuan Zhu, Meng-Yun Wu, and Dao-Qing Dai. Determining minimum set of driver nodes in protein-protein interaction networks. *BMC Bioinformatics*, 16(1), 2015.

10. D.Z. Du and P.J. Wan. *Connected Dominating Set: Theory and Applications*. Springer Optimization and Its Applications. Springer New York, 2012.

11. Johann L. Hurink and Tim Nieberg. Approximating minimum independent dominating sets in wireless networks. *Information Processing Letters*, 109(2):155–160, 2008.

12. Bernhard Korte and Jens Vygen. *Combinatorial Optimization: Theory and Algorithms*. Springer, Berlin, Heidelberg, Germany, 5th edition, 2012.

13. Masahiro Sasaki, Liang Zhao, and Hiroshi Nagamochi. Security-aware beacon based network monitoring. In *Communication Systems, 2008. ICCS 2008. 11th IEEE Singapore International Conference on*, pages 527–531, Nov 2008.

14. A. Behzad, M. Behzad, and C. E. Praeger. Fundamental Dominations in Graphs. *ArXiv e-prints*, August 2008.

15. Liang Zhao, Hiroshi Kadowaki, and Dorothea Wagner. A practical approach for finding small independent, distance dominating sets in large-scale graphs. In Rocco Aversa, Joanna Koodziej, Jun Zhang, Flora Amato, and Giancarlo Fortino, editors, *Algorithms and Architectures for Parallel Processing*, volume 8286 of *Lecture Notes in Computer Science*, pages 157–164. Springer, Berlin, Heidelberg, Germany, 2013.

16. S. Devismes, K. Heurtefeux, Y. Rivierre, A.K. Datta, and L.L. Larmore. Self-stabilizing small k-dominating sets. In *Networking and Computing (ICNC), 2011 Second International Conference on*, pages 30–39, Nov 2011.

17. M. V. Marathe, H. Breu, H. B. Hunt, III, S. S. Ravi, and D. J. Rosenkrantz. Simple heuristics for unit disk graphs. *Networks*, 25(2):59–68, 1995.

18. Ran Raz and Shmuel Safra. A sub-constant error-probability low-degree test, and a sub-constant error-probability PCP characterization of NP. In *Proceedings of the Twenty-Ninth Annual ACM Symposium on Theory of Computing*, STOC '97, pages 475–484, New York, NY, USA, 1997. ACM.

19. Magnús M. Halldórsson. Approximating the minimum maximal independence number. *Inform. Process. Lett.*, 46(4):169–172, 1993.

20. Yilin Shen, Dung T. Nguyen, and My T. Thai. Hardness complexity of optimal substructure problems on power-law graphs. In My T. Thai and Panos M. Pardalos, editors, *Handbook of Optimization in Complex Networks*, volume 57 of *Springer Optimization and Its Applications*, pages 255–277. Springer US, New York, 2012.

21. David S. Johnson. Approximation algorithms for combinatorial problems. *J. Comput. System Sci.*, 9:256–278, 1974. Fifth Annual ACM Symposium on the Theory of Computing (Austin, Tex., 1973).

22. Dhia Mahjoub and David W. Matula. Experimental study of independent and dominating sets in wireless sensor networks using graph coloring algorithms. In Benyuan Liu, Azer Bestavros, Ding-Zhu Du, and Jie Wang, editors, *Wireless Algorithms, Systems, and Applications*, volume 5682 of *Lecture Notes in Computer Science*, pages 32–42. Springer, Berlin, Heidelberg, 2009.

23. Anupama Potluri and Atul Negi. Some observations on algorithms for computing minimum independent dominating set. In Srinivas Aluru, Sanghamitra Bandyopadhyay, Umit V. Catalyurek, Devdatt P. Dubhashi, Phillip H. Jones, Manish Parashar, and Bertil Schmidt, editors, *Contemporary Computing*, volume 168 of *Communications in Computer and Information Science*, pages 57–68. Springer, Berlin, Heidelberg, 2011.

24. Masahiro Sasaki. *Algorithms and Analysis for Optimal Beacon Placement in Large-scale Network*. Master's thesis, Graduate School of Informatics, Kyoto University, Japan, 2009.

25. Jiexun Wang, Jaeseong Gim, Masahiro Sasaki, Liang Zhao, and Hiroshi Nagamochi. Efficient approximate algorithms for the beacon placement and its dual problem (abstract). In *Computational Intelligence and Software Engineering, 2009. CiSE 2009. International Conference on*, pages 1–4, Dec 2009.

26. Thomas H. Cormen, Charles E. Leiserson, Ronald L. Rivest, and Clifford Stein. *Introduction to Algorithms, Third Edition*. The MIT Press, Cambridge MA, 2009.

27. Hiroshi Eto, Fengrui Guo, and Eiji Miyano. Distance-d independent set problems for bipartite and chordal graphs. *Journal of Combinatorial Optimization*, 27(1):88–99, 2014.

28. Jure Leskovec and Andrej Krevl. SNAP Datasets: Stanford large network dataset collection. http://snap.stanford.edu/data, June 2014.

29. S. Wu, J.M. Hofman, W.A. Mason, and D.J. Watts. Who says what to whom on twitter. In *Proceedings of the 20th International Conference on World Wide Web*, page 705–714. ACM, 2011.

Chapter 6

Techniques for the Management and Querying of Big Data in Large-Scale Communication Networks

Yacine Djemaiel and Noureddine Boudriga

6.1   Introduction

The management of Big Data and its processing in an efficient manner is among the urgent needs for the research communities to ensure their correct exploitation for the different available services through interconnected complex networks. Modeling complex networks is among the needs that are discussed to represent appropriately their content and the interaction between their different components in a way that is coherent with their characteristics and constraints. There is a great need to model such networks before focusing on the processing and the management of Big Data through such networks. Adopting the representation of complex networks as a graph helps to manage them and to interrogate their contents, which is needed even by available applications-providing services on such networks or for the Big Data analytics that are performed for several fields such as climate networks and passenger train services. Through complex networks, it is needed in most cases to deal with huge volumes of data that are generated by available devices and that should be either processed or stored by the attached processing and storage resources. The processing of Big Data for complex networks is ensured by following a set of techniques considering the constraints of such networks. Even if a set of approaches have been proposed for the complex networks, several issues need to be studied to enhance the processing and the management of such networks considering both the constraints related to the handled Big Data and the complexity of such networks.

In this chapter, a modeling approach of complex networks is provided first by giving a representation of such networks. Then, the management and the processing of Big Data through complex networks are detailed. The application of the proposed models and Big Data analytics is illustrated through a set of deployed services that are provided by complex networks. Enhancements for the management and the processing of Big Data for large-scale communication networks are discussed at the end of this chapter.

6.2   Modeling Complex Networks Using Graphs

The management of Big Data for complex networks is facilitated if these latter are modeled using a structure whose management is easier. Graphs are among the structures that fulfill these needs enabling the capture of global properties of such systems where nodes represent the dynamic units of complex networks and links stand for the interaction between these units.

6.2.1   Graph structure for complex networks

A complex network can be represented as a graph, as described by Boccaletti et al. [1]. A undirected (directed) graph $G = (N, L)$ consists of two sets N and L, such that N is a nonempty set ($N \neq \emptyset$) and L is a set of unordered (ordered) pairs of elements of N. $\{n_1, n_2, \ldots, n_N\}$ are the elements of N that represent the nodes (or vertices, or points) of the graph G, while the elements of $L \equiv \{l_1, l_2, \ldots, l_K\}$ are the links or edges where N and K are, respectively, the cardinality of N and L.

In the set N, each node is usually referred to by its order i in this set. In an undirected graph, each of the links is defined by a couple of nodes i and j, denoted as (i, j) or l_{ij}. Two nodes joined by a link are considered adjacent or neighboring. A graph is connected if, for every pair of distinct nodes i and j, there is a path from i to j, otherwise it is considered disconnected. Considering this structure, it is possible to manage the interaction between nodes; the possible paths for routing may also be identified through this representation. Moreover, several additional parameters may be deduced using this structure that may be useful for applications running on such networks and for interconnected nodes, such as the number of nodes in a defined path, the set of nodes that have the same distance from a particular node that may be a gateway or a central node ensuring management and control, the number of edges connected to a node that may provide information on the set of nodes that may be discovered or accessed directly, and so on.

6.2.2 Motif concept for complex networks

A motif M, as defined in [1], is a pattern of interconnections occurring either in an undirected or in a directed graph G at a number significantly higher than in randomized versions of the graph. This concept is useful for complex networks, which enables the identification of the groups of nodes such that there is a higher density of edges within groups than between them. This feature is requested for many services running on complex networks. The identification of the significant motifs in a graph G is based on matching algorithms counting the total number of occurrences of each n-node subgraph M in the original graph G and in the randomized ones. A motif is also needed for the setting of routing strategies through complex networks and helps to enhance the delivery of network traffic to the requested destination. Moreover, this concept helps to render the network less complex by hiding the nodes that belong to a motif while keeping the interaction between motifs and preserving the topology of the complex network. The routing for complex networks is simplified using motifs since it is possible in this case to specify a routing strategy inside motifs and another strategy between motifs. A motif may also be viewed by analogy as a cluster in ad-hoc networks, which is known as a component that simplifies the routing in such networks.

6.3 Big Data Processing for Complex Networks

The processing of Big Data in complex networks requires the use of appropriate querying approaches that support the constraints of such environments and that provide a result within a tolerable processing time. Moreover, the processing of Big Data in such environments should be enhanced by applying novel approaches that enable the optimization of the time processing.

6.3.1 Optimizing processing by an asymptotic scheduling for multitask computing in Big Data platforms

Nowadays, there is a great need to efficiently process very large data sets. To fulfill this need, the research community has been interested in solving scheduling problems since this issue affects the processing time for handled Big Data by applications available through complex networks. The problem of scheduling a set of jobs across a set of machines and specifically analyzing the behavior of the system when it is overloaded and that correspond to Big Data processing is discussed in [2]. It is shown through this work that under certain conditions it can be easy to discover the best scheduling algorithm, prove its optimality, and compute its asymptotic throughput.

The observation is made on the transition period from low incoming job rate load to the very high load and back. Several algorithms experience poor performance over the transition periods. Since the asymptotically optimal algorithm makes the assumption of an infinite number of jobs, it can be used after the transition, when the job buffers are saturated. As the other scheduling algorithms are efficient under reduced load, they are combined into a single hybrid algorithm to empirically determine what the best switch point is, which leads to an asymptotic scheduling mechanism for multitask computing used in Big Data processing platforms. In this case, a finite set of jobs is given and each job consists of the execution of an undetermined number of tasks. A task represents a sequence of specific operations and its length depends on the machine on which it is scheduled to be processed. Moreover, a cluster consists of a limited number of heterogeneous processing nodes; one processing node can process a specified numbers of operations at a time and preemption is not allowed. The objective of this approach is to maximize the total number of jobs that are processed at up to a certain point. There are only three stages for each job (read, process, and write) and all of them must be scheduled on the same processing node. The scheduling problem consists in this case of pairing jobs with processing nodes in such a way that the total value is maximized. In this context, it is possible to achieve the highest possible number of completed tasks per time unit (by proving the optimality of the scheduling algorithm) when considering an infinite number of jobs.

6.3.2 Green algorithm for sampling complex networks

The processing of Big Data on complex networks may introduce a processing overhead that may engender even a denial of service. Moreover, the number of nodes and edges of complex networks may dynamically change. To reduce the processing overhead, the use of sampling algorithms can provide a green and energy-efficient way to process huge volumes of data. Using such algorithms, a small-scale sampled network may be obtained which maintains the main characteristics of the complex network (original network), especially its community structure. The community structure, as introduced in [3], depicts the aggression of nodes and the uneven distribution of edges. In most cases, nodes in the same community tend to connect closely, while nodes in different communities have fewer edges. Moreover, a community with a close community structure may get a good sampling network at a lower sampling rate, which can save more energy and time.

The sampling scheme, as detailed in [4], operates as follows: the community coefficient of every node in a network is computed; then community cluster centers of the network are selected. After that, the PageRank algorithm, introduced in [5], is used to get a rank of nodes in the network. The rank of nodes helps to find marginal nodes. The sampling starts from a community cluster center, which has the lowest rank. The selection of a lower ranked node is used to protect small communities.

6.3.3 Navigation and searching in complex networks

According to a set of experiments performed on social networks, two interesting aspects may be observed. The former is that it is possible to find relatively short paths in large-size networks. The latter is that individuals are able to find short paths even in the absence of global knowledge regarding the complex network. Two main approaches are used to construct a path from an initial node to a target node. The first approach uses local or geometric information while the second approach aims at assessing the most efficient network structure for path searching, once the way of finding a path is fixed.

6.3.3.1 Searching approach using local information

An iterative algorithm, detailed in [6], is used to find the shortest path by following a set of steps. At the initial step, an agent is put in the source vertex. The agent inspects the neighborhood of the source vertex. If the target vertex is not among those neighbors then at the next step, agents are spread to all those elements that belong to the set of neighborhoods that have not been reached by any other agent in the preceding times. In the next iterations, each new agent repeats the same procedure of verifying and spreading up to the instant in which the target vertex is attained by a given agent. This method allows the identification of the short path but it requires the presence of multi-agents and it needs $O(N)$ computation times to construct the path. If a single agent is used to construct paths between two vertices, it is needed in this case to investigate random walks on complex networks.

6.3.3.2 Network navigability

The second method aims to find the network structure that optimizes a specific search algorithm. In [6], a study of the optimal network structure related with the problem of avoiding congestion effects that arise when parallel searches are performed is given. When a single search is performed (single search problem), the optimal network is a highly polarized star-like structure. This structure is considered simple and efficient in terms of searchability due to the bounded average number of steps to find a given node, independently of the size of the system. The polarized star-like structure will be inefficient when several search processes coexist in the network, due to the limited capacity of the central node. Another approach proposed for social networks and detailed in [7] considers individuals hierarchically grouped into categories, according to their social identities. According to this approach, a tree-like hierarchy is generated that defines a social distance between any two nodes. A network is constructed by connecting two individuals with a probability which is larger for shorter computed social distances. This probability decays off exponentially for longer social distances. The message that should be communicated to a target person passes to that neighbor of the current holder who has the shortest social distance to the target.

6.4 Optimizing Big Data Management

Big Data management is among the mandatory tasks that may introduce an overhead at different levels: storage and processing. To ensure management without introducing such overhead that may prevent the management of handled Big Data, it is required to adopt appropriate representation of Big Data to deal with these issues. In this section, two approaches are discussed that help to optimize Big Data management.

6.4.1 A mark-based approach for Big Data management

Huge volumes of data are generated by running services through interconnected large-scale networks for each instant. These data may be structured, semistructured, or unstructured. This data that exceeds the capacity of commonly used hardware environments and software tools to capture, manage, and process it within a tolerable elapsed time for its user population is called Big Data, as defined in [8].

The existing schemes are unable to retrieve all the data related to generated queries, since it is not represented in an efficient manner that helps to localize the needed data by reducing the size of the structures in addition to the retrieving processing time. In some cases, a structured query language is not applicable for Big Data since a significant part of the handled data is unstructured. Moreover, the access time to the requested data may be affected by the important size of data. According to the available schemes, it is difficult to track changes to the data and to their associated security levels. Collected data from different sources may be related either due to the dependent services or the input/output relationships that may exist. However, these relationships are not preserved and considered by available management systems when handling data. To deal with such limitations of available schemes, a novel approach is proposed in [9] for representing Big Data and interrogating storage devices in an efficient manner by using extended conceptual graphs (CGs) in addition to the introduction of the mark concept. A mark, according to this approach, is considered a structure used to define a set of attributes that help to localize the data and describe it to enhance the localization and the aggregation of the needed data as a response to a query. The contribution of this model is fourfold: (1) the definition of a novel management scheme for Big Data by using a new type of CGs; (2) the use of extended CGs to represent Big Data; (3) the proposal of a structured querying language that interrogates Big Data by exploring CGs based on dynamic marks; and (4) the ability to trace interrelated data and to learn new attack strategies based on the CG content, by considering mainly the marks.

The proposed technique for the management of Big Data is inspired from the model proposed in [10] that is defined initially for video data. The proposed model is defined by: $DM = (D; Map_1; M; Map_2; C; Map_3; KB)$ where D is the set of data to be handled that may be a file, a video, stream, or a set of blocks in general that are denoted as $[b_i]$ where $i = 1, \ldots, n$. The mapping relation $\mathbf{Map_1}$ represents the correspondence between the marks and the data. This mapping is, in general, a many-to-many relationship since, in this way, a mark could be shared among data or data (e.g., if it is fragmented) could have multiple marks. This mapping allows the identification of the needed data as a response to a query since the querying language enables the identification of the marks. $\mathbf{M}$ represents the defined set of marks. A mark is represented as a structure $\mathbf{m} =$ {id_mark, source, target, user, id, concepts, predicates}, where the mark is identified by an **id_mark**. The **source** represents the application that has generated the data. The data represents the concept that is described by the target field. For each data, there is a **user** that has generated it through an application. The **id** field holds the identification of the data associated to the mark. The field **concepts** is composed of significant words that describe the data. The last field contains the set of predicates that defines the different relationships between the mark and the set of concepts in the CG that are associated to the data.

$\mathbf{Map_2}$ is a mapping relation that defines the relationships between a mark M and its concepts $\mathbf{C_1} \ldots \mathbf{C_n}$. A concept may be attached to more than one mark. This mapping is required to enhance the search and the retrieving process since it holds the mapping between the mark and the different concepts that represent the same data and that may be used as a criterion in the defined query by the user. $\mathbf{Map_3}$ is a mapping relation that associates a subset (application knowledge) of the $\mathbf{KB}$ (defined later) to M since only this knowledge is directly derived from M. It holds the relationships that may exist between the different concepts belonging to the same mark. An additional level of marking is introduced to enhance the response to a query by providing the fragment of data that is needed instead of the whole data (e.g., entries in a log file). To achieve this goal, a mark for a fragment of data is added. In this case, a set of concepts are affected for a fragment of data and a relation is defined. The definition of fragments depends on the used application and the type of data.

KB is the knowledge database that is encoded according to the CG knowledge representation scheme. It includes the different concepts, and the existing relationships between them. The links between the concepts are defined using the set of available predicates.

The marking is performed by the application that generates the data based on the KB content associated to the handled data. The set of concepts and predicates are determined by querying the KB. The generated graphs are stored in a distributed manner at each network that deploys the proposed management scheme for Big Data.

6.4.1.1 The use of conceptual graphs for Big Data

To enhance the retrieving of the required data as a response to a query, the data will be represented using a set of concepts and a set of relations between them. These concepts and the defined relations are added to the CG. Each generated mark associated to data to be stored (text file, video, etc.) holds a set of concepts that are needed to build the CG.

A CG is a finite, connected, bipartite graph composed of concepts and conceptual relations. The basic components in the knowledge base are CGs, concepts, and conceptual relations. Using a Prolog-like notation, CGs are defined as predicates of the form that is defined in [10] as follows: cg(ID,RelationList) where **ID** is a unique identifier associated with this **CG** and **RelationList** is a Prolog list that stores the conceptual relations of the specific CG. A conceptual relation is defined as: cgr(RelationName, ConceptIDs) where **ConceptIDs** is a list of concept identifiers linked by the specific conceptual relation. **RelationName** is the name of the conceptual relation. Concepts are represented as predicates of the form: cgc(ID,keywordList, Context, ConceptName) where ID is a unique identifier associated with this concept; **keywordList** is a Prolog list of identifiers of the keywords containing this concept; **Context** is either normal for the case of normal concepts or special for the case of marks; and **ConceptName** is the type-name of a normal concept or the mark label.

The predicates are used to define the relations between the different concepts. The operators ˆ and ∨ are used to represent relationships between the different concepts associated to handled data. The first operator is used as an AND between predicates defined by $predicate_1$, $predicate_2$ linking concepts $concept_1$, $concept_2$, and $concept_3$. The second operator represents an OR between predicates.

Figure 6.1 shows the proposed mark-based CG where data (e.g., D1, D2) are attached to the corresponding marks (e.g., m1, m2) that hold information needed to represent the data and that is used to ensure the querying. Each piece of data is represented through a set of concepts and relations (e.g., C1 and R1) that are defined in marks. Relations and concepts may be shared by several pieces of data such as the relation R2 that is shared for D1 and D2.

6.4.1.2 A novel structured query language for Big Data

The proposed novel structured query language for Big Data, called NSQL-DB [9], is an SQL-like querying language that interrogates available marks and explores available CGs to determine the needed data even if it is structured or not. In this section, the syntax of NSQL-DB, the retrieving and updating processes according to such language, are explained.

According to NSQL-BD, the retrieving query is formulated using the following syntax:

select concepts where (mark . field operator value | concept operator value | predicates) + .

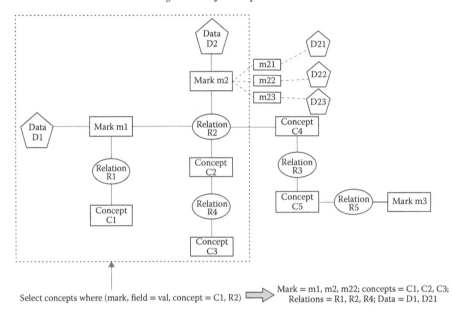

FIGURE 6.1: Illustration of the retrieving process.

The querying is performed based on a combination of several criteria including some mark field values, concept values and predicates that are evaluated for concepts. The operator "+" means that the query may include multiple criteria. The operator keyword represents the basic operators such as $=$, $\geq$, $\leq$, and so on, in addition to the proposed relations such as Boolean (AND, OR, etc.), temporal adjacency (ADJ) and interval operators (DURING, etc.). As a response to a query, a set of concepts are identified in the processed CGs that enable the identification of the related predicates and marks that are attached to the requested data.

The update of the CG content may be performed using NSQL-BD by issuing a query according to the following syntax: update mark|concept|predicate

$$\text{set concept} = \text{value}|\text{mark.field} = \text{value}|\text{predicate} = \text{value}$$

The update using the aforementioned expression may concern all components of the CG including the marks, concepts, and predicates by affecting the defined values using the set keyword. If the specified criteria do not exist, the structure is added, then the content is updated. For example, if the query specifies the following arguments for the set of keywords: mark.id = val, mark.id data = id, mark.concepts = file; session; hidden then the mark id is checked first; if it does not exist then a novel mark is generated and the provided criteria are affected to the mark fields.

As defined in [11], a meta-model is a model that describes the concepts and relationships of a certain domain. It is commonly defined by means of an object-oriented conceptual model expressed in a metamodeling language. In the case of mark-based CGs, model querying is needed. As models grow in size and complexity, the need for model persistence and model querying solutions arises to efficiently store large models and obtain information from them in an efficient, usable, and safe way.

6.4.1.3 Retrieving process

The fetching of the needed data as a response to a query is performed on the mark content instead of the huge volumes of stored data. The check is performed first for the set of mark groups. To enhance the retrieving process, a process of selection of group marks is initially performed. This step is ensured by considering the type of data that is required to be collected by considering several criteria including: the type of application that has generated the required data on a complex network (e.g., a video application, a network analysis tool, etc.), the kind of data (e.g., text, video, binary, etc.), the target of the needed data (e.g., data related to a teacher, a service, etc.), and so on.

The first step is followed by the identification of sub-marks as a response to the issued query. These "sub" marks help to identify sub-CGs that include the set of concepts and predicates which are related to the needed data or fragments of data. Figure 6.1 illustrates an example of the retrieving process for a given sample partial graph and a sample query according to the proposed NSQL-BD language.

6.4.2   Mark-based temporal conceptual graphs for enhancing Big Data management

The management of Big Data, as detailed for the aforementioned scheme (Section 6.4.1) is mainly affected by the size of the big graph data that represents the huge volumes of data. The size of this structure may increase within the size of data to be handled over time. Facing this issue, the querying time may be affected and the introduced delay may not be tolerated by running applications on a complex network. In addition, the investigation of attacks through the collected massive data could not be ensured using traditional approaches, which do not support Big Data constraints. In this context, a novel temporal conceptual graph (TCG) to represent Big Data and to optimize the size of the derived graph is proposed in [12]. The proposed scheme built on this novel graph structure enables tracing back of attacks through time using Big Data.

Generally, when dealing with Big Data, the first concern lies in determining how to manage the increasing volume of data to extract the most relevant data (useful knowledge). Several studies have been conducted and a number of analytics (such as data mining) have been applied but none of them provide a full coverage of Big Data issues and the related environments where these data will be held. Recent works have been interested in applying a (big) graph to manage and organize complex data. This technique allows parsing data into a set of interrelated clusters. It presents several benefits related to analyzing easily and storing efficiently semistructured/unstructured information, providing clear phrasing for queries about relationships, and solving new challenges for analysis (including data sizes, heterogeneity, uncertainty, and data quality) [13]. However, the applied technique lacks dynamicity since the developed graph is static and cannot evolve over time.

6.4.2.1   The proposed model

The proposed TCG introduces a time feature to the previous model introduced in Section 6.4.1, and that is defined as $TCG = (t; D(t); M(t); C(t); Map1(t); Map2(t); KB(t); Map3(t))$ where t is the instant where the TCG is generated; the remaining parameters are also defined at the same instant t.

Figure 6.2 illustrates a sample TCG according to the proposed model, generated for three instants (t_1, t_2, and t_3) in addition to the content of the CG associated to the previous scheme (Section 6.4.1), for the same instants. According to this example, the graph size is largely reduced compared to the first graph observed for the three instants. This enhancement is due to the property of the proposed TCG that includes only the nodes that have been created for the observed instant t or updated from a previous instant t' (where $t' < t$). For the provided example, $M1_{t2}$ and $M2_{t2}$ are the updated marks of $M1_{t1}$ and $M2_{t1}$ and they are linked to the TCG_{t2} as shown in Figure 6.2 using a dashed line.

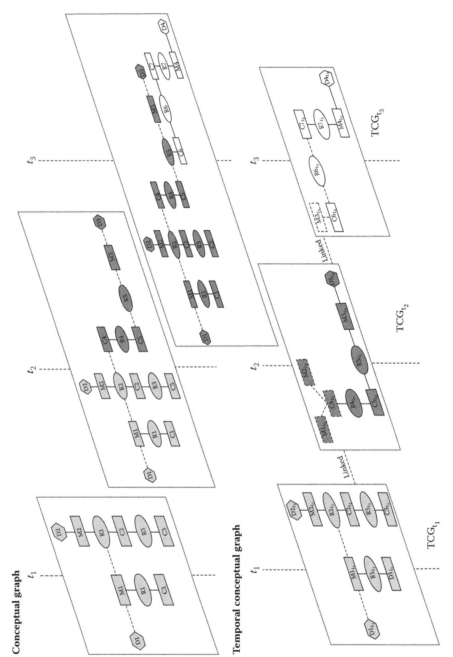

FIGURE 6.2: Illustration of the traditional and novel temporal conceptual graphs.

According to this property, the obtained graph representing Big Data at an instant t is considered a smart big graph data. This feature enables the optimization of the size of the graph without losing knowledge about the stored data in addition to the dependent data through time. Also, the querying of a TCG for an instant t or for an interval of time will be optimized compared with the interrogation of a CG. The storage of data and the management of marks are also enhanced since for each instant, a portion of the data is handled that corresponds to the modifications made at that instant.

6.4.2.2 Attack scenario reconstruction process

The reconstruction of the attack scenario is among the services that may be ensured using the aforementioned management scheme by following a set of steps that starts by fetching the set of linked TCGs starting from the detection time, denoted t_d, associated to the reported malicious event as illustrated in Figure 6.3. Based on the detection time t_d and the associated reported event (the performed malicious action), the set of data and their associated mark are extracted from the TCG_{td} by querying available marks using available mark fields, concepts, and predicates. Based on the collected data and their associated marks, a discovery of the linked TCGs is initiated backwards (for $t < t_d$) to determine the set of TCGs that are linked to the generated TCG_{td}. During this step, the set of dependent Big Data used during the attack scenario in addition to the different instants when the intruders have issued malicious actions are identified. The end of this backwards process enables the identification of the starting time for the attack and the set of dependent data among the different TCGs. The same detailed steps are also performed In the forward direction (for $t > t_d$) to identify either the end time for the performed attack or to localize the handled data and the attached marks at the current time in the case of ongoing attacks.

6.5 Processing and Monitoring of Big Data in the Cloud

The use of the cloud as a resource for the processing of Big Data is among the common tasks performed nowadays for many services that deal with these huge volumes of data and that should extract knowledge as a response to users' queries. In this section, the processing of big geo data is presented in addition to the monitoring of Big Data available through cloud infrastructures.

6.5.1 Processing of big geospatial data in the cloud

Due to the increased use of modern sensors and light detection and ranging (LiDAR) scanners, huge volumes of geospatial data are generated and should be stored and processed. This spatial information is complex and requires new processing techniques that make use of modern technology and optimization techniques. Due to these constraints, it is infeasible to process and store such data in a unique node and the data should instead be stored and processed in a distributed manner using the storage and processing resources of several interconnected nodes that form in most cases a cloud infrastructure. In [14], a modular architecture for processing of big geo data in the cloud is proposed. This architecture enables domain experts to control the processing of big geo data through a domain specific language. According to this feature, the cloud resources will be available to users from the geospatial domain. The proposed architecture is composed of a set of modules that

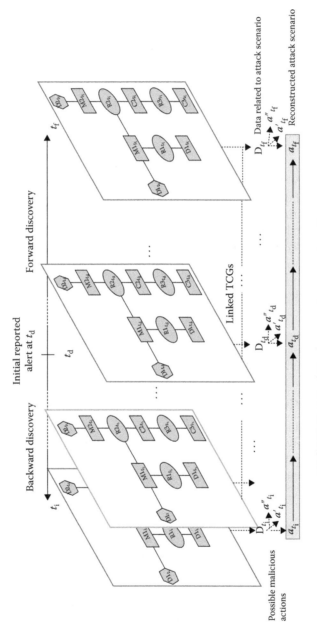

FIGURE 6.3: Attack scenario reconstruction process.

ensures the processing of big geo data in addition to the interaction with the user that is denoted in this case as a geographic information system (GIS) expert that uses the system through a web-based user interface to specify a high-level workflow using a domain-specific language. The GIS user executes this workflow that will be first parsed by a parser, then interpreted (using an interpreter), and finally processed by the job manager that ensures the querying of the catalogue service for meta-data about processing services and the data to be processed. The job manager ensures the processing by applying predefined rules to create a process chain that gives the processing services that should be executed in addition to the execution order and the involved nodes in the cloud. At the end of the execution, the job manager creates a new entry for the obtained result set in the data catalogue that will be followed by a notification sent to the user interface. To fulfill the needs of the proposed architecture, a distributed file system is used. This file system is a virtual file that spans over multiple nodes that belong to the cloud and abstracts away their heterogeneity. It provides a common interface for applications to access data on the interconnected nodes.

6.5.2 Monitoring of Big Data in the cloud

Real-time monitoring of cloud resources is crucial for several tasks including performance analysis, workload management, capacity planning, and fault detection. Applications producing Big Data make the monitoring task very hard at high sampling frequencies due to the high computational and communication overheads introduced in collecting, storing, and managing information. In this context, an adaptive algorithm for monitoring Big Data applications is proposed in [15]. This algorithm adapts the intervals of sampling and frequency of updates to data characteristics and administrator needs. The adaptivity feature enables the limitation of computational and communication costs and guarantees high reliability in capturing relevant load changes. In large data centers hosting Big Data applications, the only way to build a scalable monitoring infrastructure is to reduce the amount of monitoring data without losing its statistical properties that are at the basis of any postgathering analysis. To this purpose, two parameters are introduced. The first, parameter gain (G), is defined as one minus the ratio between the number of samples collected by the considered monitoring algorithm and the number of samples collected by the baseline monitoring algorithm that performs sampling of data at the highest possible frequency t_0. Both monitoring algorithms are supposed to operate over the same time interval that must be sufficiently long to be statistically relevant. Higher values of G denote algorithms aiming to reduce the computational and communication overhead due to monitoring. The second parameter, quality (Q), quantifies the ability of an algorithm to accurately represent load changes in system resources (e.g., load spikes).

The real-time adaptive monitoring algorithm proposed in this work consists of two phases: a training phase for the evaluation of the best parameters settings for monitoring and an adaptive monitoring phase that is the main algorithm of the proposed monitoring solution. The real-time adaptive algorithm analyzes monitored data and identifies periods of relative stability from periods of high variability. The idea of this approach is to reduce the quantity of monitored data when a resource is relatively stable, and to increase it during periods of high variability. As a consequence, the computational and communication overhead is reduced, and at the same time it is guaranteed that important system changes are not lost. This algorithm operates by dynamically setting two key variables: sampling interval t and variability Δ. The sampling interval t determines the time interval that elapses between the collection of two consecutive samples. The lower the sampling interval, the higher the amount of data to gather and to transmit. The minimum sampling interval t_m is evaluated dynamically is the lowest value that the sampling interval t can assume, and the maximum sampling interval t_M as the highest value of t.

6.6   Applications for Management of Big Data on Complex Networks

The processing of Big Data on complex networks ensures the extraction of a set of interesting knowledge that may be useful for most services exploited by users such as train services. In this section, a survey of some applications for the management of Big Data on complex networks is presented.

6.6.1   Evaluation of passenger train service plan

Train service plan evaluation remains among the challenges in railway operation. The transportation system has a complexity phenomenon that should be considered when studying such systems or trying to evaluate them. In [16], the train service plan is transferred into a train service network, which is a complex network with the characteristics of small-world and degree free. According to this model, valuable supporting information is generated for the railway operators including: a convenience degree for traveling, the transfer times, the travel time, and station-clustering coefficients.

The train service plan ensures a set of functions including the identification of the origin and destination stations of the trains, the path of the trains, the stops along the railway, and so on. The service quality provided to the public is evaluated based on the key factors that are the paths of the trains and the stops. In this way, all the stations are considered as nodes of the network and the edges are added on the network one by one according to the rules defined in [17]. If a train passes two stations and stops at the two stations, an edge will be added to the network. Moreover, if the number of edges between two nodes is greater than 1, then only one edge between the two stations is drawn and the number of edges is marked as the weight. In this case, a complex network is generated when all the trains are processed.

A set of indices of complex networks are used to evaluate the train service plan including: the shortest path, clustering coefficients, node degree, average distance, and density.

Another approach has been proposed in [18] to assess the robustness of a subway network (Beijing Subway system) in the face of random failures and malicious attacks. To achieve this goal, the topological properties of the rail transit system are quantitatively analyzed using a mathematical statistical model in addition to the development of a new weighted composite index that is used to evaluate node importance that could be used to position hub stations in a subway network.

6.6.2   The construction of complex climate networks from linear and nonlinear measures

In addition to train services, the techniques of complex network analysis have found application in climate research. The use of such networks is ensured by embedding the characteristics of climate variables such as temperature, pressure, and so on, into the topology of a complex network by appropriate linear and nonlinear measures. This approach has been adopted in [19] by developing a set of tools, denoted ClimNet, that enables the construction of large networks' adjacency matrices from climate time series while supporting functions to tune relationships to different timescales by means of symbolic ordinal analysis. The main functionality of ClimNet is to determine the adjacency matrix from a time series that integrates the obtained significant links by using the bootstrap algorithm whereas insignificant links are discarded. The obtained adjacency matrix (network) is then processed consecutively by any graph analyzing algorithm to obtain additional measures such as betweenness centrality (an indicator of a node's centrality in a network that is equal to the number of shortest paths from all vertices to all others that pass through that node) and community detection.

6.6.3 Group detection in complex networks

In complex networks, the detection of communities is of value in various applications, especially in the case of large social networks and also for epidemic outbreak prevention [20]. Communities that represent a group of nodes may correspond to people with common interests or classes with the same information signature in software networks. In [21], a propagation-based algorithm for general group detection that requires no a priori knowledge is proposed. The advantage of this scheme compared with other detection schemes is the ability to detect different types of groups by using an adequate hierarchical group refinement procedure. This feature has been tested on various real-world networks.

The proposed group detection scheme is presented for the case of simple undirected graphs. Initially, each node is labeled with a unique label. Then, at each iteration, each node adopts the label shared by most of its neighbors. Since there is in most cases numerous links within the communities, relative to the number of links toward the rest of the network, nodes that belong to communities agree on a label after only a few iterations. When the process converges, disconnected groups are classified as communities. The detection approach may be enhanced by adopting node preferences as introduced in [22]. This concept adjusts the propagation strength of the nodes and forces the label propagation process toward a desirable group.

Another issue related to the detection of communities is that most available techniques fail to determine the accurate number of nodes that belong to each community in the network. In this context, a community detection algorithm using a local community neighborhood ratio function is proposed in [23]. This scheme predicts vertex association to a specific community making use of visited node-overlapped neighbors. The processing starts by detecting local communities through iterations and using the local neighborhood ratio function; final communities are detected as a result of the merge of closely related local communities. The proposed algorithm detects local communities according to the following set of steps: it enumerates first all the vertices presented in the network and selects randomly a starting vertex and computes its proximity. After that, for each adjacent vertex pair, the relation function is computed and the related vertices are then grouped into their corresponding local communities. Each edge that belongs to the local community is labeled with its corresponding maximum vertex close relation.

6.6.4 Identification of influential nodes in complex networks based on technique for order of preference by similarity to ideal solution

Node importance is a basic measure that enables the characterization of the structure and dynamics of complex networks. Various centrality measures have been applied in complex networks to identify influential nodes. In this context, a new evaluation method of node importance in complex networks has been proposed in [24]. It is based on a technique for order performance by similarity to ideal solution, technique for order of preference by similarity to ideal solution (TOPSIS), that is, a multiple attribute decision-making technique. This technique is used to aggregate the multi-attribute to derive an evaluation of node importance of each node. This approach is not limited to one centrality measure but it considers different centrality measures. According to the proposed technique, the identification of influential nodes is performed according to a set of steps. The first step enables the construction of the network based on the connection of nodes. This step is followed by the computation of a different centrality value by computing the degree centrality, closeness centrality, and eigenvector centrality that are considered the multi-attribute in the TOPSIS application. After that, the separation from the positive ideal alternative and the negative ideal alternative is computed. A Euclidian distance is applied to compute the separation measures. At the end, the relative closeness to the ideal alternatives of nodes is determined. In this case, the alternatives having higher ideal alternatives are considered more important and should be given higher priority. The influence of the node is then identified using the aforementioned values of alternatives.

6.6.5   The design of global manufacturing networks using Big Data

The configuration of processes and resources in a network are subject in most cases of design decisions. Currently, the design and decisions are performed at the manufacturing site level, which may render decision inefficient since it is impossible in this case to take into consideration dependencies in the network. In the case of manufacturing network design, Big Data is among the main aspects that enables overall system design, the increasing validity of decisions, and high performance. In [25], an approach for applying Big Data techniques and analytics to assist the design of such networks is proposed. Several tasks are performed and some considered during the design phase include: the implementation of new products and resources, the introduction/removal of resource capacity at a site, the implementation of new manufacturing technologies, and so on. The source for the proposed approach is the production master data and the primary demand. The Big Data system that supports the decision processes requires the specification of different architecture layers. Big Data is produced based on scenarios. The creation of a scenario is conducted by modifying a subset of data representing the configuration of the manufacturing network. After the creation of such scenarios, they are analyzed, stored, and compared by the planner as a logical entity of data. The aforementioned tasks are performed at the analysis layer by an application server.

6.7   Future Trends

The inherent complexity of Big Data (including complex types, complex structures, and complex patterns) in addition to the variety and distribution through complex networks makes its representation and processing far more challenging and results in important increases in the computational complexity when compared to traditional computing models based on total data. As discussed in [26], traditional data analysis and mining tasks, such as retrieval, topic discovery, semantic analysis, and sentiment analysis, become extremely difficult when using Big Data. At present, a good understanding on addressing the complexity of Big Data is still missing. For instance, there is a lack of deep understanding on the inherent relationship between data complexity and computational complexity of Big Data, as well as domain-oriented Big Data processing methods that are applicable in complex environments including complex networks. Based on these constraints, there is still a need to design highly efficient computational models and methods for solving problems using Big Data that are able to process and store them in complex networks.

An additional fundamental problem related to the use of Big Data in complex networks is how to formulate or quantitatively describe the essential characteristics of the complexity of Big Data. The study of complexity theory of Big Data will help in understanding essential characteristics of complex patterns in Big Data, simplify its representation, obtain better knowledge abstraction, and guide the design of computing models and algorithms on Big Data. As a consequence, we need to establish the theory and models of data distribution under multimodal interrelationships.

Based on the key features of Big Data, notably multiple sources, huge volume, and fast-changing in addition to the complexity of the network where these data are stored and processed, it is difficult for traditional computing methods such as machine learning to support the processing, analysis, and computation of Big Data in an efficient manner. Such computations cannot simply rely on analysis tools and iterative algorithms used in traditional approaches for handling small amounts of data. Assumptions made in traditional computations based on independent and identical distribution of data and adequate sampling for generating reliable statistics are not applicable. When solving problems involving Big Data, we need to reexamine and investigate its computability, computational complexity, and algorithms.

To address the computational complexity of Big Data applications, we need to focus on the whole life cycle of Big Data applications to study data-centric computing paradigms based on the characteristics of Big Data. We also need to develop algorithms for distributed and streaming computing and form a Big Data–oriented computing framework where communication, storage, and computing are well integrated and optimized.

Due to the great size of handled Big Data, we need to explore existing reduction-based computing methods where Big Data is reduced without losing knowledge.

The design of system architectures, computing frameworks, processing modes, and benchmarks for highly energy-efficient Big Data processing platforms is the key issue to be addressed in system complexity. Solving these problems can lay the principles for designing, implementing, testing, and optimizing Big Data processing systems. Their solutions will form an important foundation for developing hardware and software system architectures with energy-optimized and efficient distributed storage and processing.

The evaluation and optimization of energy efficiency of Big Data processing systems is a great research challenge. We need to comprehensively measure a variety of energy efficiency factors, including system throughput, parallel processing capabilities, job calculation accuracy, and energy consumption per unit. It is required to conduct fundamental research on performance evaluation, distributed system architecture, streaming computing framework, and online data processing, while taking into account features of value sparsity and weak access locality.

The modeling of complex networks as a graph helps to manage and process Big Data in an efficient manner but it is still insufficient due to the dynamicity of the interconnected nodes through such networks in addition to the great size of the graph. In this context, it is required to enhance the representation of the complex network by another structure that considers the aforementioned constraints.

The traditional approaches that enable the storage of Big Data are not applicable for complex networks; we need to design a storage approach that may enhance the size of the Big Data and handle sufficient data on the interconnected nodes to prevent the loss of data or an incoherent fragment that renders the overall Big Data not useful. Using cloud infrastructures to ensure the storage and processing of Big Data has been revolutionizing the IT industry and has interested the research community recently. As mentioned in [27], cloud analytics solutions need to consider the multiple cloud deployment models adopted by enterprises, where a cloud can be private, public, or hybrid by combining both clouds, and additional resources from a public cloud can be provided as needed to a private cloud.

The design of energy management systems using Big Data is among the challenges for complex networks such as a smart electricity grid. This network enables a two-way flow of power and data between suppliers and consumers to facilitate the power flow optimization in terms of economic efficiency, reliability, and sustainability [28]. The challenge in such networks is how to take advantage of the users' participation to reduce the cost of power.

6.8   Conclusion

The management and processing of Big Data through complex networks is among the interesting topics that have been studied by research communities and that lead to a set of techniques and models that deal with the set of issues in relation with this kind of network. Such models have been used to process and store Big Data in complex networks and to ensure Big Data analytics in addition to a set of services that exploit these schemes for different fields. The modeling of such networks is among the topics that is investigated in this chapter to illustrate how to design such networks by considering the properties of such networks and their complexity. Even if several approaches have been proposed dealing with the processing and management of Big Data in complex networks, several issues still need to be studied to enhance the representation, management, and processing of Big Data through complex networks.

References

1. S. Boccaletti, V. Latora, Y. Moreno, M. Chavez, and D.-U. Hwang. Complex networks: Structure and dynamics. *Phys. Rep.* 424 (4–5): 175–308, 2006.

2. S. Frent and F. Pop. Asymptotic scheduling for many task computing in Big Data platforms. *Inf. Sci.* 319: 71–91, 2015.

3. M.E. Newman. Fast algorithm for detecting community structure in networks. *Phys. Rev.* E 69 (6) 066133, 2004.

4. C. Tong, Y. Lian, J. Niu, Z. Xie, and Y. Zhang. A novel green algorithm for sampling complex networks. *J. Netw. Comput. Appl.* 59: 55–62. 2015.

5. L. Page, S. Brin, R. Motwani, and T. Winograd. *The PageRank Citation Ranking: Bringing Order to the Web.* San Francisco, CA: Stanford InfoLab, 1999.

6. S. Boccaletti, V. Latora, Y. Moreno, M. Chavez, and D.U. Hwang. Complex networks: Structure and dynamics. *Phys. Rep.* 424: 175–308, 2006.

7. J. Watts, P.S. Dodds, and M.E.J. Newman. Identity and search in social networks. *Science* 296: 1302–1305, 2002.

8. I. Franks, ed. *Taming the Big Data Tidal Wave: Finding Opportunities in Huge Data Streams with Advanced Analytics.* Hoboken, NJ: John Wiley & Sons, 2012.

9. Y. Djemaiel, N. Essaddi, and N. Boudriga. Optimizing Big Data management using conceptual graphs: A mark-based approach. *Proceedings of the 17th International Conference on Business Information Systems*, BIS 2014, May 22–23, pp. 1–12, Larnaca, Cyprus.

10. F. Kokkoras, H. Jiang, I.P. Vlahavas, A.K. Elmagarmid, E.N. Houstis, and W.G. Aref. Smart videotext: A video data model based on conceptual graphs. *Multimedia Syst.* 8 (4): 328–338, 2002.

11. J.E. Pagan and J.G. Molina. Querying large models efficiently. *Inf. Softw. Technol.* 56 (6): 586–622, 2014.

12. Y. Djemaiel, B.A. Fessi, and N. Boudriga. A mark based-temporal conceptual graphs for enhancing Big Data management and attack scenario reconstruction. *Proceedings of the 18th International Conference on Business Information Systems*, BIS 2015, June 24–26, pp. 62–73, Poznań, Poland.

13. J. Riedy, D.A. Bader, and D. Ediger. *Streaming Graph Analytics for Massive Graphs.* Atlanta, GA: Georgia Institute of Technology, College of Computing, 2012.

14. M. Krämer and I. Senner. A modular software architecture for processing of big geospatial data in the cloud. *Comput. Graph.* 49: 69–81, 2015.

15. M. Andreolini, M. Colajanni, M. Pietri, and S. Tosi. Adaptive, scalable and reliable monitoring of Big Data on clouds. *J. Parallel Distrib. Comput.* 79: 67–79, 2015.

16. X. Meng, Y. Qin, and L. Jia. Comprehensive evaluation of passenger train service plan based on complex network theory. *Measurement* 58: 221–229, 2014.

17. X. Meng, L. Jia, and J. Xie. Complex characteristic analysis of passenger train flow network. *Proceeding of CCDC2010*, pp. 2533–2536, 2010.

18. Y. Yang, Y. Liu, M. Zhou, F. Li, and C. Sun. Robustness assessment of urban rail transit based on complex network theory: A case study of the Beijing Subway. *Saf. Sci.* 79: 149–162, 2015.

19. J.I. Deza and H. Ihshaish. The construction of complex networks from linear and nonlinear measures—Climate networks. *Proc. Comput. Sci.* 51: 404–412, 2015.

20. J. Ruan and W. Zhang. An efficient spectral algorithm for network community discovery and its applications to biological and social networks. *Proceedings of the IEEE International Conference on Data Mining*, pp. 643–648, 2007.

21. L. Šubelj and M. Bajec. Group detection in complex networks: An algorithm and comparison of the state of the art. *Phys. A Stat. Mech. Appl.* 397: 144–156, 2014.

22. X.Y. Leung, P. Hui, P. Liò, and J. Crowcroft. Towards real-time community detection in large networks. *Phys. Rev.* E 79 (6): 066107, 2009.

23. J. Eustace, X. Wang, and Y. Cui. Community detection using local neighborhood in complex networks. *Phys. A Stat. Mech. Appl.* 436 (C): 665–677, 2015.

24. Y. Du, C. Gao, Y. Hu, S. Mahadevan, and Y. Deng. A new method of identifying influential nodes in complex networks based on TOPSIS. *Phys. A Stat. Mech. Appl.* 399 (C): 57–69, 2014.

25. P. Golzer, L. Simon, P. Cato, and M. Amberg. Designing global manufacturing networks using Big Data. *Procedia* 33: 191–196, 2015.

26. X. Jin, B.W. Wah, X. Cheng, and Y. Wang. Significance and challenges of Big Data research. *Big Data Res.* 2 (2): 59–64, 2015.

27. M.D. Assuncao, R.N. Calheiros, S. Bianchi, M.A.S. Netto, and R. Buyya. Big Data computing and clouds: Trends and future directions. *J. Parallel Distrib. Comput.* 79: 3–15, 2015.

28. P.D. Diamantoulakis, V.M. Kapinas, and G.K. Karagiannidis. Big Data analytics for dynamic energy management in smart grids. *Big Data Res.* 2 (3): 94–101, 2015.

Chapter 7

Large Random Matrices and Big Data Analytics

Robert Caiming Qiu

7.1   Introduction

The growth of "Big Data" is changing the paradigm, especially in cases in which massive amounts of data are distributed across locations. Massive data analysis is not the province of any one field, but is rather a thoroughly interdisciplinary enterprise. Solutions to massive data problems will require an intimate blending of ideas from computer science and statistics, with essential contributions also needed from applied and pure mathematics, from optimization theory, and from various engineering areas, notably signal processing and information theory. In general, by bringing interdisciplinary perspectives to bear on massive data analysis, it will be possible to discuss trade-offs that arise when one jointly considers the computational, statistical, scientific, and human-centric constraints that frame a problem [1].

There is no standard definition for Big Data. In this chapter, following the methodology of [1], we roughly represent Big Data in terms of high-dimensional data vectors. Geometrically, these are points in a high-dimensional vector space. Due to uncertainty, it is natural to consider the p-dimensional random vector $\mathbb{R}^p$ or $\mathbb{C}^p$, where p is a large integer. Here p scalar-valued random variables are taken into account simultaneously—both spatially and temporally. Now, we are given n such samples of dimension p, $\mathbf{x}_1, \mathbf{x}_2, ..., \mathbf{x}_n \in \mathbb{C}^p$. Our aim is to study the collective statistical properties of the new matrix-valued random variable $\mathbf{X} = [\mathbf{x}_1, \mathbf{x}_2, ..., \mathbf{x}_n] \in \mathbb{C}^{p \times n}$. The central idea is to treat $\mathbf{X}$ as a whole, rather than multiple samples of random vectors. This conceptual straightforwardness offers many new insights. We assume that the size $p \times n$ of the random matrix is arbitrary. This assumption makes the traditional multivariate analysis [2] invalid, where the dimension p is very small compared with n, the number of data samples. This simplification allows us to obtain more explicit results. Mathematically, we focus on the scalar-valued random variables!

In the novel framework of Big Data, the mathematical foundation is the marriage of statistics and applied matrix analysis. In addition, convex optimization is also required since the functions of random matrix $\mathbf{X}$ such as eigenvalues are convex. Random matrices are relevant to statistics, computer science, and theoretical physics. It seems natural to use random matrices as a novel paradigm to model large data sets, as we did in [3–5].

If we know the statistics of each individual scalar random variable $X_{ij}, i = 1, ..., p$, $j = 1, ..., n$, now can we "transport" the probabilistic structure from the sample space Ω onto the target space E. A random "variable" is not a variable in the analytical sense [6], but a function! Functions of random matrices are the fundamental objects of study in this chapter.

In previous studies [3–5], we use the unifying power of random matrix models to tie together wireless communication, sensing, and smart grid. This chapter adds to this line of work from a novel prospect: we restrict ourselves by studying the trace function $\mathrm{Tr}\, f\,(\mathbf{X})$ or $\mathrm{Tr}\, f\,(\mathbf{XX}^H)$. This restriction enables us to obtain more explicit results by focusing on the scalar-valued random variables. If the eigenvalues of one certain random matrix $\mathbf{X}$ are denoted by $\lambda_1, ..., \lambda_p$, the sum of these eigenvalues

$$S_p = f\,(\lambda_1) + \cdots + f\,(\lambda_p) \tag{7.1}$$

are of central interest in the context of this chapter. Our target is different from the context of classical probability [6,7]

$$S_n = Y_1 + \cdots + Y_n \tag{7.2}$$

where $Y_i, i = 1, ..., n$ are independent scalar-valued random variables in that Equation 7.1 involves highly-dependent random variables $f(\lambda_i), i = 1, ..., n$. This fundamental difference offers many novel results. When $f(\cdot)$ is a convex function, we have

$$\frac{f(\lambda_1) + \cdots + f(\lambda_p)}{p} \leqslant f\left(\frac{\lambda_1 + ... + \lambda_p}{p}\right)$$

Example 1 (Data modeling for power grid)   *For each node, we observe a sequence of T samples that is represented with a random (column) vector $\mathbf{x}_i$, for the ith node. For a number N of nodes, we represent N random vectors with a data matrix*

$$\mathbf{X} = (\mathbf{x}_1, \mathbf{x}_2, ..., \mathbf{x}_N) \in \mathbb{C}^{T \times N}$$

Often we deal with Gaussian random matrix $\mathbf{G}$ with the entries G_{ij} Gaussian random variables. The entries are independent. The G is a non-Hermitian random matrix, which is a lot more difficult than the Hermitian case.

To mine the massive data, we usually study two types of Hermitian random matrices $\frac{1}{2}(\mathbf{G} + \mathbf{G}^H)$ and $\mathbf{G}\mathbf{G}^H$. Similarly, we can study the general case

$$\frac{1}{2}(\mathbf{X} + \mathbf{X}^H) \quad and \quad \mathbf{X}\mathbf{X}^H$$

Our fundamental technique arises from the simple observation that the two cases (hypotheses) for $\mathbf{G}$ and $\mathbf{X}$ are different! The key is to discover the statistical metrics that differentiate the two cases. We call this anomaly detection, where the case $\mathbf{G}$ is normal.

Modern research topics are of interest:

1. What is the smallest size of N such that the random matrix paradigm is still valid?

2. What happens if the entries of $\mathbf{X}$ are non-Gaussian?

□

Our whole chapter is motivated by the previous example. Our analysis is guided by the study of linear eigenvalue statistics, Equation 7.1, in analogy with the sums of independent random variables, Equation 7.2 (Section 7.2). We highlight the similarities and differences of the two equations. Almost all results are taken from the recent literature on random matrix theory. It appears that what might be novel is our view of large random matrices as basic building blocks in systematically modeling a variety of applications, in particular power grids and wireless networks.

7.2   Sums of Independent Random Variables

The aim is to set the stage for the comparison with the sums of eigenvalue functions of (large) random matrices $\mathbf{X}$ of size $p \times n$. The current trend is to study (1) finite n, (2) non-Gaussian entries $X_{ij}, i = 1, ..., p, j = 1, ..., n$. For the applications of special interest to us (e.g., large power grids), we ask the natural question: How small is the value n required such that the random matrix model is still valid for the data sets?

One of the fundamental results of probability theory is the strong law of large numbers. Let $(\xi_i)_{i \geqslant 1}$ be independent and identically distributed (i.i.d.) and defined on the same space. Let $\mu = \mathbb{E}\xi_i$ and $\sigma^2 = \sigma_{\xi_i}^2 < \infty$. Let $S_n = \sum\limits_{i=1}^{n} \xi_i$.

Then

$$\lim_{n \to \infty} \frac{S_n}{n} = \lim_{n \to \infty} \frac{1}{n} \sum_{i=1}^{n} \xi_i = \mu \quad \text{almost surely and in } L^2$$

The central limit theorem gives a rate of convergence of $\sqrt{n}$ for the strong law of large numbers. Let $\xi_1, \xi_2, ..., \xi_n$ be a sequence of independent random variables with finite second moments. Let $m_k = \mathbb{E}\xi_k$, $\sigma_k^2 = \operatorname{Var} \xi_k > 0$, $S_n = \xi_1 + ... + \xi_n$, $D_n^2 = \sum\limits_{k=1}^{n} \sigma_k^2$, and let $F_k(x)$ be the distribution function of the random variables ξ_k. Let us suppose that the Lindeberg condition is satisfied: for every $\varepsilon > 0$

$$\frac{1}{D_n^2} \sum_{k=1}^{n} \int_{\{x : |x - m_k| > \varepsilon D_n\}} (x - m_k)^2 dF_k(x) \to 0, \quad n \to \infty \tag{7.3}$$

Then

$$\frac{S_n - \mathbb{E}S_n}{\sqrt{\operatorname{Var} S_n}} \xrightarrow{d} \mathcal{N}(0, 1) \tag{7.4}$$

where d denotes convergence in distribution.

Let $Z_n = (S_n - \mathbb{E}S_n) / \sqrt{D_n}$ and $F_{Z_n}(x) = \mathbb{P}(Z_n \leqslant x)$. Then

$$F_{Z_n}(x) \to \Phi(x), \quad n \to \infty$$

where $\Phi(x) = \frac{1}{\sqrt{2\pi}} \int_{-\infty}^{x} e^{-t^2/2} dt$. Since $\Phi(x)$ is continuous, the convergence here is actually uniform:

$$\sup_{x \in \mathbb{R}} |F_{Z_n}(x) - \Phi(x)| \to 0, \quad n \to \infty \tag{7.5}$$

In particular, it follows that

$$\mathbb{P}(S_n \leqslant x) - \Phi\left(\frac{x - \mathbb{E}S_n}{D_n}\right) \to 0, \quad n \to \infty$$

It is also natural to raise the question of the *rate of convergence* in Equation 7.5. In essence this is equivalent to asking for a rate of convergence result for the central limit theorem. In the case when the numbers $\xi_1, \xi_2, ...$ are independent and uniformly distributed with the finite theid moment $\mathbb{E}|\xi_1|^3 < \infty$, this question is answered by the *Berry–Esseen inequality*:

$$\sup_{x \in \mathbb{R}} |F_{Z_n}(x) - \Phi(x)| \leqslant C \frac{\mathbb{E}|\xi_1 - \mathbb{E}\xi_1|^3}{\sigma^3 \sqrt{n}} \tag{7.6}$$

where $\sigma^2 = \mathbb{E}|\xi_1 - \mathbb{E}\xi_1|^2 < \infty$, and the absolute constant C satisfies the inequality

$$1/(2\pi) \leqslant C \leqslant 0.8$$

Let $\xi_1, \xi_2, ..., \xi_n$ be independent identically distributed (i.i.d) random variables with zero mean and variance σ^2, finite third moment $\mathbb{E}|\xi_1|^3 < \infty$; it is known that the following *nonuniform* inequality holds: for all $x \in \mathbb{R}$,

$$|F_n(x) - \Phi(x)| \leqslant C \frac{\mathbb{E}|\xi_1|^3}{\sigma^3 \sqrt{n}} \cdot \frac{1}{(1 + |x|)^3} \tag{7.7}$$

where $F_n(x) = \mathbb{P}(S_n \leqslant x)$ with $S_n = \frac{\xi_1 + \cdots + \xi_n}{\sigma \sqrt{n}}, n \geqslant 1$. For example, we can test this inequality using Bernoulli random varibles.

Let $\mathbf{X}_1, ..., \mathbf{X}_n$ be a sequence of independent *random vectors* (with values in $\mathbb{R}^p$) with mean zero and (finite) covariance matrix $\mathbf{\Sigma}$. Then Equation 7.3 shows that for all $x \in \mathbb{R}$

$$\frac{\mathbf{X}_1 + \cdots + \mathbf{X}_n}{\sqrt{n}} \xrightarrow{d} \mathcal{N}(0, 1) \tag{7.8}$$

We define characteristic functions as $\varphi_n(t) = \int_{-\infty}^{\infty} e^{itx} dF_n(x)$, $t \in \mathbb{R}$. There is a one-to-one correspondence between distribution functions and characteristic functions (Fourier transforms φ_n). So, we can study the properties of distribution functions by using the corresponding Fourier transforms. It is a fortunate fact that weak convergence $F_n \xrightarrow{w} F$ of distribution is equivalent to pointwise convergence $\varphi_n \to \varphi$ of the corresponding Fourier transforms.

7.3 Circular Law and the Single Ring Law for Non-Hermitian Random Matrix Models

The phenomenological nature of random matrix models may be regarded as a drawback. But this feature provides, on the other hand, the model-free frameworks that make the approach valid for a wide variety of systems having different microscopic natures or origins. These frameworks have a certain amount of "robustness" of the random matrix models. It is believed that a "sufficiently large" number of them should have no dependence or rather weak dependence on the random matrix ensemble used. This belief of universality partly explains the fact that the major ideas of the random matrix models are based on the Gaussian ensemble and the circular ensemble [8].

In classical probability, the law of large numbers and the central limit theorems are the core. Theorems for Wigner's semicircle law and Marchenko–Pastur law can be viewed as random matrix analogues of the law of large numbers from classical probability theory. Thus a central limit theorem for or fluctuations of linear eigenvalue statistics is a natural second step in studies of the eigenvalue distribution of any ensemble of random matrices.

We define the i.i.d. matrix as follows. Let ξ be a random variable. We say $\mathbf{X}_N$ is an i.i.d. random matrix of size N with atom variable ξ if $\mathbf{X}_N$ is an $N \times N$ matrix whose entries are i.i.d. copies of ξ.

Theorem 1 (Circular law [9]) *Let ξ be a complex random variable with mean zero and unit variance. For each $N \geq 1$, let $\mathbf{X}_N$ be an i.i.d. random matrix of size N with atom variable ξ. Then, for any bounded and continuous function $f : \mathbb{C} \to \mathbb{C}$,*

$$\int_{\mathbb{C}} f(z) d\mu_{\frac{1}{\sqrt{N}} \mathbf{X}_N}(z) \to \frac{1}{\pi} \int_{\mathbb{U}} f(z) d^2 z$$

almost surely as $N \to \infty$ *where* $\mathbb{U}$ *is the unit disk in the complex plane* $|z| \leqslant 1$ *and* $d^2 z = dxdy$, *with* $z = x + iy$.

The single ring theorem (or law), by Guionnet, Krishnapur, and Zeitouni [10], describes the empirical distribution of the eigenvalues of a large generic matrix with prescribed singular values, that is, an $N \times N$ matrix of the form $\mathbf{A} = \mathbf{UTV}$, with $\mathbf{U}; \mathbf{V}$ some independent Haar-distributed unitary matrices and $\mathbf{T}$ a deterministic matrix whose singular values are the ones prescribed. More precisely, under some technical hypotheses, as the dimension N tends to infinity, if the empirical distribution of the singular values of $\mathbf{A}$ converges to a compactly supported limit measure Θ on the real line, then the empirical eigenvalues distribution of $\mathbf{A}$ converges to a limit measure μ on the complex plane which depends only on Θ. The limit measure μ is rotationally invariant in $\mathbb{C}$ and its support is the annulus $S := \{z \in \mathbb{C}; a \leqslant |z| \leqslant b\}$ with $a, b \geq 0$ such that

$$a^{-2} = \int x^{-2} d\Theta(x) \text{ and } b^2 = \int x^2 d\Theta(x) \qquad (7.9)$$

There is no eigenvalue outside the annulus (the support of the limiting spectral distribution).

7.4   Fluctuations of Linear Eigenvalue Statistics

The central idea of this chapter is to use massive random matrices as a new paradigm for Big Data analytics. In particular, we give a tutorial review of linear eigenvalue statistics in random matrix theory. This chapter is based on the unified framework that is first initiated by L.A. Pastur [11] in the context of random matrices. This framework is based on two facts: integration by parts and the Poincare–Nash inequality. Integration by parts is valid for both Gaussian [12, Eq. (2.1.39)] and non-Gaussian [12, Eq. (18.1.19)] random matrices. The Poincare–Nash inequality is valid for Gaussian [12, Eq. (2.1.45)] and general matrix models [13, Eq(1.37)].

For a random matrix $\mathbf{A} \in \mathbb{C}^{n \times n}$ whose eigenvalues $\lambda_i, i = 1, ..., n$, are known as strong correlated random variables, the central object of statistical study is the so-called linear eigenvalue statistics defined as

$$\text{Tr} f(\mathbf{A}) = \sum_{i=1}^{n} f(\lambda_i) \qquad (7.10)$$

where Tr represents the trace of a square matrix and $f(\cdot)$ is a test function that assumes certain smoothness. Define the empirical eigenvalue distribution

$$\mu_n(dx) = \frac{1}{n} \sum_{i=1}^{n} \delta(x - \lambda_i) dx$$

where $\lambda_i, i = 1, 2, ..., n$ are eigenvalues of $\mathbf{X}_n$. The following functions are of special interest in the context of this chapter:

1. Linear function $f(x) = x$.

2. Radius $f(z) = |z|, \quad z \in \mathbb{C}$.

3. Moments $f(x) = x^k, k \in \mathbb{N}$.

4. Generalized variance $f(x) = \log x$.

5. Mutual information $f(x) = \log(1 + x)$.

6. Likelihood ratio test (LRT) $f(x) = \log x - x - 1$.

7. Stieltjes transform $f(x) = \frac{1}{w-x}$, $\operatorname{Im} w \neq 0, w \in \mathbb{C}$.

8. Exponentials $f(x) = e^{jtx}, t \in \mathbb{R}$.

9. Analytical function $f(z) = \sum\limits_{n=0}^{\infty} a_n z^n, \quad z \in \mathbb{C}$.

Let us use some examples to illustrate the fluctuations of linear eigenvalue statistics.

7.4.1 Moments $f(x) = x^k, k \in \mathbb{N}$

Consider an ensemble $\mathbf{X}_n = \frac{1}{\sqrt{n}}(A_{ij})_{i,j=1}^n$ of random $n \times n$ Hermitian matrices with matrix elements A_{ij} of mean zero and unit variance. A classical result of Wigner [14] states that if $A_{ij}, 1 \leq i \leq j \leq n$, are i.i.d. and satisfy a moment bound, then

$$\lim_{n \to \infty} \frac{1}{n} \mathbb{E} \operatorname{Tr}\left(\mathbf{X}_n^k\right) = \int_{-2}^{2} x^k \frac{1}{2\pi} \sqrt{4 - x^2} dx \tag{7.11}$$

namely the limit exists and is equal to the kth moment of the semicircle law.

Define the empirical eigenvalue distribution

$$\mu_n(dx) = \frac{1}{n} \sum_{i=1}^{n} \delta(x - \lambda_i) dx$$

where $\lambda_i, i = 1, 2, ..., n$ are eigenvalues of $\mathbf{X}_n$. Convergence of Equation 7.11 implies that μ_n converges weakly, in expectation, to the semicircle law $\frac{1}{2\pi} \sqrt{4 - x^2} I(|x| \leq 2) dx$. Hence Wigner's result on the limit of empirical distribution is in some sense analogous to the law of large numbers of classical probability theory (Section 7.2), even though it is linked to the central limit theorem of free probability [15].

The monic Chebyshev polynomials of the first kind are defined as through the trigonometric identity

$$T_m(2\cos(\theta)) = 2\cos(m\theta)$$

The main result of Johansson et al. [16] is that, for a Wigner matrix with Gaussian entries of variance σ^2, the monic rescaled Chebyshev polynomials

$$T_m(x, \sigma^2) = \sigma^{2m} T_m\left(\frac{x}{\sigma^2}\right), \quad m \geq 1$$

diagonalize the covariance matrix. That is

$$\left\{\operatorname{Tr}\left(T_m\left(\mathbf{X}_n, \sigma^2\right)\right) - \mathbb{E} \operatorname{Tr}\left(T_m\left(\mathbf{X}_n, \sigma^2\right)\right)\right\}_{m \geq 1}$$

converges to a family of independent Gaussian random variables.

7.4.2 Generalized variance $f(x) = \log x$

Condition 1: Let $X_{ij}, i = 1, 2, ..., p, j = 1, 2, ..., n$, be standard Gaussian random variables with zero mean and unit variance $\mathcal{N}(0, 1)$.

We denote the data matrix by $\mathbf{X} \in \mathbb{C}^{p \times n}$. The sample covariance matrix, a random matrix, is defined as $\mathbf{S}_p^{(n)} = \mathbf{X}\mathbf{X}^H$, where H represents the conjugate and transpose (or Hermitian) of a matrix. $\mathbf{S}_p^{(n)}$ is the $p \times p$ nonnegative random matrix. Using a matrix equality, we obtain

$$\log \det \left(\mathbf{S}_p^{(n)} \right) = \sum_{i=1}^{p} \log (\lambda_i)$$

where $\lambda_i, i = 1, ..., p$ are the nonnegative eigenvalues.

We consider the asymptotic regime

$$p \to \infty, n \to \infty, \text{ but } p/n \to y, \ 0 < y < 1 \tag{7.12}$$

Theorem 2 (Central limit theorem) *Under **Condition** 1 and Equation 7.12, $\left(1/(n-1)_p \right)$* $\det \left(\mathbf{S}_p^{(n)} \right)$ *converges in distribution to a log-normal distribution,* $(n-1)_p = (n-1)(n-2) \cdots$ $(n-p)$. *The parameters are*

$$\frac{1}{\sqrt{-2 \log (1 - y)}} \left[\log \det \left(\mathbf{S}_p^{(n)} \right) - \log (n-1)_p \right]$$

is asymptotically $\mathcal{N}(0, 1)$.

Theorem 3 (Estimates for expectation and variance) *If* $1 \leq p < n$, *then*

$$-\frac{1}{3n} \frac{1 - y^\star}{y^\star} \leqslant \mathbb{E} \left[\log \det \left(\mathbf{S}_p^{(n)} \right) - \log (n-1)_p \right] \leqslant \frac{1}{n} \frac{1 - y^\star}{y^\star}$$

$$-\frac{2}{n} \frac{y^\star}{1 - y^\star} \leqslant \text{Var} \left[\log \det \left(\mathbf{S}_p^{(n)} \right) - \log (n-1)_p \right] + 2 \log (1 - y^\star) \leqslant \frac{3}{n} \frac{y^\star}{1 - y^\star}$$

where $y^\star = \frac{p}{n}$.

Theorem 4 (Berry–Esseen type inequality for convergence rate) *Let* $G_p^{(n)}(x)$ *be the distribution function of*

$$\frac{\log \det \left(\mathbf{S}_p^{(n)} \right) - \mathbb{E} \log \det \left(\mathbf{S}_p^{(n)} \right)}{\sqrt{\text{Var} \log \det \left(\mathbf{S}_p^{(n)} \right)}} = \frac{\sum_{j=1}^{p} (\log U_j - \mathbb{E} \log U_j)}{\sqrt{\sum_{j=1}^{p} \text{Var} \log U_j}}$$

where $U_1, U_2, ..., U_p$ *are independent* χ^2 *random variables with* $n - p + 1, n - p + 2, ..., n$, *respectively. Then*

$$\sup_{-\infty < x < \infty} \left| G_p^{(n)}(x) - \Phi(x) \right| \leqslant \frac{1}{\sqrt{n}} \cdot \frac{C}{\sqrt{(1 - y^\star) \log (1/ (1 - y^\star))} - y^\star/n}$$

where $y^\star = p/n$, *and* C *is an absolute constant,* $1 \leq p < n$, *and* $\Phi(x)$ *is the standard Gaussian distribution.*

Theorem 4 can be applied to study how fast the generalized variance converges to the asymptotic limit. This is interesting in anomaly detection in a power grid where a number N of nodes are observed across a time window of T data samples and analyzed jointly [17], leading to a use of large random matrix $\mathbf{X}$ of $N \times T$. If we use the generalized variance as our function of choice, the results in this subsection are relevant.

7.4.3 Regularized likelihood ratio test $f(x) = x - \log x - 1$

Here we study the inference of a covariance matrix from high-dimensional data. To be specific, we propose a regularized likelihood ratio statistic to test a high-dimensional covariance matrix in one sample. In estimation, the sample covariance matrix $\mathbf{S}$ often becomes poorly conditioned and the eigenvalues spread out when p increases. For this reason, many estimators for a large-scale covariance matrix are proposed in the literature. Among many, the linear shrinkage estimator, defined by

$$\hat{\boldsymbol{\Sigma}}_{\text{lin}} = \alpha \mathbf{S} + (1 - \alpha)\,\mathbf{I} \qquad (7.13)$$

is the most popular and widely used for many multivariate procedures as an alternative to the sample covariance matrix $\mathbf{S}$. The tuning parameter α is assumed to be a constant in (0,1).

We consider the inference of a large scale covariance matrix whose dimension p is large compared to the sample size n. To be specific, we are interested in testing whether the covariance matrix equals a given known matrix, $\mathcal{H}_0 : \boldsymbol{\Sigma} = \boldsymbol{\Sigma}_0$, where $\boldsymbol{\Sigma}_0$ is a given matrix and is assumed to be $\mathbf{I}_p$ without loss of generality. The LRT statistic to test $\mathcal{H}_0 : \boldsymbol{\Sigma} = \mathbf{I}$ vs. $\mathcal{H}_1 : \boldsymbol{\Sigma} \neq \mathbf{I}$ is defined by

$$\text{LRT} = \text{Tr}(\mathbf{S}) - \log \det(\mathbf{S}) - \text{Tr}(\mathbf{I}_p) = \sum_{i=1}^{p} (\ell_i - \log \ell_i - 1)$$

where $\mathbf{S}$ is the sample covariance matrix and ℓ_i is the ith largest eigenvalue of the sample covariance matrix. Its asymptotic null distribution for finite p, which is the chi-square distribution with degrees of freedom $p(p+1) = 2$, is not accurate any more, if p increases. Its correct asymptotic distribution is computed by Bai et al. [18] for the case p/n approaches $\gamma \in (0, 1)$ and both n and p increase. They further numerically show that their asymptotic normal distribution defines a valid procedure for testing $\mathcal{H}_0 : \boldsymbol{\Sigma} = \mathbf{I}$. The results of Bai et al. [18] are refined by Jiang et al. [19]<who include the asymptotic null distribution for the case $p/n = 1$. Though the correction of the null distribution, the sample covariance, is known to have redundant variability when p is large, it still remains a general question that the LRT is asymptotically optimal for the testing problem in the n, p large scheme.

Here, we study to improve the "corrected" LRT using the regularized covariance matrix estimator. In detail, we consider a modification of the LRT, denoted by regularized LRT (rLRT), that is defined by

$$\text{rLRT} = \text{Tr}\left(\hat{\boldsymbol{\Sigma}}\right) - \log \det\left(\hat{\boldsymbol{\Sigma}}\right) - \text{Tr}(\mathbf{I}_p) = \sum_{i=1}^{p} (\lambda_i - \log \lambda_i - 1) \qquad (7.14)$$

where $\hat{\boldsymbol{\Sigma}}$ is a regularized covariance matrix and λ_i is the ith largest eigenvalue of $\hat{\boldsymbol{\Sigma}}$. Here, we consider the regularization via linear shrinkage (Equation 7.13). The linearly shrunken sample covariance matrix (simply shrinkage estimator) is known to reduce expected estimation loss of the sample covariance matrix. [20]

Many multivariate statistical procedures are based on $F^{\mathbf{S}}$, the empirical spectral distribution of the sample covariance matrix. Specifically, of interest here is a family functional eigenvalues which is also called linear spectral statistics (LSS) or linear eigenvalue statistics:

$$\hat{\theta} = \frac{1}{p} \sum_{i=1}^{p} g(\ell_i) = \int g(x) dF^{\mathbf{S}}(x)$$

for a function $g(x)$ with some complex-analytic conditions. We define $\varphi(x) = \alpha x + (1 - \alpha)$ and $g(x) = \varphi(x) - \log\{\varphi(x)\} - 1$. Then we can write

$$\mathrm{rLRT} = p \int g(x) dF^{\mathbf{S}_n}$$

The $F^{\gamma,\delta_1}(x)$ denotes the Marchenko–Pasture law. This law is used as an approximation for rLRT.

The central limit theorem for the LRT function of the sample covariance matrix is used in [21] for sensing.

7.4.4 Non-Hermitian random matrices $\mathbf{AA}^H - \mathbf{A}^H\mathbf{A}$

We study the functions of non-Hermitian random matrices. Let $\mathbf{A}_n, n = 1, 2, \ldots$ be an $n \times n$ matrix, whose entries a_{ij} for $i, j = 1, \ldots, n$ are *independent* complex random variables on a probability space $(\Omega, \mathcal{A}, \mathbb{P})$ with the *same* distribution function $F_a(z) = \mathbb{P}\{\mathrm{Re}\, a \leqslant \mathrm{Re}\, z, \mathrm{Im}\, a \leqslant Imz\}$, where z is complex. We assume that *all moments* $\mathbb{E}|a|^k, k = 1, 2, \ldots$ exist and $\mathbb{E}a = 0$, $\mathbb{E}|a|^2 = 1$. In [22], Wegmann studied the Hermitian matrices such as

$$\mathbf{A}^r(\mathbf{A}^H)^r, \mathbf{A}^r + (\mathbf{A}^H)^r, i\left(\mathbf{A}^r - (\mathbf{A}^H)^r\right), \mathbf{A}^r(\mathbf{A}^H)^r \pm (\mathbf{A}^H)^r\mathbf{A}^r$$

where $r = 1, 2, \ldots$.

The commutator $\mathbf{D}(\mathbf{A}) := \mathbf{AA}^H - \mathbf{A}^H\mathbf{A}$ in a sense measures the nonnormality of a matrix $\mathbf{A}$. We obtain $\lim_{n\to\infty} \|\mathbf{A}_n\|_F^2 / n^2 = 1$, and $\lim_{n\to\infty} \|\mathbf{D}(\mathbf{A}_n)\|_F^2 / n^3 = 2$, with probability 1 for random matrices and so

$$\lim_{n\to\infty} n \|\mathbf{D}(\mathbf{A}_n)\|_F^2 / \|\mathbf{A}_n\|_F^4 = 2 \quad \text{probability almost surely}$$

where $\| \cdot \|_F$ represents the Frobenius norm.

Let $\lambda_i^{(n)}$ be the eigenvalue of $\mathbf{A}_n$. It is shown in [22] that

$$\lim_{n\to\infty} \sup \sum_{i=1}^n \left|\lambda_i^{(n)}\right|^2 / \|\mathbf{A}_n\|_F^2 \leqslant 0.8228 \cdots, \quad \text{probability almost surely}$$

Our motivation to study the functions of random matrices arises from the belief that any abnormal behavior in power grids or wireless networks will "break" the normal system status that is modelled as large random data formed from N sensors across T data samples. In this case we have $N = T$, tending to infinity.

7.5 Central Limit Theorem for Hermitian Random Matrices

Theorems for Wigner's semicircle law and the Marchenko–Pastur law can be viewed as random matrix analogues of the law of large numbers from classical probability theory. Thus a central limit theorem for or fluctuations of linear eigenvalue statistics is a natural second step in studies of the eigenvalue distribution of any ensemble of random matrices. Here we only give the result for a sample covariance matrix. This section is motivated for applications in smart grids [23].

For each $n \geq 1$, let $\mathbf{A}_n = \frac{1}{n}\mathbf{X}_n^H\mathbf{X}_n$ be a real sample covariance matrix of size n, where $\mathbf{X}_n = \{X_{ij}\}_{1 \leqslant i,j \leqslant n}$, and $\{X_{ij} : 1 \leqslant i, j \leqslant n\}$ is a collection of real independent random variables with zero mean and unit variance. The eigenvalues are ordered such that $\lambda_1(\mathbf{A}_n) \leqslant \lambda_2(\mathbf{A}_n) \leqslant \cdots \leqslant \lambda_n(\mathbf{A}_n)$. The test function f from the space $\mathcal{H}_s$ has the norm

$$\|f\|_s^2 = \int (1 + 2|\omega|)^{2s}|F(\omega)|^2 d\omega$$

for some $s > 3/2$, where $F(\omega)$ is the Fourier transform of f defined by

$$F(\omega) = \frac{1}{\sqrt{2\pi}} \int e^{j\omega t} f(t)dt$$

We note that if f is a real-valued function with $f \in \mathcal{H}_s$ for some $s > 3/2$, then both f and its derivative f' are continuous and bounded almost everywhere [24]. In particular, this implies that f is Lipschitz.

Suppose that $\mathbb{E}[X_{ij}^4] = m_4$ for all $1 \leq i, j \leq n$ and all $n \geq 1$. Assume there exists $\varepsilon > 0$ such that

$$\sup_{n \geqslant 1} \sup_{1 \leqslant i,j \leqslant n} \mathbb{E}|X_{ij}|^{4+\varepsilon} < \infty$$

Let f be a real-valued function with $\|f\|_s < \infty$ for some $s > 3/2$. Then

$$\sum_{i=1}^{n} f(\lambda_i(\mathbf{A}_n)) - \mathbb{E}\sum_{i=1}^{n} f(\lambda_i(\mathbf{A}_n)) \to \mathcal{N}(0, v^2[f]) \qquad (7.15)$$

in distribution as $n \to \infty$, where the variance $v^2[f]$ is a function of f defined by

$$v^2[f] = \frac{1}{2\pi^2} \int_0^4 \int_0^4 \left(\frac{f(x) - f(y)}{x - y}\right)^2 \frac{(4 - (x - 2)(y - 2))}{\sqrt{4 - (x - 2)^2}\sqrt{4 - (y - 2)^2}} dxdy$$
$$+ \frac{m_4 - 3}{4\pi^2}\left(\int_0^4 \frac{x - 2}{\sqrt{4 - (x - 2)^2}}dx\right)^2 \qquad (7.16)$$

For Big Data, we are interested in the performance of algorithms at different scales of matrix sizes n. The variance of the linear eigenvalue statistics does not grow to infinity in the limit $n \to \infty$ for sufficiently smooth test functions. This points to very effective cancellations between different terms of sum and a rigidity property [25] for the distribution of the eigenvalues.

We also refer to Najim and Yao [26] for a recent result. Consider a $N \times n$ matrix

$$\mathbf{Y}_n = \frac{1}{\sqrt{n}}\mathbf{\Sigma}_n^{1/2}\mathbf{X}_n$$

where $\mathbf{\Sigma}_n$ is a nonnegative definite Hermitian matrix and $\mathbf{X}_n$ is a random matrix with i.i.d. real or complex standardized entries. The fluctuations of the linear statistics of the eigenvalues,

$$\mathrm{Tr}\, f(\mathbf{Y}_n\mathbf{Y}_n^H) = \sum_{i=1}^{N} f(\lambda_i), \qquad \lambda_i \text{ eigenvalues of } \mathbf{Y}_n\mathbf{Y}_n^H$$

are shown to be Gaussian, in the regime where both dimensions of matrix $\mathbf{Y}_n$ go to infinity at the same pace and in the case where f is an analytic function. The main improvement with respect to (w.r.t.) Bai and Silverstein's CLT [27] lies in the fact that Najim and Yao [26] consider general entries with finite fourth moment, but whose fourth cumulant is nonnull, that is, whose fourth moment may differ from the moment of a (real or complex) Gaussian random variable. As a consequence, extra terms proportional to

$$|\nu|^2 = \left|\mathbb{E}(X_{11}^n)^2\right|^2 \qquad \kappa = \mathbb{E}|X_{11}^n|^4 - |\nu|^2 - 2$$

appear in the limiting variance and in the limiting bias, which not only depend on the spectrum of matrix $\boldsymbol{\Sigma}_n$ but also on its eigenvectors.

7.6 Central Limit Theorem for Non-Hermitian Random Matrices

In this section, we present an exhaustive treatment of the central limit theorems for non-Hermitian random matrices. This type of problem is far more difficult than the counterparts of Hermitian random matrices. Non-Hermitian matrices are more basic than Hermitian matrices in the sense that the latter can be always obtained using the former. We are motivated by applications in power grids and wireless systems.

7.6.1 Gaussian fluctuations for the circular law

Universality says that if we replace the large random matrix with its Gaussian counterpart, the result remains the same. This is the reason why most results are obtained using Gaussian random matrices. In this section, we focus on the random matrix with Gaussian random variables.

A fundamental non-Hermitian ensemble of random matrix theory is that of $N \times N$ matrices with independent complex Gaussian entries of mean zero and variance $1/N$. This model is typically attributed to Ginibre [28] who derived the joint distribution of eigenvalues z_k lying in the complex plane $\mathbb{C}$. That object is given by:

$$dP_N(z_1, z_2, ..., z_N) = \frac{1}{Z_N} \prod_{1 \leqslant \ell < k \leqslant N} |z_\ell - z_k|^2 \mu_N(dz_1) \cdots \mu_N(dz_N) \qquad (7.17)$$

in which $\mu_N(dz) = e^{-N|z|^2} d\operatorname{Re}(z) d\operatorname{Im}(z)$ and Z_N is the appropriate normalizer. A straightforward large deviation analysis will take you from Equation 7.17 to the circular law: for $N \to \infty$ the measure $\frac{1}{N} \sum_{i=1}^{N} \delta_{z_i}$ tends weakly to the uniform measure on the disk —z— 1. The goal here is to describe the associated fluctuations. In the non-Hermitian setting, Forrester [29] was the first who studied this type of fluctuation question.

Our study is centered around linear eigenvalue statistics of the form $X(f) = \sum_{i=1}^{N} f(z_i)$, and the behavior of their covariances:

$$\operatorname{Cov}(X(f), X(g)) = \int_{\mathbb{C}^N} X(f) X(g) \, dP_N - \int_{\mathbb{C}^N} X(f) dP_N \int_{\mathbb{C}^N} X(g) dP_N$$

as $N \to \infty$. In particular, if the statistic is one of either the moduli, $f(z_k) = f(|z_k|)$, or the angles, $f(z_k) = f(\arg(z_k))$, we obtain exact formulas for the covariance at finite N which are amenable to a precise asymptotic analysis.

Theorem 5 [30] *For statistics of the moduli, let f and g be bounded over $\mathbb{R}^+$. Then*

$$
\begin{aligned}
\operatorname{Cov}\left(X\left(f\right),X\left(g\right)\right) &= \sum_{\ell=1}^{N}\left\{\mathbb{E}\left[f\left(\sqrt{\tfrac{1}{N}s_\ell}\right)g\left(\sqrt{\tfrac{1}{N}s_\ell}\right)\right]\right. \\
&\left.- \mathbb{E}\left[f\left(\sqrt{\tfrac{1}{N}s_\ell}\right)\right]\mathbb{E}\left[g\left(\sqrt{\tfrac{1}{N}s_\ell}\right)\right]\right\}
\end{aligned}
\tag{7.18}
$$

where $s_\ell = \eta_1 + \cdots + \eta_\ell$ for η_ℓ is a sequence of independent exponential random variables of mean one.

Theorem 6 [30] *Let $f\left(z\right) = f\left(|z|\right)$ and $g\left(z\right) = g\left(|z|\right)$ lie in $C^{2,\delta}$ for $r = |z| \in (0, 1+\varepsilon)$ with some $\delta > 0$ and $\varepsilon > 0$ and otherwise be bounded. Then, as $N \to \infty$, $\operatorname{Cov}\left(X(f), X(g)\right) = \frac{1}{2}\int_0^1 \frac{df(r)}{dr}\frac{dg(r)}{dr}r\,dr + o\left(1\right)$. Moreover, the centered (but unnormalized) random variable $X\left(f\right) - \mathbb{E}\left[X\left(f\right)\right]$ converges to a mean zero Gaussian with the corresponding variance.*

One motivation for the previous two theorems is the statistics in a power grid [17,23] where the moduli statistics is of interest.

Rider and Silverstein [31] give a version of the central limit theorem for a complex Gaussian random matrix. Let ξ be a complex random variable with mean zero, unit variance, and which satisfies (i) $\mathbb{E}\left[\xi^2\right] = 0$, (ii) $\mathbb{E}\left[|\xi|^k\right] \leq k^{\alpha k}$ for $k > 2$ and some $\alpha > 0$, (iii) $\operatorname{Re}\xi$ and $\operatorname{Im}\xi$ have a bounded joint density. For each $N \geq 1$, let $\mathbf{X}_N$ be an i.i.d. random matrix with atom variable ξ. Consider test functions $f_1, ..., f_k$ analytic in a neighborhood of the disk $|z| \leq 4$ and otherwise bounded. Then, as $N \to \infty$, the *random vector*

$$
\left(\operatorname{Tr} f_j\left(\frac{1}{\sqrt{N}}\mathbf{X}_N\right) - N f_j\left(0\right)\right)_{j=1}^k
$$

converges in distribution to a mean-zero, multivariate Gaussian vector $(\mathcal{G}\left(f_1\right), \mathcal{G}\left(f_2\right), ..., \mathcal{G}\left(f_k\right))$ with covariances

$$
\mathbb{E}\left[\mathcal{G}\left(f_1\right)\overline{\mathcal{G}\left(f_1\right)}\right] = \frac{1}{\pi}\int_{\mathbb{U}}\frac{d}{dz}f_i\left(z\right)\overline{\frac{d}{dz}f_i\left(z\right)}d^2z
$$

in which $\mathbb{U}$ is the unit disk $|z| \leq 1$ and $d^2z = dx\,dy$, with $z = x + iy$, the bar standards for the complex conjugate.

A version of the previous result was proven by Nourdin and Peccati [32] when $\mathbf{X}_N$ is a real i.i.d. random matrix and $f_1, f_2, ..., f_k$ are polynomials. Let X be a centered real random variable, having unit variance and with finite moments of all orders, that is, $\mathbb{E}\left(X\right) = 0, E\left(X^2\right) = 1$ and $E|X|^k < \infty$ for every $k \geq 3$. We consider a doubly indexed collection $\mathcal{X} = \{X_{ij} : i, j \geqslant 1\}$ of i.i.d. copies of X. For every integer $N > 2$, we denote by $\mathbf{X}_N$ the $N \times N$ random matrix

$$
\mathbf{X}_N = \left\{\frac{X_{ij}}{\sqrt{N}} : i, j = 1, ..., N\right\}
\tag{7.19}
$$

Theorem 7 [32] *Fix $m \geq 1$, as well as integers $1 \leq k_1 < \cdots < k_m$. Then, the following holds:*

1. *As $N \to \infty$,*

$$
\left(\operatorname{Tr}\left(\mathbf{X}_N^{k_1}\right) - \mathbb{E}\operatorname{Tr}\left(\mathbf{X}_N^{k_1}\right), ..., \operatorname{Tr}\left(\mathbf{X}_N^{k_m}\right) - \mathbb{E}\operatorname{Tr}\left(\mathbf{X}_N^{k_m}\right)\right) \xrightarrow{Law} \left(Z_{k_1}, ..., Z_{k_m}\right)
\tag{7.20}
$$

where $\mathbf{Z} = \{Z_k : k \geqslant 1\}$ denotes a collection of real independent centered Gaussian random variables such that, for every $k \geqslant 1, E\left(Z_k^2\right) = k$.

2. *Write $\beta = E|X|^3$. Suppose that the function $\varphi : \mathbb{R}^m \to \mathbb{R}$ is three times differentiable and that its partial derivatives up to the order three are bounded by some constant $B < \infty$. Then, there exists a finite constant $C = C(\beta, B, m, k_1, ..., k_m)$, not depending on N, such that*

$$\left| \mathbb{E}\left[\varphi\left(\frac{\mathrm{Tr}\left(\mathbf{X}_N^{k_1}\right) - \mathbb{E}\,\mathrm{Tr}\left(\mathbf{X}_N^{k_1}\right)}{\sqrt{\mathrm{Var}\,\mathrm{Tr}\left(\mathbf{X}_N^{k_1}\right)}}, ..., \frac{\mathrm{Tr}\left(\mathbf{X}_N^{k_m}\right) - \mathbb{E}\,\mathrm{Tr}\left(\mathbf{X}_N^{k_m}\right)}{\sqrt{\mathrm{Var}\,\mathrm{Tr}\left(\mathbf{X}_N^{k_m}\right)}} \right) \right] \right.$$
$$\left. - \mathbb{E}\left[\varphi\left(\frac{Z_{k_1}}{k_1}, ..., \frac{Z_{k_1}}{k_m} \right) \right] \right| \leqslant C \frac{1}{N^{1/4}} \tag{7.21}$$

Consider the Ginibre ensemble, that is the $n \times n$ random matrix in which all entries are independent complex Gaussians of mean zero and variance $1/n$. The eigenvalues $z_1, z_2, ..., z_n$ form a point process. Compared to n points dropped independently and uniformly on the unit disk $\mathbb{U}$, the Ginibre points are clearly more regular. For example, their sum, the matrix trace, is complex Gaussian with variance 1, while for the independent points the variance is $\frac{2}{3}n$. See [33] for details.

Theorem 8 (Noise limit for eigenvalues [33]) *Let $f : \mathbb{C} \to \mathbb{R}$ possess continuous partial derivatives in a neighborhood of $\mathbb{U}$, and grow at most exponentially at infinity. Then, as $n \to \infty$, the distribution of the random variable $\sum_{k=1}^{n} (f(z_k) - \mathbb{E}f(z_k))$ converges to a zero-mean Gaussian random variable with variance $\sigma_f^2 + \tilde{\sigma}_f^2$, where is the squared Dirichlet (H^1) norm of f on the unit disk, and*

$$\tilde{\sigma}_f^2 = \frac{1}{2} \|f\|_{H^{1/2}(\partial\mathbb{U})}^2 = \frac{1}{2} \sum_{k \in \mathbb{Z}} |k| \left| \hat{f}(k) \right|^2$$

is the $H^{\frac{1}{2}}$-norm on the unit circle $\partial\mathbb{U}$, where $\hat{f}(k) = \frac{1}{2\pi} \int_0^{2\pi} f(\theta)e^{-ik\theta}d\theta$ is the kth Fourier coefficient of f restricted to $|z| = 1$.

Theorem 9 (The Gaussian free field limit [GFF] [33]) *For the Ginibre characteristic polynomial $p_n(z) = \prod_{k=1}^{n}(z - z_k)$, let*

$$h_n(z) = \log|p_n(z)| - \mathbb{E}\log|p_n(z)|$$

Then h_n converges weakly without normalization to h^*, the planar GFF conditioned to be harmonic outside the disk. More precisely, for functions f we have

$$\int h_n(z) f(z)d^2z \Rightarrow \int h^*(z) f(z)d^2z$$

in distribution, and the same holds for the joint distribution for the integrals against multiple test functions $f_1, ..., f_k$.

The planar GFF [34] is a model to study the scaling limit of uniformly random (discrete) $\mathbb{R}^2 \to \mathbb{R}$ surfaces. It has not previously been connected with any matrix model.

We are guided by the heuristics that the abnormal conditions of power grids will perturb the beautiful properties of the Ginibre ensemble.

7.6.2   Gaussian fluctuations for the single ring law

Benaych-George and Rochet [35] studied the analytic test functions in the single ring law. We consider a non-Hermitian random matrix $\mathbf{A}$ whose distribution is invariant under the left and right actions of the unitary group. The so-called single ring law states that the empirical eigenvalue distribution of $\mathbf{A}$ converges to a limit measure supported by a ring S. We study the joint weak convergence, as $N \to \infty$, of (scalar-valued) random variables of the type

$$\mathrm{Tr}\left(f\left(\mathbf{A}\right)\mathbf{M}\right)$$

for f an analytic function on the single ring S whose Laurent series expansion has null constant term and $\mathbf{M}$ a deterministic $N \times N$ matrix satisfying some limit conditions. These limit conditions allow us to consider both:

- Fluctuations, around their limits as predicted by the single ring law, of LSS of $\mathbf{A}$ (for $\mathbf{M} = \mathbf{I}$):

$$f\left(\mathbf{A}\right) = \sum_{i=1}^{N} f\left(\lambda_i\right)$$

  where $\lambda_1, ..., \lambda_N$ denote the eigenvalues of $\mathbf{A}$

- Finite rank projections of $f\left(\mathbf{A}\right)$ (for $\mathbf{M} = \sqrt{N} \times$ (a matrix with bounded rank)), like the matrix entries of $f\left(\mathbf{A}\right)$

When nonnormal matrices are concerned, functional calculus makes sense only for analytic functions: if one denotes by $\lambda_1, ..., \lambda_N$ the eigenvalues of a non-Hermitian matrix $\mathbf{A}$, one can estimate $\sum_{i=1}^{N} f\left(\lambda_i\right)$ out of the numbers $\mathrm{Tr}\,\mathbf{A}^K$ or $\mathrm{Tr}\left(\left(z\mathbf{I} - \mathbf{A}\right)^{-1}\right)$ only when f is analytic. For a $\mathcal{C}^2$ test function f, one relies on the explicit joint distribution of the λ_i's or on Girko's so-called Hermitization technique, which expresses the empirical spectral measure of $\mathbf{A}$ as the Laplacian of the function $z \mapsto \log |\det\left(z\mathbf{I} - \mathbf{A}\right)|$ (see [36,37]).

We can prove that for $\mathbf{A} = \mathbf{UTV}$, an $N \times N$ matrix of the type introduced previously, and $f\left(z\right) = \sum_{k \in \mathbb{Z}} a_k z^k$, an analytic function on a neighborhood of the limit support S of the empirical eigenvalue distribution of $\mathbf{A}$, the random variable

$$\mathrm{Tr}\,f\left(\mathbf{A}\right) - N a_0$$

converges in distribution, as $N \to \infty$, to a centered complex Gaussian random variable with a given covariance matrix. This is a first step in the study of the noise in the single ring theorem. The random variable

$$\mathrm{Tr}\,f\left(\mathbf{A}\right) - N a_0 = \sum_{i=1}^{N} \left(f\left(\lambda_i\right) - \mathbb{E}f\left(\lambda_i\right)\right)$$

does not need to be renormalized to have a limit in distribution, which reflects the eigenvalue repulsion phenomenon (indeed, would the λ_i's have been i.i.d., this random variable would have had order $\sqrt{N}$, e.g., see Section 7.2).

7.7   No Eigenvalue Outside the Support of the Limit Spectral Distribution

For a large random matrix, a unique phenomenon occurs: there is no eigenvalue outside the support of the limit spectral distribution. This special property is powerful in applications such as anomaly detection in a power grid [17,23].

For Hermitian random matrices, Bai and Silverstein [38] state the theorem and use it in separating eigenvalues [39].

For the single ring theorem of the large non-Hermitian random matrix, Guionnet and Zeitouni [40] give the answer. For the Ginibre matrix, see [41].

We focus on the family of $N \times N$ random matrices with all entries independent and distributed as complex Gaussian of mean zero and variance $1/N$. The well-known circular law states that $\frac{1}{N} \sum_{i=1}^{N} \delta_{z_i}$ converges weakly to the uniform measure on the unit disc $D = \{z : |z| \leqslant 1\}$. Therefore, identifying an N-dependent disc which scales to D and captures all the eigenvalues is equivalent to studying the spectral radius. The main result is the following.

Theorem 10 [41] *Let R_N be the spectral radius for the matrix $\mathbf{X}_N$. That is, $R_N = \max_{1 \leqslant k \leqslant N} |z_k|$. Then, using Equation 7.17,*

$$\lim_{N \to \infty} P_N \left[\sqrt{4N\gamma_N} \left(R_N - 1 - \sqrt{\frac{\gamma_N}{4N}} \right) \leqslant x \right] = \exp\left[-\exp\left(-x \right) \right] \qquad (7.22)$$

where $\gamma_N = \log \frac{N}{2\pi} - 2 \log \log N$. In other words, R_N is well approximated by

$$R_N \cong 1 + \sqrt{\frac{\gamma_N}{4N}} - \frac{1}{\sqrt{4N\gamma_N}} \log Z$$

with Z an exponential random variable of mean 1.

Theorem 11 [41] *We have the following almost sure statement for the behavior of R_N. For any $\delta > 0$ and $M < \infty$,*

$$\mathrm{Prob}\left[1 + \frac{M}{N} \leqslant R_N \leqslant 1 + (2 + \delta) \sqrt{\frac{\log N}{N}} \right] = 1$$

as $N \to \infty$.

7.8   Berry–Esseen Type Inequalities for Large Random Matrices

We often need Berry–Esseen inequalities such as Equations 7.6 and 7.7 in the theory of sums of independent random variables. In analogy, we have results for large random matrices.

Following [42], we obtain optimal bounds of order $O\left(1/n\right)$ for the rate of convergence to the semicircle law and to the Marchenko–Pastur law for the expected spectral distribution functions of random matrices from the Gaussian unitary ensemble (GUE) and Laguerre unitary ensemble (LUE), respectively. Let $g(x)$ and $G(x)$ denote the density and the distribution function of the standard semicircle law respectively, that is

$$g(x) = \frac{1}{2\pi}\sqrt{4-x^2}I_{\{|x|\leqslant 2\}}, G(x) = \int_{-\infty}^{x} g(t)\,dt$$

where $I_{\{B\}}$ is the indicator of an event B. The Kolmogorov distance between $\mathbb{E}F_n(x)$ and $G(x)$, that is

$$\Delta_n = \sup_x |\mathbb{E}F_n(x) - G(x)|$$

where $F_n(x)$ is the empirical spectral distribution function defined by $F_n(x) = \frac{1}{n}\sum_{i=1}^{n} I_{\{\lambda_i\leqslant x\}}$, with $\lambda_1 \leqslant \cdots \leqslant \lambda_n$ the eigenvalues of the GUE and LUE. In the case of GUE the following representation for the expected spectral density holds:

$$p_n(x) = \sqrt{\frac{1}{2n}}\sum_{k=1}^{n-1} \varphi_k^2\left(x\sqrt{\frac{n}{2}}\right)$$

where $\varphi_k(x)$ denote the Hermitian orthogonal functions. There exist absolute positive constants, γ and C, such that, for $\left[-2+\gamma n^{-2/3}, 2-\gamma n^{-2/3}\right]$,

$$|p_n(x) - g(x)| \leqslant \frac{C}{n(4-x^2)} \tag{7.23}$$

For the LUE, we have similar bounds. For applications in the power grid, n is the nodes of the power grid; we are thus able to study how the performance of anomaly detection scales with the number of nodes n.

As shown in [43], the Kolmogorov distance between the spectral distribution function of a random covariance matrix $\frac{1}{p}\mathbf{X}\mathbf{X}^T$, where $\mathbf{X}$ is an $n \times p$ matrix with independent entries and the distribution function of the Marchenko–Pastur law, is of order $O(1/\sqrt{n})$.

Let $\mathbf{X} = (X_{jk})_{j,k=1}^{n}$ denote a Hermitian random matrix with entries X_{jk}, which are independent for $1 \leq j \leq k \leq n$. We consider the rate of convergence of the empirical spectral distribution function of the matrix $\mathbf{X}$ to the semi-circular law assuming that $\mathbb{E}X_{jk} = 0$, $\mathbb{E}X_{jk}^2 = 1$, and that the distributions of the matrix elements X_{jk} have a uniform subexponential decay in the sense that there exists a constant κ such that for any $1 \leq j \leq k \leq n$ and any $t \geq 1$ we have

$$\Pr\{|X_{jk}| > t\} \leqslant \frac{1}{\kappa}\exp(-t^\kappa)$$

It is shown in [44] that the Kolmogorov distance between the empirical spectral distribution of the Wigner matrix $\mathbf{W} = \frac{1}{\sqrt{n}}\mathbf{X}$ and the semicircular law is of order $O\left(n^{-1}\log^b n\right)$ with some positive constant $b > 0$.

Let $\mathbf{X} = (X_{jk})$ denote an n p random matrix with entries Xjk, which are independent for $1 \leqslant j \leqslant n, 1 \leqslant k \leqslant p$. We consider the rate of convergence of the empirical spectral distribution function of the matrix $\mathbf{S} = \frac{1}{p}\mathbf{X}\mathbf{X}^*$ to the Marchenko–Pastur law. We assume that $\mathbb{E}X_{jk} = 0, \mathbb{E}X_{jk}^2 = 1$, and that the distributions of the matrix elements X_{jk} have a uniformly sub-exponential decay in the sense that there exists a constant $\kappa > 0$ such that for any $n, p \geq 1$ and $1 \leqslant j \leqslant n, 1 \leqslant k \leqslant p$ and any $t \geq 1$ we have

$$\Pr\{|X_{jk}| > t\} \leqslant \frac{1}{\kappa}\exp\left(-t^\kappa\right)$$

It is shown in [45] that the Kolmogorov distance between the empirical spectral distribution of the sample covariance matrix $\mathbf{S}$ and the Marchenko–Pastur distribution is of order $O\left(n^{-1}\log^{4+\frac{4}{\kappa}}n\right)$ with high probability.

7.9 Gaussian Tools for Massive Data Analysis

We often need to calculate the expectation of the trace function. For Gaussian random variables, two fundamental analytic tools are the Nash–Poincare inequality and the integration by parts formula. We are motivated to illustrate how these analytical tools are used for linear eigenvalue statistics.

7.9.1 Expected moments of random matrices with complex Gaussian entries

Recall that for ξ in $\mathbb{R}$ and $\sigma^2 \in \,]0, \infty[\,$, $\mathcal{N}\left(\xi, \frac{1}{2}\sigma^2\right)$ denotes the Gaussian distribution with mean ξ and variance σ^2. The normalized trace is defined as

$$\mathrm{tr}_n = \frac{1}{n}\mathrm{Tr}$$

The first class, denoted Hermitian Gaussian random matrices or HGRM(n,σ^2), is a class of Hermitian $n \times n$ random matrices $\mathbf{A} = (a_{ij})$, satisfying that the entries $\mathbf{A} = (a_{ij}), 1 \leqslant i \leqslant j \leqslant n$, form a set of $\frac{1}{2}n(n+1)$ independent, Gaussian random variables, which are complex valued whenever $i < j$, and fulfill that

$$\mathbb{E}(a_{ij}) = 0, \text{ and } \mathbb{E}\left(|a_{ij}|^2\right) = \sigma^2, \text{ for all } i, j$$

The case $\sigma^2 = \frac{1}{2}$ gives the normalization used by Wigner [46] and Mehta [47], while the case $\sigma^2 = \frac{1}{n}$ gives the normalization used by Voiculescu [15]. We say that $\mathbf{A}$ is a standard Hermitian Gaussian random $n \times n$ matrix with entries of variance σ^2, if the following conditions are satisfied:

1. The entries $a_{ij}, 1 \leqslant i \leqslant j \leqslant n$, form a set of $\frac{1}{2}n(n+1)$ independent, complex valued random variables.

2. For each i in $\{1, 2, ..., n\}$, a_{ii} is a real valued random variable with distribution $\mathcal{N}\left(0, \frac{1}{2}\sigma^2\right)$.

3. When $i \leq j$, the real and imaginary parts $\mathrm{Re}\,(a_{ij})\,,\mathrm{Im}\,(a_{ij})$, of a_{ij} are independent, identically distributed random variables with distribution $\mathcal{N}\left(0, \frac{1}{2}\sigma^2\right).$

4. When $j > i$, $a_{ij} = \bar{a}_{ji}$, where $\bar{d}$ is the complex conjugate of d.

We denote by HGRM(n,σ^2) the set of all such random matrices. If $\mathbf{A}$ is an element of HGRM(n,σ^2), then

$$\mathbb{E}\left(|a_{ij}|^2\right) = \sigma^2, \text{ for all } i, j$$

The distribution of the real valued random variable a_{ii} has density

$$x \mapsto \frac{1}{\sqrt{2\pi\sigma^2}} \exp\left(-\frac{x^2}{2\sigma^2}\right) \quad x \in \mathbb{R}$$

(w.r.t.) Lebesgue measure on $\mathbb{R}$, whereas, if $i \leq j$, the distribution of the complex valued random variable a_{ij} has density

$$z \mapsto \frac{1}{\sqrt{2\pi\sigma^2}} \exp\left(-\frac{z^2}{2\sigma^2}\right), \quad z \in \mathbb{C}$$

w.r.t. Lebesgue measure on $\mathbb{C}$. For any complex matrix $\mathbf{H}$, we have that

$$\mathrm{Tr}\left(\mathbf{H}^2\right) = \sum_{i=1}^{n} h_{ii} + 2\sum_{i<j} |h_{ij}|^2$$

The distribution of an element $\mathbf{A}$ of HGRM(n,σ^2) has the density

$$\mathbf{H} \mapsto \frac{1}{\sqrt{2\pi\sigma^2}} \exp\left(-\frac{1}{2\sigma^2}\mathrm{Tr}\left(\mathbf{H}^2\right)\right), \quad \mathbf{H} \in \mathbb{C}^{n \times n}$$

The second class, denoted Gaussian random matrices or GRM(m,n,σ^2), is a class of $m \times n$ random matrices $\mathbf{B} = (b_{ij})\,, 1 \leqslant i \leqslant m, 1 \leqslant j \leqslant n$. This class forms a set of mn independent complex-valued, Gaussian random variables, satisfying that

$$\mathbb{E}\,(b_{ij}) = 0, \text{ and } \mathbb{E}\left(|b_{ij}|^2\right) = \sigma^2, \text{ for all } i, j$$

We say $\mathbf{B}$ is a standard Gaussian random matrix of $m \times n$ with entries of variance σ^2, if real valued random variables $\mathrm{Re}\,(b_{ij})\,,\mathrm{Im}\,(b_{ij})\,, 1 \leqslant i, j \leqslant n$, form a family of $2mn$ i.i.d. random variables, with distribution $\mathcal{N}\left(0, \frac{1}{2}\sigma^2\right)$. This class starts with Wishart [48] and Hsu [49].

We are interested in the explicit formulas for the mean values $\mathbb{E}\,(\mathrm{Tr}\,[\exp\,(s\mathbf{A})])$ in the Wigner case, and

$$\mathbb{E}\,(\mathrm{Tr}\,[\mathbf{B}^*\mathbf{B}\exp\,(s\mathbf{B}^*\mathbf{B})])$$

in the Wishart case, as functions of a complex parameter s. A new and entirely analytical treatment of these problems is given by the classical work of Haagerup and Thorbjornsen [50] which we follow closely for this section.

7.9.2 Hermitian Gaussian random matrices HGRM(n, σ^2)

We need the confluent hyper-geometric function [51, Vol.1, p. 248] $(a, c, x) \mapsto \Phi\,(a, c, x)$ that is defined as

$$\Phi\,(a, c, x) = \sum_{n=0}^{\infty} \frac{(a)_n x^n}{(c)_n n!} = 1 + \frac{a}{c}\frac{x}{1} + \frac{a(a+1)}{c(c+1)}\frac{x^2}{2} + \cdots$$

for a, c, x in $\mathbb{C}$, such that $c \notin \mathbb{Z}\backslash\mathbb{N}$. In particular, if $a \in \mathbb{Z}\backslash\mathbb{N}$, then $(x) \mapsto \Phi(a, c, x)$ is a polynomial in x of degree $-a$, for any permitted c. For any nonnegative integer n, and any complex number w, we apply the notation

$$(w)_n = \begin{cases} 1, & \text{if } n = 0 \\ w(w + 1)(w + 2) \cdots (w + n - 1), & \text{if } n \in \mathbb{N} \end{cases}$$

For any element $\mathbf{A}$ of $\text{HGRM}(n, \sigma^2)$ and any $s \in \mathbb{C}$, we have that

$$\mathbb{E}\left(\text{Tr}\left[\exp(s\mathbf{A})\right]\right) = n \cdot \exp\left(\frac{\sigma^2 s^2}{2}\right) \cdot \Phi\left(1 - n, 2; -\sigma^2 s^2\right)$$

If $(\mathbf{X}_n)$ is a sequence of random matrices, such that $\mathbf{X}_n \in \text{HGRM}\left(n, \frac{1}{n}\right)$ for all n in $\mathbb{N}$, then for any $s \in \mathbb{C}$, we have that

$$\lim_{n \to \infty} \mathbb{E}\left(\text{Tr}\left[\exp(s\mathbf{X}_n)\right]\right) = \frac{1}{2\pi} \int_{-2}^{2} \exp(sx)\sqrt{4 - x^2} dx$$

and the convergence is uniform on compact subsets of $\mathbb{C}$. Further we have that for the kth moment of $\mathbf{X}_n$

$$\lim_{n \to \infty} \mathbb{E}\left(\text{tr}_n\left[\mathbf{X}_n^k\right]\right) = \frac{1}{2\pi} \int_{-2}^{2} x^k \sqrt{4 - x^2} dx$$

and in general, for every continuous bounded function $f : \mathbb{R} \mapsto \mathbb{C}$,

$$\lim_{n \to \infty} \mathbb{E}\left(\text{tr}_n\left[f(\mathbf{X}_n)\right]\right) = \frac{1}{2\pi} \int_{-2}^{2} f(x)\sqrt{4 - x^2} dx$$

Often, a recursion formula is efficient in calculation. Let $\mathbf{A}$ be an element of $\text{HGRM}(n, 1)$, and for integer k define:

$$C(k, n) = \mathbb{E}\left(\text{Tr}\left[\mathbf{A}^{2k}\right]\right)$$

Then the initial values are $C(0, n) = n, C(1, n) = n^2$, and for fixed n in $\mathbb{N}$, the numbers $C(k, n)$ satisfy the recursion formula

$$C(k + 1, n) = n \cdot \frac{4k + 2}{k + 2} \cdot C(k, n) + \frac{k\left(4k^2 - 1\right)}{k + 2} \cdot C(k - 1, n), \qquad k \geqslant 1 \qquad (7.24)$$

We can further show that $C(k, n)$ has the following form [52]:

$$C(k, n) = \sum_{i=0}^{\left[\frac{k}{2}\right]} a_i(k) n^{k+1-2i}, \qquad k \in \mathbb{N}_0, n \in \mathbb{N} \qquad (7.25)$$

Here the notation $\mathbb{N}_0$ denotes the integer that does not include zero, in contrast with $\mathbb{N}$. The coefficients $a_i(k), i, k \in \mathbb{N}_0$ are determined by the following recursive formula:

$$a_i(k) = 0, \quad i \geqslant \left[\frac{k}{2}\right] + 1$$

$$a_0(k) = \frac{1}{k + 1}\binom{2k}{k}, \quad k \in \mathbb{N}_0$$

$$a_i(k + 1) = \frac{4k + 2}{k + 2} \cdot a_i(k) + \frac{k\left(4k^2 - 1\right)}{k + 2} \cdot a_{i-1}(k - 1), \quad k, i \in \mathbb{N}$$

A list of the numbers of $a_i(k)$ is given in [52, p. 459].

From Equations 7.24 and 7.25, we can get [50] for any $\mathbf{A}$ in $\text{HGRM}(n, 1)$,

$$\mathbb{E}\left(\text{Tr}\left[\mathbf{A}^2\right]\right) = n^2$$
$$\mathbb{E}\left(\text{Tr}\left[\mathbf{A}^4\right]\right) = 2n^3 + n$$
$$\mathbb{E}\left(\text{Tr}\left[\mathbf{A}^6\right]\right) = 5n^4 + 10n^2$$
$$\mathbb{E}\left(\text{Tr}\left[\mathbf{A}^8\right]\right) = 14n^5 + 70n^3 + 21n$$
$$\mathbb{E}\left(\text{Tr}\left[\mathbf{A}^{10}\right]\right) = 42n^6 + 420n^4 + 483n^2$$

and so on. If we replace the $\mathbf{A}$ mentioned previously by an element $\mathbf{X}$ of $\text{HGRM}(n, \frac{1}{n})$, and Tr by tr_n, then we have to divide the previously mentioned numbers by n^{k+1}. Finally, for $\mathbf{X}$ of $\text{HGRM}(n, \frac{1}{n})$, we have

$$\mathbb{E}\left(\text{tr}_n\left[\mathbf{A}^2\right]\right) = 1$$
$$\mathbb{E}\left(\text{tr}_n\left[\mathbf{A}^4\right]\right) = 2 + \frac{1}{n^2}$$
$$\mathbb{E}\left(\text{tr}_n\left[\mathbf{A}^6\right]\right) = 5 + \frac{10}{n^2}$$
$$\mathbb{E}\left(\text{tr}_n\left[\mathbf{A}^8\right]\right) = 14 + \frac{70}{n^2} + \frac{21}{n^4}$$
$$\mathbb{E}\left(\text{tr}_n\left[\mathbf{A}^{10}\right]\right) = 42 + \frac{420}{n^2} + \frac{483}{n^4}$$

and so on. The constant term in $\mathbb{E}\left(\text{tr}_n\left[\mathbf{A}^{2k}\right]\right)$ is

$$a_0\left(k\right) = \frac{1}{k+1}\left(\begin{array}{c} 2k \\ k \end{array}\right) = \frac{1}{2\pi}\int_{-2}^{2} x^{2k}\sqrt{4 - x^2}dx$$

in concordance with Wigner's semicircle law.

7.9.3 Hermitian Gaussian random matrices $\text{GRM}(m, n, \sigma^2)$

We first define the function $\varphi_k^\alpha\left(x\right)$ as

$$\varphi_k^\alpha\left(x\right) = \left[\frac{k!}{\Gamma\left(k + \alpha + 1\right)} x^\alpha \exp\left(-x\right)\right]^{1/2} \cdot L_k^\alpha\left(x\right), \quad k \in \mathbb{N}_0 \tag{7.26}$$

where $L_k^\alpha(x)_{k\in\mathbb{N}_0}$ is the sequence of generalized Laguerre polynomials of order α, that is,

$$L_k^\alpha\left(x\right) = \left(k!\right)^{-1} x^{-\alpha} \exp\left(x\right) \cdot \frac{d^k}{dx^k}\left(x^{k+\alpha} \exp\left(-x\right)\right), \quad k \in \mathbb{N}_0$$

Here $\Gamma\left(x\right)$ is the gamma function.

Now we can state a corollary from [50]. Let $\mathbf{B}$ be an element of $\text{GRM}(m, n, 1)$, let $\varphi_k^\alpha\left(x\right), \alpha \in]0, \infty[, k \in \mathbb{N}_0$ be the functions introduced in Equation 7.26, and let $f :]0, \infty[\mapsto \mathbb{R}$ be a Borel function.[1] If $m \geq n$, we have that

$$\mathbb{E}\left(\text{Tr}\left[f\left(\mathbf{B}^*\mathbf{B}\right)\right]\right) = \int_0^\infty f(x)\left[\sum_{i=0}^{n-1}\left(\varphi_k^{m-n}\left(x\right)\right)^2\right]dx$$

If $m \leq n$, we have that

$$\mathbb{E}\left(\text{Tr}\left[f\left(\mathbf{B}^*\mathbf{B}\right)\right]\right) = (n - m)f(0) + \int_0^\infty f(x)\left[\sum_{i=0}^{n-1}\left(\varphi_k^{n-m}\left(x\right)\right)^2\right]dx$$

---

[1]A map $f : X \mapsto Y$ between two topological spaces is called Borel (or Borel measurable) if $f^{-1}(A)$ is a Borel set for any open set A.

We need to define the hyper-geometric function F.

$\mathbf{Q}$ of $m \times n$ is an element from $\mathrm{GRM}(m, n, \frac{1}{n})$. Denote $c = \frac{m}{n}$. We have

$$\mathbb{E}\left(\mathrm{tr}_n\left[\mathbf{Q}^*\mathbf{Q}\right]\right) = c$$

$$\mathbb{E}\left(\mathrm{tr}_n(\mathbf{Q}^*\mathbf{Q})^2\right) = c^2 + c$$

$$\mathbb{E}\left(\mathrm{tr}_n(\mathbf{Q}^*\mathbf{Q})^3\right) = \left(c^3 + 3c^2 + c\right) + cn^{-2}$$

$$\mathbb{E}\left(\mathrm{tr}_n(\mathbf{Q}^*\mathbf{Q})^4\right) = \left(c^4 + 6c^3 + 6c^2 + c\right) + \left(5c^2 + 5c\right)n^{-2}$$

$$\mathbb{E}\left(\mathrm{tr}_n(\mathbf{Q}^*\mathbf{Q})^5\right) = \left(c^5 + 10c^4 + 20c^3 + 10c^2 + c\right) + \left(15c^3 + 40c^2 + 15c\right)n^{-2} + 8cn^{-4}$$

$$(7.27)$$

In general, $\mathbb{E}\left(\mathrm{tr}_n(\mathbf{Q}^*\mathbf{Q})^k\right)$ is a polynomial of degree $\left[\frac{k-1}{2}\right]$ in n^{-2}, for fixed c.

7.10 An Integration by Parts Formula for Functions of Gaussian Random Variables

Lemma 1 *Let z^* stand for the conjugate of a complex number z. Let $\mathbf{x} = [X_1, ..., X_N]^T$ be a complex Gaussian random vector whose law is determined by*

$$\mathbb{E}\left[\mathbf{x}\right] = \mathbf{0}, \mathbb{E}\left[\mathbf{x}\mathbf{x}^T\right] = \mathbf{0}, \mathbb{E}\left[\mathbf{x}\mathbf{x}^H\right] = \mathbf{\Sigma}$$

or $\mathbf{x} \sim \mathcal{CN}\left(\mathbf{0}, \mathbf{\Sigma}\right)$. Let

$$f = f\left(X_1, ..., X_N, X_1^*, ..., X_N^*\right)$$

be a $\mathbb{C}^1$ complex function, polynomially bounded together with its derivatives, then

$$\mathbb{E}\left[X_i f\left(\mathbf{x}\right)\right] = \sum_{j=1}^{N} \mathbf{\Sigma}_{ij} \mathbb{E}\left[\frac{\partial f\left(\mathbf{x}\right)}{\partial X_j^*}\right] \tag{7.28}$$

This formula, Equation 7.28, relies on an integration by parts and thus is referred to as the integration by parts formula for *Gaussian vectors*. It is widely used in mathematical physics and has been used in random matrix theory.

Remark 1 (Integration by parts formula for functionals of matrices with i.i.d. entries) *Let $f\left(\mathbf{W}\right)$ be a $\mathcal{C}^1$ complex function of the elements of $\mathbf{W}$ and $\mathbf{W}^*$, polynomially bounded together with its derivatives, where $\mathbf{W}$ has i.i.d. entries $W_{ij} \sim \mathcal{CN}\left(0, 1\right)$. Then,*

$$\mathbb{E}\left[W_{ij} f\left(\mathbf{W}\right)\right] = \mathbb{E}\left[\frac{\partial f\left(\mathbf{W}\right)}{\partial X_{ij}^*}\right] \tag{7.29}$$

$\square$

7.11 Poincare–Nash Inequality

Lemma 2 *Let* $\mathbf{x}$ *and* f *be as Section 7.10 and let*

$$\nabla_{\mathbf{x}} f = \left(\frac{\partial f}{\partial X_1}, ..., \frac{\partial f}{\partial X_N} \right)^T \text{ and } \nabla_{\mathbf{x}*} f = \left(\frac{\partial f}{\partial X_1^*}, ..., \frac{\partial f}{\partial X_N^*} \right)^T$$

Then the following inequality holds true:

$$\text{Var}\left(f\left(\mathbf{x} \right) \right) \leqslant \left[\left(\nabla_{\mathbf{x}} f\left(\mathbf{x} \right) \right)^T \boldsymbol{\Sigma} \left(\nabla_{\mathbf{x}} f\left(\mathbf{x} \right) \right)^* \right] + \mathbb{E}\left[\left(\nabla_{\mathbf{x}*} f\left(\mathbf{x} \right) \right)^H \boldsymbol{\Sigma} \, \nabla_{\mathbf{x}} f\left(\mathbf{x} \right) \right] \qquad (7.30)$$

Remark 2 (Poincare–Nash Inequality for functionals of matrices with i.i.d. Gaussian entries) *Let* $f\left(\mathbf{W} \right)$ *be a function of the elements of* $\mathbf{W}$ *and* $\mathbf{W}^*$ *as in Remark 1, where* $\mathbf{W} \in \mathbb{C}^{N \times n}$ *has i.i.d. Gaussian entries* $W_{ij} \sim \mathcal{CN}\left(0, 1 \right)$. *Then,*

$$\text{Var}\left(f\left(\mathbf{W} \right) \right) \leqslant \sum_{i=1}^{N} \sum_{j=1}^{n} \mathbb{E}\left[\left| \frac{\partial f\left(\mathbf{W} \right)}{\partial W_{ij}} \right|^2 + \left| \frac{\partial f\left(\mathbf{W} \right)}{\partial W_{ij}^*} \right|^2 \right] \qquad (7.31)$$

$\square$

Lemma 3 (Identities for complex derivatives) *Let* $\mathbf{H} \in \mathbb{C}^{N \times K}$. *Then*

$$\frac{\partial H_{pq}}{\partial H_{ij}^*} = 0, \qquad \frac{\partial H_{pq}}{\partial H_{ij}} = \delta_{ip}\delta_{jq}, \qquad \frac{\partial \left[\mathbf{H}\mathbf{H}^H \right]_{pq}}{\partial H_{ij}^*} = \delta_{iq}H_{pj}, \qquad \frac{\partial \left[\mathbf{H}\mathbf{H}^H \right]_{pq}}{\partial H_{ij}} = \delta_{ip}H_{qj}^*,$$

$$\frac{\partial \left[\mathbf{H}^H\mathbf{H} \right]_{pq}}{\partial H_{ij}^*} = \delta_{jp}H_{iq}, \qquad \frac{\partial \left[\mathbf{H}^H\mathbf{H} \right]_{pq}}{\partial H_{ij}} = \delta_{jq}H_{ip}^*$$

Moreover, denote $\mathbf{Q} = \left(\frac{1}{K}\mathbf{H}\mathbf{H}^H + x\mathbf{I}_N \right)^{-1}$ *and* $\tilde{\mathbf{Q}} = \left(\frac{1}{K}\mathbf{H}^H\mathbf{H} + x\mathbf{I}_N \right)^{-1}$ *for some real number* $x > 0$. *Then,*

$$\frac{\partial Q_{pq}}{\partial H_{ij}^*} = -\frac{1}{K}\left[\mathbf{Q}\mathbf{H} \right]_{pj}Q_{iq}, \qquad \frac{\partial Q_{pq}}{\partial H_{ij}} = -\frac{1}{K}\left[\mathbf{H}^H\mathbf{Q} \right]_{jq}Q_{pi}$$

$$\frac{\partial \tilde{Q}_{pq}}{\partial H_{ij}^*} = -\frac{1}{K}\tilde{Q}_{pi}\left[\mathbf{H}\tilde{\mathbf{Q}} \right]_{iq}, \qquad \frac{\partial \tilde{Q}_{pq}}{\partial H_{ij}} = -\frac{1}{K}\tilde{Q}_{jq}\left[\tilde{\mathbf{Q}}\mathbf{H}^H \right]_{pi}$$

The Poincare–Nash inequality turns out to be *extremely useful* to deal with variances of various quantities of interest related to random matrices.

Example 2 (A random matrix with a separable variance profile) *Consider the* $N \times n$ *matrix* $\mathbf{Y}_n$ *of the form*

$$\mathbf{Y}_n = \mathbf{D}^{1/2}\mathbf{G}_n\tilde{\mathbf{D}}^{1/2} \qquad (7.32)$$

where

$$\mathbf{D} = \text{diag}\left(d_1, ..., d_N \right) \text{ and } \tilde{\mathbf{D}} = \text{diag}\left(\tilde{d}_1, ..., \tilde{d}_n \right)$$

are, respectively, $N \times N$ and $n \times n$ diagonal matrices, and

$$\mathbf{G}_n = \mathbf{V}_n^H \mathbf{W}_n \tilde{\mathbf{U}}_n$$

has i.i.d. entries with complex Gaussian distribution $\mathcal{CN}(0,1)$ since $\mathbf{V}_n$ and $\tilde{\mathbf{U}}_n$ are deterministic unitary matrices. Since every individual entry of $\mathbf{Y}_n$ has the form $Y_{ij} = \sqrt{d_i \tilde{d}_j} X_{ij}$, we can call $\mathbf{Y}_n$ a random matrix with a separable variance profile.

Consider $\mathbf{Y}_n$ defined in Equation 7.32. When $\mathbf{x}$ is the vector of the stacked columns of matrix $\mathbf{Y}$, that is, $\mathbf{x} = [Y_{11}, ..., Y_{Nn}]^T$, we have

$$\mathbb{E}[Y_{ij} f(\mathbf{Y})] = d_i \tilde{d}_i \mathbb{E}\left[\frac{\partial f(\mathbf{Y})}{\partial \bar{Y}_{ij}}\right] \tag{7.33}$$

while Equation 7.30 becomes

$$\text{Var}(f(\mathbf{Y})) \leqslant \sum_{i=1}^{N} \sum_{j=1}^{n} d_i \tilde{d}_i \mathbb{E}\left[\left|\frac{\partial f(\mathbf{Y})}{\partial Y_{ij}}\right|^2 + \left|\frac{\partial f(\mathbf{Y})}{\partial \bar{Y}_{ij}}\right|^2\right] \tag{7.34}$$

□

Every function $\varphi : \mathbb{R}^{2n} \mapsto \mathbb{C}$ can be written as

$$\varphi(\mathbf{x}, \mathbf{y}) = \tilde{\varphi}\left(\frac{\mathbf{z} + \mathbf{z}^*}{2}, \frac{\mathbf{z} - \mathbf{z}^*}{2}\right) = \tilde{\varphi}(\mathbf{z}, \mathbf{z}^*)$$

with $\mathbf{z} = \mathbf{x} + i\mathbf{y} \in \mathbb{C}^n$. We also define the classical differential operators

$$\frac{\partial}{\partial z} = \frac{1}{2}\left(\frac{\partial}{\partial x} - i\frac{\partial}{\partial y}\right), \text{ and} \frac{\partial}{\partial z^*} = \frac{1}{2}\left(\frac{\partial}{\partial x} + i\frac{\partial}{\partial y}\right)$$

Lemma 4 *Let $z_1 = x_1 + iy_1, ..., z_n = x_n + iy_n$ be n i.i.d. $\mathcal{N}(0,1)$ random variables defined on $\mathbb{R}^{2n}$ with polynomially bounded partial derivatives. Then, if $\mathbf{x} = (x_1, ..., x_n)^T$ and $\mathbf{y} = (y_1, ..., y_n)^T$, it holds that*

$$\text{Var}[\varphi(\mathbf{x}, \mathbf{y})] = \text{Var}[\tilde{\varphi}(\mathbf{z}, \mathbf{z}^*)] \leqslant \sigma^2 \sum_{i=1}^{n}\left(\mathbb{E}\left|\frac{\partial \tilde{\varphi}(\mathbf{z}, \mathbf{z}^*)}{\partial z_i}\right| + \mathbb{E}\left|\frac{\partial \tilde{\varphi}(\mathbf{z}, \mathbf{z}^*)}{\partial z_i^*}\right|\right)$$

So far, we have presented the Poincare–Nash inequality for a Gaussian matrix. For general matrix models, we have a similar inequality.

Theorem 12 (Poincare–Nash inequality for general matrix models [53]) *If we have a probability distribution of Hermitian random matrix $\mathbf{M}_n$*

$$p_n(\mathbf{M}_n)\, d\mathbf{M}_n = \frac{1}{Z_n} \exp\{-n \operatorname{Tr} V(\mathbf{M}_n)\}\, d\mathbf{M}_n \tag{7.35}$$

where

$$d\mathbf{M}_n = \prod_{j=1}^{n} dM_{jj} \prod_{1 \leqslant j \leqslant k \leqslant n} (d\operatorname{Re} M_{jk})(d\operatorname{Im} M_{jk})$$

obeying conditions

$$V(t) \geqslant (2 + \varepsilon)\ln|t|, \quad \varepsilon > 0, \quad t \to \infty$$

and

$$|V(\lambda) - V(\mu)| \leqslant C(L)|\lambda - \mu|^{\gamma}, \quad \gamma > 0, \quad |\lambda|, |\mu| \leqslant L$$

then for any smooth bounded function $\varphi : \mathbb{R} \to \mathbb{C}$ *with bounded derivative we have the bound for Rth classical cumulant of* $\frac{1}{N} \operatorname{Tr} \varphi (\mathbf{M}_n)$

$$\left| k_r \left(\frac{1}{n} \operatorname{Tr} \varphi (\mathbf{M}_n) \right) \right| \leqslant C_\varphi r! a^r \frac{1}{n^r}, \quad a > 0, \quad , r \geqslant 2 \tag{7.36}$$

where the constant C_φ *depends only on the* L^∞*-norm of* φ *and* φ'.

For example, consider $\varphi (\lambda) = (\lambda - z)^{-1}$; we have

$$|\varphi (\lambda)| \leqslant \frac{1}{|\operatorname{Im} z|}, \quad |\varphi' (\lambda)| \leqslant \frac{1}{|\operatorname{Im} z|^2}$$

and thus

$$k_2 (a) = \operatorname{Var} \{a\}, \quad \mathbb{E} \left\{ |a - \mathbb{E} \{a\}|^4 \right\} = k_4 (a) + 3 k_2^2 (a)$$

7.12 Analysis of Linear Eigenvalue Statistic

Linear eigenvalue statistics is defined as

$$\rho_n (\lambda) = \frac{1}{n} \sum_{i=1}^{n} \delta \left(\lambda - \lambda_i^{(n)} \right) \tag{7.37}$$

where

$$\lambda_1^{(n)}, ..., \lambda_n^{(n)} \tag{7.38}$$

are the random eigenvalues of some random matrix, and δ is the Dirac delta function. We want to find the following moments:

$$\mathbb{E} \{\rho_n (\lambda)\} \text{ as } n \to \infty, \quad \mathbb{E} \{\rho_n^2 (\lambda)\} \text{ as } n \to \infty \tag{7.39}$$

where

$$\rho_n^2 (\lambda) = \frac{1}{n^2} \sum_{i=1}^{n} \sum_{j=1}^{n} \delta \left(\lambda - \lambda_i^{(n)} \right) \delta \left(\lambda - \lambda_j^{(n)} \right) = \frac{1}{n^2} \sum_{i=1}^{n} \delta^2 \left(\lambda - \lambda_i^{(n)} \right)$$
$$+ \sum_{i \neq j}^{n} \delta \left(\lambda - \lambda_i^{(n)} \right) \delta \left(\lambda - \lambda_j^{(n)} \right)$$

We find that summand $\delta^2 \left(\lambda - \lambda_i^{(n)} \right)$ of the first sum on the right-hand side (r.h.s.) is not well defined since it is often said that the square of delta function is infinity.

To avoid this we must "smooth" the delta function:

$$\frac{1}{n} \sum_{i=1}^{n} u \left(\lambda - \lambda_i^{(n)} \right) = \int u (\lambda - \mu) \rho_n (\lambda) d\mu$$

Motivated by this, we deal with normalized linear eigenvalue statistics, defined for any sufficiently smooth test:

$$\xi_n [\varphi] = \frac{1}{n} \sum_{i=1}^{n} \varphi \left(\lambda_i^{(n)} \right) = \int \varphi (\mu) \rho_n (\lambda) d\mu \tag{7.40}$$

7.12.1 The self-averaging property of the eigenvalue density

We consider the variance

$$\text{Var}\left\{\xi_n\left[\varphi\right]\right\} = \mathbb{E}\left\{\left|\xi_n\left[\varphi\right]\right|^2\right\} - \left|\mathbb{E}\left\{\xi_n\left[\varphi\right]\right\}\right|^2 \tag{7.41}$$

of a linear eigenvalue statistic of Wishart matrices. This is the simplest measure of the fluctuations of $\xi_n\left[\varphi\right]$ around its expectation $\mathbb{E}\left\{\xi_n\left[\varphi\right]\right\}$.

$$n \to \infty, p \to \infty, p/n \to c \in (1, \infty). \tag{7.42}$$

We have the bound

$$\text{Var}\left\{\xi_n\left[\varphi\right]\right\} \leqslant \frac{4}{\beta n^2 p} \text{Tr}\,\boldsymbol{\Sigma}^2 \left(\max_{\lambda \in \mathbb{R}} \left|\varphi'\left(\lambda\right)\right|\right) \tag{7.43}$$

where $\beta = 1$ for real symmetric and $\beta = 2$ for Hermitian Wishart matrices.

We want to keep the spectrum bounded for n, p of Equation 7.42 rather than escaping to infinity, so we assume that in the limit of Equation 7.42

$$\max_n \frac{1}{n} \text{Tr}\,\boldsymbol{\Sigma}^2 \leqslant C < \infty \tag{7.44}$$

Similarly, a stronger condition is necessary to obtain Equation 7.42. Assuming this and

$$\max_{\lambda \in \mathbb{R}} \left|\varphi'\left(\lambda\right)\right| < \infty \tag{7.45}$$

we find from Equation 7.43 that

$$\text{Var}\left\{\xi_n\left[\varphi\right]\right\} = O\left(1/n^2\right) \tag{7.46}$$

Note that if the eigenvalues in Equation 7.38 of $\boldsymbol{\Xi}$ were i.i.d. random variables, then the variance of their linear statistics would be equal to $\frac{1}{n}\text{Var}\left\{\varphi\left(\lambda_1\right)\right\}$, that is, $O\left(1/n\right)$ for any φ such that $\text{Var}\left\{\varphi\left(\lambda_1\right)\right\} < \infty$. Thus, Equation 7.46 exhibits strong dependence between eigenvalues of Wishart random matrices. In other words, there must be *very effective cancellation* in the sum of strongly correlated eigenvalues, which can be viewed as a global manifestation of the regularity of eigenvalue distribution [16, p. 152].

The order of $O\left(1/n^2\right)$ in Equation 7.46 is the case if the test functions satisfy Equation 7.45. Consider, for instance, the Heaviside function

$$\varphi_H\left(\lambda\right) = \begin{cases} \varphi_0, & \lambda \in [a, b] \\ 0, & \text{otherwise} \end{cases}$$

as in Equation 7.40. In this case $\xi_n\left[\varphi\right]$ is just the relative number $\xi_n\left(a, b\right)$ of eigenvalues of $\boldsymbol{\Xi}$ falling in the interval $[a, b]$. We can show that at least for $\boldsymbol{\Sigma} = a^2 \mathbf{I}_n$, we have

$$\text{Var}\left\{\xi_n\left(a, b\right)\right\} = \frac{\log n}{\pi^2 n^2}\left(1 + o\left(1\right)\right), \quad n \to \infty \tag{7.47}$$

The fact that the fluctuations of linear eigenvalue statistics around their mean are $O\left(1/n^2\right)$ as in Equation 7.46 implies that for large systems the density is effectively deterministic. This property is known as the *self-averaging property* in random matrix theory and the theory of disordered systems [47]. The property is similar to the law of large numbers of probability theory, according to which if $X_1, ..., X_n$ are i.i.d. random variables, then the random variable

$$S_n = \frac{1}{n}\sum_{i=1}^{n} X_i$$

tends to the nonrandom limit $\mathbb{E}\{X_1\}$ as $n \to \infty$. The simplest manifestation of this is the form of variance

$$\text{Var}\{S_n\} := \mathbb{E}\left\{|S_n|^2\right\} - |\mathbb{E}\{S_n\}|^2 = \frac{1}{n}\text{Var}\{X_1\} = \frac{1}{n}\left(\mathbb{E}\left\{|X_1|^2\right\} - |\mathbb{E}\{X_1\}|^2\right)$$

of S_n. Indeed, it follows from the previously mentioned that if $\mathbb{E}\left\{|X_1|^2\right\}$ is finite, then

$$\text{Var}\{S_n\} = O\left(1/n\right), \quad n \to \infty$$

that is, the fluctuations of S_n are negligible for large n.

7.13 Asymptotic Density of the Eigenvalues

Let Ξ be an $n \times n$. We define the density of the random eigenvalues as a real symmetric or Hermitian Wishart matrix with p degrees of freedom and an $n \times n$ matrix Σ, that is, $\Xi \sim \mathcal{W}_n(p, \Sigma)$ or

$$\Xi = \frac{1}{p}\Sigma^{1/2}\mathbf{X}\mathbf{X}^T\Sigma^{1/2} \tag{7.48}$$

Here $\mathbf{X} = \{X_{ij}\}_{i,j=1}^{n,p}$ is the $n \times p$ matrix, whose entries are the standard Gaussian random variables, determined by the relations

$$\mathbb{E}\{X_{ij}\} = 0, \qquad \mathbb{E}\{X_{i_1 j_1} X_{i_2 j_2}\} = \delta_{i_1,i_2}\delta_{j_1,j_2} \tag{7.49}$$

and Σ is positive definite.

Consider the linear statistics in Equation 7.40 of eigenvalues of Ξ, corresponding to a real or complex valued test function φ. It can be shown that

$$\lim_{n\to\infty, p\to\infty, p/n\to c\in[1,\infty)} \mathbb{E}\{\xi_n[\varphi]\} = \int \varphi(\lambda)\rho(\lambda)d\lambda \tag{7.50}$$

where ρ is the limiting eigenvalue density of Ξ, which can be found by solving a functional equation for its Stieltjes transform

$$f(z) = \int \frac{1}{\lambda - z}\rho(\lambda)\,d\lambda, \quad \text{Im } z \neq 0 \tag{7.51}$$

Given f, one can find ρ from the inversion formula

$$\rho(\lambda) = \frac{1}{\pi}\text{Im } f(\lambda + i0) \tag{7.52}$$

Assuming that the limiting eigenvalue density of Σ exists and denoting it ν, we can write the equation for f of Equation 7.51 as

$$f(z) = c\int \frac{\nu(\sigma)}{\sigma((c-1)f(z) - zf(z)) - cz} \tag{7.53}$$

The equation has to be solved in the class of functions analytic for $\text{Im } z \neq 0$ and such that

$$\text{Im } f(z)\,\text{Im } z > 0, \quad \text{Im } z \neq 0 \tag{7.54}$$

7.14 The Marchenko–Pastur (MP) Density

Equations 7.53 and 7.54 are hard to solve in closed form for an arbitrary ν. Numerical methods are normally necessary to solve the equation and obtain the limiting density of the random eigenvalues from Equation 7.52. We consider the simplest case where eigenvalues are asymptotically "flat," hence

$$\nu\left(\sigma\right) = \delta\left(\sigma - a^2\right)$$

For this case the matrix $\boldsymbol{\Sigma}$ becomes the unit matrix times a^2, that is $\boldsymbol{\Sigma} = a^2 \mathbf{I}_n$. Using this, Equation 7.53 reduces to the quadratic equation

$$a^2 z f^2\left(z\right) + \left(cz + a^2\left(1 - c\right)\right) f\left(z\right) + c = 0 \tag{7.55}$$

whose solution, satisfying Equation 7.54, is

$$f\left(z\right) = \frac{-\left(cz + a^2\left(1 - c\right)\right) + \sqrt{\left(cz + a^2\left(1 - c\right)\right)^2 - 4a^2 cz}}{2a^2 z} \tag{7.56}$$

where the branch of the square root is defined by its asymptotic $cz + O\left(1\right), z \to \infty$.

Substituting $f(z)$ from Equation 7.56 into Equation 7.52 yields that, for $c > 1$

$$\rho\left(\lambda\right) = \frac{c}{2\pi a^2 \lambda} \begin{cases} \sqrt{\left(a_+ - \lambda\right)\left(\lambda - a_-\right)} & \lambda \in [a_-, a_+] \\ 0, & \lambda \notin [a_-, a_+] \end{cases} \tag{7.57}$$

where

$$a_\pm = a^2 \left(1 \pm c^{-1/2}\right)^2$$

Equation 7.57 is now known as the Marchenko–Pastur density, which was originally derived in [54]. The significance of the work in [54] lies also in that they provide the analytical tool of this class of problems.

7.15 Derivation of Variance Bound Using the Nash–Poincare Inequality

We present the derivation of Equation 7.43, based on the Nash–Poincare inequality. For technical simplicity, we consider the Hermitian analog of the Wishart matrices, often called the Laguerre Ensemble. The result for the real symmetric, that is, genuine Wishart matrices, Equation 7.53, can be obtained analogously. Consider random matrices (see Equation 7.53) of the form

$$\boldsymbol{\Xi} = \frac{1}{p} \boldsymbol{\Sigma}^{1/2} \mathbf{X} \mathbf{X}^H \boldsymbol{\Sigma}^{1/2} \tag{7.58}$$

where $\boldsymbol{\Sigma}$ is positive definite. Here $\mathbf{X} = \{X_{ij}\}_{i,j=1}^{n,p}$ is the $n \times p$ matrix, whose entries are the standard complex Gaussian random variables, determined by the relations

$$\mathbb{E}\left\{X_{ij}\right\} = 0, \quad \mathbb{E}\left\{X_{i_1 j_1} X^*_{i_2 j_2}\right\} = \delta_{i_1, i_2} \delta_{j_1, j_2} \tag{7.59}$$

where the $*$ stands for the complex conjugate of a complex number.

We follow the four steps to obtain Equation 7.43.

(1) Given the standard complex Gaussian random variables

$$\mathbb{E}\left\{Z_i\right\} = 0, \quad \mathbb{E}\left\{Z_i Z_j^*\right\} = \delta_{ij}$$

and a differentiable function Φ of $2N$ complex variables, consider the random variable

$$\Psi = \Phi\left(Z_1, ..., Z_N, Z_1^*, ..., Z_N^*\right)$$

Then its variance admits the bound

$$\mathrm{Var}\left\{\Psi\right\} := \mathbb{E}\left\{|\Psi|^2\right\} - |\mathbb{E}\left\{\Psi\right\}|^2 \leqslant \sum_{i=1}^{N} \mathbb{E}\left(\left|\frac{\partial \Phi}{\partial Z_i}\right|^2 + \left|\frac{\partial \Phi}{\partial Z_i^*}\right|^2\right) \qquad (7.60)$$

known as the Nash–Poincare inequality (see e.g., [55]).

(2) Given a Hermitian matrix $\mathbf{A}(t)$ depending on a parameter t and a function $\varphi : \mathbb{R} \to \mathbb{C}$, consider the matrix function $\varphi\left(\mathbf{A}\left(t\right)\right).$ Then we have

$$\frac{d}{dt}\,\mathrm{Tr}\,\varphi\left(\mathbf{A}\left(t\right)\right) = \mathrm{Tr}\left[\varphi'\left(\mathbf{A}\left(t\right)\right)\mathbf{A}'\left(t\right)\right] \qquad (7.61)$$

It follows from Equation 7.58 that the entries $\{\Xi_{ij}\}_{i,j=1}^{n}$ are

$$\Xi_{ij} = \frac{1}{p}\sum_{k=1}^{n}\sum_{\ell=1}^{n}\sum_{\alpha=1}^{p} R_{ik} X_{k\alpha} X_{k\alpha}^* R_{\ell j} \qquad (7.62)$$

where $\mathbf{R} = \mathbf{\Sigma}^{1/2}$. Besides, it follows from the spectral theorem for Hermitian matrices and Equation 7.40 that

$$\xi_n\left[\varphi\right] = \frac{1}{n}\,\mathrm{Tr}\,\varphi\left(\mathbf{\Xi}\right)$$

Take in Equation 7.60 $\frac{1}{n}\,\mathrm{Tr}\,\varphi\left(\mathbf{\Xi}\right)$ as Ψ, and $\{X_{k\alpha}\}_{k,\alpha=1}^{n,p}$ as $\{Z_i\}_{i=1}^{N}$, hence $N = np$. This gives

$$\mathrm{Var}\left\{\mathrm{Tr}\,\varphi\left(\mathbf{\Xi}\right)\right\} \leqslant \frac{1}{n^2}\sum_{\alpha=1}^{p}\sum_{k=1}^{n}\mathbb{E}\left(\left|\frac{\partial\,\mathrm{Tr}\,\varphi\left(\mathbf{\Xi}\right)}{\partial X_{k\alpha}}\right|^2 + \left|\frac{\partial\,\mathrm{Tr}\,\varphi\left(\mathbf{\Xi}\right)}{\partial Z_{k\alpha}^*}\right|^2\right) \qquad (7.63)$$

Take now in Equation 7.61 $X_{k\alpha}$ as t, and $\frac{1}{p}\mathbf{R}\mathbf{X}\mathbf{X}^H\mathbf{R}$ as $\mathbf{A}$, and use the formulas (see Equation 7.62)

$$\frac{\partial}{\partial X_{k\alpha}}\left(\mathbf{R}\mathbf{X}\mathbf{X}^H\mathbf{R}\right)_{lm} = R_{ik}\left(\mathbf{X}^H\mathbf{R}\right)_{\alpha\ell}, \quad \frac{\partial}{\partial X_{\ell\alpha}^*}\left(\mathbf{R}\mathbf{X}\mathbf{X}^H\mathbf{R}\right)_{k\ell} = R_{jk}\left(\mathbf{X}^H\mathbf{R}\right)_{k\alpha} \qquad (7.64)$$

After some algebraic manipulation we obtain

$$\mathrm{Var}\left\{\mathrm{Tr}\,\varphi\left(\mathbf{\Xi}\right)\right\} \leqslant \frac{2}{n^2 p}\mathbb{E}\left\{\mathrm{Tr}\,\mathbf{\Xi}\varphi'\left(\mathbf{\Xi}\right)\mathbf{\Sigma}[\varphi'\left(\mathbf{\Xi}\right)]^*\right\} = \frac{2}{n^2 p}\mathbb{E}\left\{\mathrm{Tr}\left[\varphi'\left(\mathbf{\Xi}\right)\right]^*\varphi'\left(\mathbf{\Xi}\right)\mathbf{\Sigma}^{1/2}\mathbf{\Xi}\mathbf{\Sigma}^{1/2}\right\}$$

(3) Now we use the inequality $|\mathrm{Tr}\,\mathbf{AB}| \leqslant \|\mathbf{A}\|\,\mathrm{Tr}\,\mathbf{B}$, valid for any matrix $\mathbf{A}$ ($\|\mathbf{A}\|$ is the Euclidian norm of $\mathbf{A}$) and a positive definite $\mathbf{B}$. Choosing $\mathbf{A} = [\varphi'\left(\mathbf{\Xi}\right)]^*\varphi'\left(\mathbf{\Xi}\right)$ and $\mathbf{B} = \mathbf{\Sigma}^{1/2}\mathbf{\Xi}\mathbf{\Sigma}^{1/2}$, we obtain

$$\left|\mathrm{Tr}\,\mathbf{\Xi}\varphi'\left(\mathbf{\Xi}\right)\mathbf{\Sigma}[\varphi'\left(\mathbf{\Xi}\right)]^*\right| \leqslant \left\|\varphi'\left(\mathbf{\Xi}\right)\right\|^2 \mathrm{Tr}\,\mathbf{\Xi}\mathbf{\Sigma}$$

(4) The inequality $|\psi(\Xi)| \leqslant \max_{x \in \mathbb{R}} |\psi(x)|$, valid for a Hermitian Ξ and any function ψ and implying that

$$\text{Var}\{\xi_n[\varphi]\} \leqslant \frac{2}{n^2 p}\left(\max_{x \in \mathbb{R}}|\varphi'(x)|\right)^2 \mathbb{E}\{\text{Tr}\,\Xi\Sigma\} \tag{7.65}$$

We obtain from Equation 7.59 that

$$\mathbb{E}\{\text{Tr}\,\Xi\Sigma\} = \text{Tr}\,\Sigma^2$$

Plugging this in Equation 7.65 and using Equation 7.44, we obtain Equation 7.43 with the complex case $\beta = 2$. The real case $\beta = 1$ is similar.

7.16 Derivation of Asymptotic Density Using the Integration by Parts Formula

We present the derivation of Equations 7.50 through 7.57, based on the integration by parts formula. Let $\mathcal{N}_n$ be the normalized counting measure of eigenvalues $\left\{\lambda_i^{(n)}\right\}_{i=1}^n$, defined for any interval Δ of spectral axis as

$$\mathcal{N}_n(\Delta) = \#\left\{i = 1, ..., n : \lambda_i^{(n)} \in \Delta\right\}/n \tag{7.66}$$

The measure $\mathcal{N}_n$ has density ρ_n of Equation 7.37. According to the spectral theorem, we obtain for Hermitian matrices that the Stieltjes transform of $\mathcal{N}_n$ is

$$g_n(z) = \int \frac{1}{\lambda - z}d\mathcal{N}_n(\lambda) = \frac{1}{n}\sum_{i=1}^n \frac{1}{\lambda_i^{(n)} - z} = \frac{1}{n}\text{Tr}(\Xi - z\mathbf{I}_n)^{-1}, \quad \text{Im}\,z \neq 0 \tag{7.67}$$

This function of a complex variable z plays a fundamental role in the whole theory of random matrices. The linear relation is pointed out. Our goal is to prove the limit

$$f(z) = \lim_{n,p \to \infty, p/n \to c \in [1,\infty)} \mathbb{E}[g_n(z)]$$

satisfies Equation 7.53, when Ξ is defined in Equation 7.58. Indeed, this relation and the general properties of the Stieltjes transform (see e.g., [56, Section 59]) imply Equations 7.50 and 7.57.

By using the identity

$$(\mathbf{A} - z\mathbf{I}_n)^{-1} = -z^{-1} + z^{-1}(\mathbf{A} - z\mathbf{I}_n)^{-1}$$

and Equation 7.58, we write

$$\begin{aligned} f_n(z) = \mathbb{E}[g_n(z)] &= -z^{-1} + (nz)^{-1}\mathbb{E}\left[\text{Tr}\,\Xi(\Xi - z\mathbf{I}_n)^{-1}\right]\\ &= -z^{-1} + (npz)^{-1}\mathbb{E}\left[\text{Tr}\,\mathbf{X}^H\Sigma\mathbf{X}\left(\frac{1}{p}\mathbf{X}^H\Sigma\mathbf{X} - z\mathbf{I}_n\right)^{-1}\right]\\ &= -z^{-1} + (nz)^{-1}\mathbb{E}[\text{Tr}\,\mathbf{K}] \end{aligned} \tag{7.68}$$

where we used the cyclicity of trace ($\text{Tr}\,\mathbf{ABC} = \text{Tr}\,\mathbf{BCA}$) and denoted

$$\mathbf{Q} = \mathbf{X}^H\Sigma\mathbf{X}\mathbf{G}, \quad \mathbf{G} = \left(\frac{1}{p}\mathbf{X}^H\Sigma\mathbf{X} - z\mathbf{I}_n\right)^{-1} \tag{7.69}$$

We need now the key formula, valid for the standard Gaussian complex variables $(\xi_1, ..., \xi_q, \xi_1^*, ..., \xi_q^*)$

$$\mathbb{E}\left[\xi_i \Phi\left(\xi_1, ..., \xi_q, \xi_1^*, ..., \xi_q^*\right)\right] = \mathbb{E}\left[\frac{\partial}{\partial \xi_i^*} \Phi\left(\xi_1, ..., \xi_q, \xi_1^*, ..., \xi_q^*\right)\right], \quad i = 1, ..., q \qquad (7.70)$$

which can be shown by integration by parts, and Equation 7.64

$$\frac{\partial G_{\alpha\beta}}{\partial X_{\ell\alpha}^*} = \frac{1}{p} G_{\alpha\alpha} F_{\ell\beta}, \quad \mathbf{F} = \mathbf{\Sigma X G}$$

for $\mathbf{G}$ of Equation 7.69. Note that we treat $\mathbf{X}$ and $\mathbf{X}^H$ as independent quantities, according to Equations 7.59 and 7.70.

Now we write

$$\mathbb{E}\left[Q_{jk}\right] = \frac{1}{p} \sum_{\alpha=1}^{p} \sum_{\beta=1}^{p} \sum_{\ell=1}^{n} \mathbb{E}\left[\Sigma_{j\ell} X_{\ell\alpha} G_{\alpha\beta} X_{k\beta}^*\right]$$

and taking $X_{\ell\alpha}$ as ξ_i and $G_{\alpha\beta} X_{k\beta}^*$ as Φ in Equation 7.70, we obtain the matrix relation

$$\mathbb{E}\left[\mathbf{Q}\right] = h_n\left(z\right) \mathbf{\Sigma} - h_n\left(z\right) \mathbf{\Sigma}\mathbb{E}\left[\mathbf{Q}\right] - \mathbf{\Sigma}\mathbb{E}\left[\tilde{s}_n\left(z\right)\mathbf{Q}\right]$$

where

$$s_n\left(z\right) = \frac{1}{p} \operatorname{Tr} \mathbf{G}, \quad h_n\left(z\right) = \mathbb{E}\left[s_n\left(z\right)\right], \quad \tilde{s}_n\left(z\right) = s_n\left(z\right) - h_n\left(z\right)$$

The relation implies

$$\mathbb{E}\left[\mathbf{Q}\right] = h_n\left(z\right) \mathbf{\Sigma}(1 + h_n\left(z\right) \mathbf{\Sigma})^{-1} - \mathbf{\Sigma}(1 + h_n\left(z\right) \mathbf{\Sigma})^{-1}\mathbb{E}\left[\tilde{s}_n\left(z\right)\mathbf{Q}\right] \qquad (7.71)$$

Substituting Equation 7.71 into 7.68, we obtain that

$$f_n\left(z\right) = -z^{-1} + (nz)^{-1} h_n\left(z\right) \operatorname{Tr}\left\{\mathbf{\Sigma}(1 + h_n\left(z\right) \mathbf{\Sigma})^{-1}\right\} - z^{-1}$$

$$\times \mathbb{E}\left[\tilde{s}_n\left(z\right) \frac{1}{n} \operatorname{Tr}\left\{\mathbf{\Sigma}(1 + h_n\left(z\right) \mathbf{\Sigma})^{-1}\mathbf{Q}\right\}\right] \qquad (7.72)$$

It follows from Equation 7.43 with $\varphi\left(\lambda\right) = (\lambda - z)^{-1}, \operatorname{Im} z \neq 0$, that

$$\operatorname{Var}\left[s_n\left(z\right)\right] := \mathbb{E}\left[\left|\tilde{s}_n\left(z\right)\right|^2\right] \leqslant \frac{2C}{pn|\operatorname{Im} z|^2}$$

where C is given in Equation 7.44.

This allows us to prove that the third term on the r.h.s. of Equation 7.72 is $O\left(1/n\right)$ (even $O\left(1/n^2\right)$). This and the spectral theorem for the positive definite correlation matrix $\mathbf{\Sigma}$ give

$$f_n\left(z\right) = -z^{-1} + (nz)^{-1} h_n\left(z\right) \sum_{i=1}^{n} \frac{\sigma_i}{1 + h_n\left(z\right)} + O\left(n^{-1}\right)$$

where $\{\sigma_i\}_{i=1}^{n}$ are eigenvalues of $\mathbf{\Sigma}$. So, assuming that there is the limiting distribution ν of σ's, in the limit we obtain Equation 7.42,

$$f\left(z\right) = -\frac{1}{z} + \frac{h\left(z\right)}{z} \int \frac{\sigma\nu\left(\sigma\right) d\sigma}{1 + h\left(z\right)} \qquad (7.73)$$

Note that for $p \geq n$ the $p \times p$ matrix $\mathbf{X}^H \mathbf{\Sigma X}$ has $p - n$ zero eigenvalues and the rest of them coincide with those $\mathbf{\Sigma}^{1/2} \mathbf{X X}^H \mathbf{\Sigma}^{1/2}$. This implies

$$h_n\left(z\right) = -\frac{1}{z}\frac{p - n}{p} + \frac{n}{p} f_n\left(z\right)$$

hence, the limiting relation

$$h\left(z\right) = -\frac{1}{z}\left(1 - c\right) + cf\left(z\right) \tag{7.74}$$

It is easy to find that Equations 7.73 and 7.74 imply 7.53.

7.17 Variance of the Trace Function of a Gaussian Random Matrix

The trace function is a (scalar-valued) linear function, thought of as an average of the diagonal entries. Its variance is of independent interest. Gaussian random matrices are considered in this section.

Theorem 13 (Hoydis, Couillet, and Piantanida [57]) *Let* $\mathbf{G} \in \mathbb{C}^{p \times n}$ *have i.i.d. entries* $G_{ij} \sim \mathcal{CN}\left(0, 1\right)$ *and let* $\mathbf{A} \in \mathbb{C}^{n \times p}$ *and* $\mathbf{B} \in \mathbb{C}^{p \times p}$. *Then,*

1. $\mathrm{Var}\left[\mathrm{Tr}\,\mathbf{AG}\right] = \mathrm{Tr}\,\mathbf{AA}^{H}$.

2. $\mathrm{Var}\left[\mathrm{Tr}\,\mathbf{BGG}^{H}\right] \leqslant 2n\,\mathrm{Tr}\,\mathbf{BB}^{H}$.

Proof 1

$$\mathrm{Var}\left[\mathrm{Tr}\,\mathbf{AG}\right] = \mathbb{E}\left[\left(\mathrm{Tr}\,\mathbf{AG}\right)^{2}\right] = \sum_{i=1}^{p}\sum_{j=1}^{n}\left|\mathbf{A}_{ij}\right|^{2} = \mathrm{Tr}\,\mathbf{AA}^{H}$$

Part (2) can be proven relying on Lemmas 2 and 3:

$$
\begin{aligned}
\mathrm{Var}\left[\mathrm{Tr}\,\mathbf{BGG}^{H}\right] &\leqslant \sum_{i=1}^{p}\sum_{j=1}^{n}\mathbb{E}\left[\left|\frac{\partial\,\mathrm{Tr}\,\mathbf{BGG}^{H}}{\partial G_{ij}}\right|^{2} + \left|\frac{\partial\,\mathrm{Tr}\,\mathbf{BGG}^{H}}{\partial G_{ij}^{*}}\right|^{2}\right] \\
&= \sum_{i=1}^{p}\sum_{j=1}^{n}\mathbb{E}\left[\left|\sum_{k=1}^{p}\sum_{\ell=1}^{p}B_{k\ell}\frac{\partial[\mathbf{GG}^{H}]_{k\ell}}{\partial G_{ij}}\right|^{2} + \left|\sum_{k=1}^{p}\sum_{\ell=1}^{p}B_{k\ell}\frac{\partial[\mathbf{GG}^{H}]_{k\ell}}{\partial G_{ij}^{*}}\right|^{2}\right] \\
&= \sum_{i=1}^{p}\sum_{j=1}^{n}\mathbb{E}\left[\left|\sum_{\ell=1}^{p}B_{\ell i}G_{\ell j}^{*}\right|^{2} + \left|\sum_{k=1}^{p}B_{ik}G_{kj}^{*}\right|^{2}\right] \\
&= \sum_{i=1}^{p}\sum_{j=1}^{n}\left(\sum_{\ell=1}^{p}\left|B_{\ell i}\right|^{2} + \sum_{k=1}^{p}\left|B_{ik}\right|^{2}\right) \\
&= 2n\,\mathrm{Tr}\,\mathbf{BB}^{H}
\end{aligned}
$$

□

The sample covariance (or noncentered Wishart) random matrix $\mathbf{S} = \frac{1}{n}\mathbf{GG}^{H} \in \mathbb{C}^{p \times p}$ is applied to obtain

$$\mathrm{Var}\left[\mathrm{Tr}\,\mathbf{BS}\right] = \mathrm{Var}\left[\mathrm{Tr}\,\mathbf{B}\frac{1}{n}\mathbf{GG}^{H}\right] \leqslant \frac{2}{n}\,\mathrm{Tr}\,\mathbf{BB}^{H} \tag{7.75}$$

The centered Wishart matrix $\mathbf{W} = \frac{1}{n}\mathbf{GG}^{H} - \mathbf{I}_{p} \in \mathbb{C}^{p \times p}$ is applied to obtain

$$\mathrm{Var}\left[\mathrm{Tr}\,\mathbf{BW}\right] = \mathrm{Var}\left[\mathrm{Tr}\,\mathbf{B}\frac{1}{n}\mathbf{GG}^{H} - \mathrm{Tr}\,\mathbf{B}\right] = \mathrm{Var}\left[\mathrm{Tr}\,\mathbf{B}\frac{1}{n}\mathbf{GG}^{H}\right] \leqslant \frac{2}{n}\,\mathrm{Tr}\,\mathbf{BB}^{H} \tag{7.76}$$

where the second equality and the inequality follow from the following:

$$\mathrm{Var}\left[c\right] = 0, \quad \mathrm{Var}\left[c + X\right] = \mathrm{Var}\left[X\right], \quad \mathrm{Var}\left[cX\right] = c^{2}\,\mathrm{Var}\left[X\right]$$

for a random variable X and a constant c. Note either $\mathbf{S} = \frac{1}{n}\mathbf{G}\mathbf{G}^H$ or its centered version $\mathbf{W} = \frac{1}{n}\mathbf{G}\mathbf{G}^H - \mathbf{I}_p$ gives the same variance bound. When $\mathbf{B} = \mathbf{I}_p$, then $\operatorname{Tr}\mathbf{B}\mathbf{B}^H = p$. When n and p are large, say $n = 640$ and $p = 256$, we obtain $2p/n = 0.8$. Classical statistical methods only consider the case of fixed p, and large n, such that $2p/n \to 0$, resulting in vanishing variance $\operatorname{Var}[\operatorname{Tr}\mathbf{B}\mathbf{W}] \to 0$.

Using Theorem 13, we obtain

$$\operatorname{Var}[\operatorname{Tr}\mathbf{A}\mathbf{G}] = \operatorname{Tr}\mathbf{A}\mathbf{A}^H \text{ and } \operatorname{Var}\left[\operatorname{Tr}\mathbf{B}\mathbf{G}\mathbf{G}^H\right] \leqslant 2n\operatorname{Tr}\mathbf{B}\mathbf{B}^H \qquad (7.77)$$

where $\mathbf{G} \in \mathbb{C}^{n \times p}$ have i.i.d. Gaussian entries and $\mathbf{A} \in \mathbb{C}^{p \times n}$ and $\mathbf{B} \in \mathbb{C}^{n \times n}$ are two complex matrices. Equation 7.77 can be used to bound the variance.

Consider the matrix hypothesis testing problem

$$\mathcal{H}_0 : \mathbf{X}$$
$$\mathcal{H}_1 : \mathbf{X} + \mathbf{Y}$$

where $\mathbf{X}$ and $\mathbf{Y}$ are two random matrices. We are interested in the special case when $\mathbf{X} = \mathbf{G}$ is a Gaussian random matrix. This problem is equivalent to the following:

$$\mathcal{H}_0 : \mathbf{T}\mathbf{G}$$
$$\mathcal{H}_1 : \mathbf{T}(\mathbf{G} + \mathbf{Y})$$

where $\mathbf{T}$ is a nonrandom test matrix. We test whether if the variances for two hypotheses are different. In other words,

$$\mathcal{H}_0 : \operatorname{Var}[\operatorname{Tr}(\mathbf{T}\mathbf{G})]$$
$$\mathcal{H}_1 : \operatorname{Var}[\operatorname{Tr}(\mathbf{T}\mathbf{G}) + \operatorname{Tr}(\mathbf{T}\mathbf{Y})]$$

where the (linear) trace function is used to convert two matrix-valued random variables into two scalar-valued random variables. It follows that

$$\operatorname{Var}[X + Y] \leq \left(\sqrt{\operatorname{Var}[X]} + \sqrt{\operatorname{Var}[Y]}\right)^2$$

where X and Y may or may not be independent. We obtain

$$\mathcal{H}_0 : \operatorname{Var}[\operatorname{Tr}(\mathbf{T}\mathbf{G})] = \operatorname{Tr}\mathbf{T}\mathbf{T}^H$$
$$\begin{aligned}\mathcal{H}_1 : \operatorname{Var}[\operatorname{Tr}(\mathbf{T}\mathbf{G}) + \operatorname{Tr}(\mathbf{T}\mathbf{Y})] &\leqslant \left(\sqrt{\operatorname{Var}[\operatorname{Tr}(\mathbf{T}\mathbf{G})]} + \sqrt{\operatorname{Var}[\operatorname{Tr}(\mathbf{T}\mathbf{Y})]}\right)^2 \\ &= \left(\sqrt{\operatorname{Tr}\mathbf{T}\mathbf{T}^H} + \sqrt{\operatorname{Var}[\operatorname{Tr}(\mathbf{T}\mathbf{Y})]}\right)^2\end{aligned}$$

The second line uses Part 1 of Theorem 13.

7.18 Resolvent Matrices and Log-Determinant

We assume

$$\frac{n}{K} = \beta + o\left(n^{-2}\right), \quad \frac{N}{K} = c + o\left(n^{-2}\right)$$

for some constants $\beta, c > 0$ as $n \to \infty$. We write as $n \xrightarrow{\beta, c} \infty$.

We need to define a function

$$\theta\left(t\right) = \frac{c-1}{2t} - \frac{1}{2} + \frac{\sqrt{(1-c+t)+4ct}}{2t} \tag{7.78}$$

for $t \in \mathbb{R} > 0$. We also define

$$\tilde{\theta}\left(t\right) = \theta\left(t\right) - \frac{c-1}{2t}$$

The function $m\left(z\right) = \frac{1}{c}\theta\left(-z\right)$ for $\mathbb{C}\backslash\mathbb{R}_+$ corresponds to the Stieltjes transform of the Marchenko–Pastur law. We list some properties of $\theta(t)$:

1. $\theta\left(t\right) > \frac{c}{\left(1+\sqrt{c}\right)^2+t} > 0$

2. $\theta\left(t\right) < \frac{c}{t}$

3. $\theta\left(t\right) = \frac{c}{1-c+t(1+\theta(t))}$

4. $\frac{\theta(t)}{1+\theta(t)} = c - t\theta\left(t\right)$

5. $\frac{1}{1+\theta(t)} = 1 - c + t\theta\left(t\right)$

6. $\theta'\left(t\right) = -\frac{\theta(t)(1+\theta(t))}{1-c+t(1+2\theta(t))}$

The following theorem is important since we can approximate the expectation of the trace function of the resolvent matrix using a *deterministic* function $\theta\left(t\right)$. More important, the approximation error is on the order of $1/(t^2n^2)$.

Theorem 14 *The random matrix* $\mathbf{H} \in \mathbb{C}^{N \times K}$ *has i.i.d. entries* $H_{ij} \sim \mathcal{CN}\left(0, 1\right)$. *For* $t \in \mathbb{R} > 0$, *let* $\mathbf{Q}\left(t\right) = \left(\frac{1}{K}\mathbf{H}\mathbf{H}^H + t\mathbf{I}_N\right)^{-1}$ *and* $\tilde{\mathbf{Q}}\left(t\right) = \left(\frac{1}{K}\mathbf{H}^H\mathbf{H} + t\mathbf{I}_N\right)^{-1}$. *Then, as* $n \xrightarrow{\beta,c} \infty$,

$$\mathbb{E}\left[\frac{1}{K}\operatorname{Tr}\mathbf{Q}\left(t\right)\right] = \theta\left(t\right) + \mathcal{O}\left(\frac{1}{t^2n^2}\right), \quad \mathbb{E}\left[\frac{1}{K}\operatorname{Tr}\tilde{\mathbf{Q}}\left(t\right)\right] = \tilde{\theta}\left(t\right) + \mathcal{O}\left(\frac{1}{t^2n^2}\right)$$

Proof 2 *The proof follows from a direct adaption of [12, Theorem 7.2.2], see also [58, Theorem 3 and Proposition 5] for a more complex matrix model. As opposed to these works, we maintain the dependence on* $t \in \mathbb{R}$ *in the bounds. Let* $\psi_n\left(t\right) \triangleq \mathbb{E}\left[\frac{1}{K}\operatorname{Tr}\mathbf{Q}\left(t\right)\right]$. *Then,*

$$
\begin{aligned}
\psi_n\left(t\right) &\stackrel{(a)}{=} \frac{c}{t} - \frac{1}{t}\sum_{i,j,k}\mathbb{E}\left[H_{ij}H_{kj}^*Q_{ki}\left(t\right)\right] \\
&\stackrel{(b)}{=} \frac{c}{t} - \frac{1}{t}\psi_n\left(t\right) + \frac{1}{t}\mathbb{E}\left[\frac{1}{K}\operatorname{Tr}\mathbf{Q}\left(t\right)\frac{\mathbf{H}^H\mathbf{H}}{K}\frac{1}{K}\operatorname{Tr}\mathbf{Q}\left(t\right)\right] \\
&\stackrel{(c)}{=} \frac{c}{t} - \frac{1}{t}\psi_n\left(t\right) + \frac{c}{t}\psi_n\left(t\right) - \frac{1}{t}\mathbb{E}\left[\left(\frac{1}{K}\operatorname{Tr}\mathbf{Q}\left(t\right)\right)^2\right] \\
&\stackrel{(d)}{=} \frac{c}{t} - \frac{1}{t}\psi_n\left(t\right) + \frac{c}{t}\psi_n\left(t\right) - \psi_n(t)^2 + \varepsilon_n
\end{aligned}
\tag{7.79}
$$

with $|\varepsilon_n| \leqslant \frac{2c}{t^4K^2}$. *(b) follows from an application of integration by parts formula 7.28 together with the derivation rules in Lemma 3. (d) results by using*

$$
\begin{aligned}
|\varepsilon_n| \leqslant \operatorname{Var}\left[\frac{1}{K}\operatorname{Tr}\mathbf{Q}\left(t\right)\right] \leqslant \frac{2}{K^2}\sum_{i,j}\mathbb{E}\left[\left|\frac{\partial\operatorname{Tr}\mathbf{Q}(t)}{\partial H_{ij}^*}\right|^2\right] &= \frac{2}{K^2}\sum_{i,j}\mathbb{E}\left[\left|\frac{1}{K}\left[\mathbf{Q}(t)^2\mathbf{H}\right]_{ij}\right|^2\right] \\
&= \frac{2}{K^2}\mathbb{E}\left[\operatorname{Tr}\mathbf{Q}(t)^2\mathbf{H}^H\mathbf{H}\mathbf{Q}(t)^2\right]
\end{aligned}
$$

from Remark 1, and then the fact that

$$\frac{2}{K^2}\mathbb{E}\left[\operatorname{Tr}\mathbf{Q}(t)^2\mathbf{H}^H\mathbf{H}\mathbf{Q}(t)^2\right] \leqslant \frac{2}{t^4 K^4}\mathbb{E}\left[\operatorname{Tr}\mathbf{H}\mathbf{H}^H\right] = \frac{2N}{t^4 K^3}$$

By Equation 7.79 and Property 3, which can be written $\theta(t) = \frac{c}{t} - \frac{1}{t}\theta(t) + \frac{c}{t}\theta(t) - \theta(t)^2$,
we obtain

$$\begin{aligned}\psi_n(t) - \theta(t) &= (\theta(t) - \psi_n(t))\frac{c-1}{2t} + \theta(t)^2 - \psi_n(t)^2 + \varepsilon_n \\ &= (\theta(t) - \psi_n(t))\left(\frac{c-1}{2t} + \theta(t) + \psi_n(t)\right) + \varepsilon_n\end{aligned} \tag{7.80}$$

which follows respectively from Property 3. Then from $t\psi_n(t) > 0$ and $\theta(t) > 0$ (Property 1),
and finally from $t\theta(t)/c < 1$ (Property 2).

Gathering the terms in $\theta(t) - \psi_n(t)$ of 7.80, according to this reasoning, we obtain

$$|\psi_n(t) - \theta(t)| = \left|\frac{\varepsilon_n}{1 + \frac{c-1}{2t} + \theta(t) + \psi_n(t)}\right| < \varepsilon_n$$

This terminates the proof of the first part. For the second part, it is sufficient to notice that
$\frac{1}{K}\operatorname{Tr}\tilde{\mathbf{Q}}(t) = \frac{1}{K}\operatorname{Tr}\tilde{\mathbf{Q}}(t) - \frac{c-1}{2t}$. $\square$

Now we are in a position to approximate the log-determinant using a determinstic function with the error of $1/n^2$

Theorem 15 *The random matrix $\mathbf{H} \in \mathbb{C}^{N \times K}$ has i.i.d. entries $H_{ij} \sim \mathcal{CN}(0,1)$. Let $\sigma^2 > 0$. Then, as $n \xrightarrow{\beta,c} \infty$,*

$$\mathbb{E}\left[\frac{1}{K}\log\det\left(\mathbf{I}_N + \frac{1}{\sigma^2}\frac{1}{K}\mathbf{H}\mathbf{H}^H\right)\right] = C(\sigma^2) + \mathcal{O}\left(\frac{1}{n^2}\right)$$

where

$$C(\sigma^2) = \log(1 + \theta(\sigma^2)) + c\log\left(1 + \frac{1}{\sigma^2(1 + \theta(\sigma^2))}\right) - \frac{\theta(\sigma^2)}{1 + \theta(\sigma^2)}$$

and $\theta(t)$ is defined in Equation 7.78.

Proof 3 *The approach is taken from [58, Theorem 1]. The properties of the special function $\theta(t)$ are used. We refer to [57] for the details. $\square$*

7.19 Fast Decay Rate of the Linear Eigenvalue Statistics

On the almost sure location of the singular values of certain Gaussian block-Hankel large random matrices [59].

At discrete-time n, we have

$$\mathbf{y}_n = \mathbf{A}\mathbf{s}_n + \mathbf{v}_n$$

where $\mathbf{A}$ is an $M \times K$ matrix, $\mathbf{s}_n$ is a $K \times 1$ column vector, and $\mathbf{v}_n$ is an additive Gaussian white noise with zero mean and covariance matrix $\mathbb{E}\left[\mathbf{v}_n\mathbf{v}_n^H\right] = \sigma^2\mathbf{I}_M$. We denote by $\mathbf{Y}_N = [\mathbf{y}_1, ..., \mathbf{y}_N]$ the $M \times N$ observation matrix (assuming $M \leq N$), which can be easily written as

$$\mathbf{Y}_N = \mathbf{A}\mathbf{S}_N + \mathbf{V}_N \tag{7.81}$$

where $\mathbf{S}_N = [\mathbf{s}_1, ..., \mathbf{s}_N]$ and $\mathbf{V}_N = [\mathbf{v}_1, ..., \mathbf{v}_N]$. After the normalization

$$\mathbf{X}_N = \frac{1}{\sqrt{N}}\mathbf{Y}_N, \mathbf{B}_N = \frac{1}{\sqrt{N}}\mathbf{A}\mathbf{S}_N, \mathbf{W}_N = \frac{1}{\sqrt{N}}\mathbf{V}_N$$

we obtain

$$\mathbf{X}_N = \mathbf{B}_N + \mathbf{W}_N \qquad (7.82)$$

where $\mathbf{W}_N$ is a complex Gaussian white noise matrix with i.i.d. entries that have zero mean and variance σ^2/N.

The support of a particular function ϕ is denoted as $\mathrm{supp}\,(\varphi)$. $\mathcal{C}^\infty\,(\mathbb{R},\mathcal{E})$ and $\mathcal{C}_c^\infty\,(\mathbb{R},\mathcal{E})$ are, respectively, the set of all smooth functions (smooth compactly supported functions) taking values in the set $\mathcal{E} \in \mathbb{R}$.

Lemma 5 *Let $\varphi \in \mathcal{C}_c^\infty\,(\mathbb{R},\mathbb{R})$. Then,*

$$\mathbb{E}\left[\frac{1}{M}\,\mathrm{Tr}\,\varphi\left(\mathbf{X}_N\mathbf{X}_N^H\right)\right] = \int_{\mathcal{S}_N}\varphi\,(\lambda)\,d\mu_N\,(\lambda) + \mathcal{O}\left(\frac{1}{N^2}\right) \qquad (7.83)$$

Moreover, if φ is constant on each cluster of $\mathcal{S}_N$ for all large N, then

$$\mathrm{Var}\left[\frac{1}{M}\,\mathrm{Tr}\,\varphi\left(\mathbf{X}_N\mathbf{X}_N^H\right)\right] = \mathcal{O}\left(\frac{1}{N^4}\right) \qquad (7.84)$$

Proof 4 *The ideas are developed by Haagerup and Thorbjornsen [60] and used extensively in Capitaine and Donati-Martin [61] and Capitaine et al. [62] where the deformed Wigner matrices are considered.*

We first recall Lemmas 6 and 7. It follows that

$$\mathbb{E}\left[\frac{1}{M}\,\mathrm{Tr}\,\varphi\left(\mathbf{X}_N\mathbf{X}_N^H\right)\right] = \frac{1}{\pi}\limsup_{y\downarrow 0}\left(\int_{\mathbb{R}+}\varphi\,(x)\,\mathbb{E}\left[\frac{1}{M}\,\mathrm{Tr}\,\mathbf{Q}_N\,(x+iy)\right]dx\right) \qquad (7.85)$$

as well as

$$\int_{\mathcal{S}_N}\varphi\,(\lambda)\,d\mu_N\,(\lambda) = \frac{1}{\pi}\limsup_{y\downarrow 0}\left(\int_{\mathbb{R}+}\varphi\,(x)\,\mathbb{E}\left[\frac{1}{M}\,\mathrm{Tr}\,\mathbf{T}_N\,(x+iy)\right]dx\right) \qquad (7.86)$$

Subtracting Equation 7.85 from 7.86, and using 7.88, we get

$$\mathbb{E}\left[\frac{1}{M}\,\mathrm{Tr}\,\varphi\left(\mathbf{X}_N\mathbf{X}_N^H\right)\right] - \int_{\mathcal{S}_N}\varphi\,(\lambda)\,d\mu_N\,(\lambda) = \frac{1}{N^2}\lim_{y\downarrow 0}\frac{1}{\pi}\,\mathrm{Im}\left(\int_{\mathbb{R}+}\varphi\,(x)\,\chi_N\,(x+iy)\,dx\right) \qquad (7.87)$$

Using Lemma 7 leads directly to Equation 7.83. To prove Equation 7.84, we rely on the Poincare inequality. Indeed, we have

$$\mathbb{E}\left[\frac{1}{M}\,\mathrm{Tr}\,\varphi\left(\mathbf{X}_N\mathbf{X}_N^H\right)\right] \leqslant \frac{\sigma^2}{N}\sum_{i=1}^{M}\sum_{j=1}^{N}\mathbb{E}\left[\left|\frac{\partial}{\partial W_{i,j}}\left\{\frac{1}{M}\,\mathrm{Tr}\,\varphi\left(\mathbf{X}_N\mathbf{X}_N^H\right)\right\}\right|^2\right.$$
$$\left. + \left|\frac{\partial}{\partial W_{i,j}^*}\left\{\frac{1}{M}\,\mathrm{Tr}\,\varphi\left(\mathbf{X}_N\mathbf{X}_N^H\right)\right\}\right|^2\right]$$

Applying Lemma 8 gives

$$\frac{\partial}{\partial W_{i,j}}\left\{\frac{1}{M}\,\mathrm{Tr}\,\varphi\left(\mathbf{X}_N\mathbf{X}_N^H\right)\right\} = \frac{1}{M}\left[\mathbf{X}_N^H\varphi'\left(\mathbf{X}_N\mathbf{X}_N^H\right)\right]_{j,i}$$

$$\frac{\partial}{\partial W_{i,j}^*}\left\{\frac{1}{M}\,\mathrm{Tr}\,\varphi\left(\mathbf{X}_N\mathbf{X}_N^H\right)\right\} = \frac{1}{M}\left[\varphi'\left(\mathbf{X}_N\mathbf{X}_N^H\right)\mathbf{X}_N\right]_{i,j}$$

Therefore,

$$\mathrm{Var}\left[\frac{1}{M}\,\mathrm{Tr}\,\varphi\left(\mathbf{X}_N\mathbf{X}_N^H\right)\right] \leqslant \frac{C}{N^2}\mathbb{E}\left[\frac{1}{M}\,\mathrm{Tr}\,\varphi'\left(\mathbf{X}_N\mathbf{X}_N^H\right)^2\mathbf{X}_N\mathbf{X}_N^H\right]$$

with $C > 0$ a constant independent of N. Function $h(\lambda) = \lambda\varphi'(\lambda)$ belongs to $\mathcal{C}_c^\infty(\mathbb{R},\mathbb{R})$ and thus Equation 7.83 gives

$$\mathbb{E}\left[\frac{1}{M}\operatorname{Tr}\varphi'(\mathbf{X}_N\mathbf{X}_N^H)^2\mathbf{X}_N\mathbf{X}_N^H\right] = \int_{\mathcal{S}_N}\lambda\varphi'(\lambda)^2 d\mu_N(\lambda) + \mathcal{O}\left(\frac{1}{N^2}\right)$$

Since ϕ is constant on each cluster of $\mathcal{S}_N$ by assumption, $\operatorname{supp}(h) \cap \mathcal{S}_N = \emptyset$ for all N and thus

$$\mathbb{E}\left[\frac{1}{M}\operatorname{Tr}\varphi'(\mathbf{X}_N\mathbf{X}_N^H)^2\mathbf{X}_N\mathbf{X}_N^H\right] = \mathcal{O}\left(\frac{1}{N^2}\right)$$

which shows 7.84. $\square$

Lemma 6 *For $z \in \mathbb{C}\backslash\mathbb{R}$, it holds that*

$$\mathbb{E}[\hat{m}(z)] = m_N(z) + \frac{1}{N^2}\chi_N(z) \tag{7.88}$$

for all N, where $\chi_N(z)$ is homomorphic on $\mathbb{C}\backslash\mathbb{R}$ and satisfies $|\chi_N(z)| \leqslant P_1(|z|)P_2\left(\frac{1}{|\operatorname{Im} z|}\right)$, where P_1 and P_2 are polynomials.

Proof 5 *We see Vallet, Loubaton, and Mestre [63] for a proof.*

$\square$

Lemma 7 *Let $\varphi \in \mathcal{C}_c^\infty(\mathbb{R},\mathbb{R})$ be independent of N and (h_N) be a sequence of holomorphic function on $\mathbb{C}\backslash\mathbb{R}$ satisfying $|h_N(z)| \leqslant P_1(|z|)P_2\left(\frac{1}{|\operatorname{Im} z|}\right)$. Then, we have*

$$\limsup_{y\downarrow 0}\int_\mathbb{R}|\varphi(x)|h_N(x+iy)\,dx \leqslant C < \infty$$

with C a constant independent of N.

Proof 6 *This lemma is proven in Haagerup and Thorbjornsen [60] and generalized in Capitaine and Donati-Martin [61].*

$\square$

Lemma 8 *[60] Let N be a positive integer, let I be an open interval in $\mathbb{R}$, and let $t \mapsto a(t) : I \to \mathcal{M}_N(\mathbb{C})_{\mathrm{sa}}$ be a C^1-function. Consider further a function ϕ in $C^1(\mathbb{R})$. Then the function*

$$t \mapsto \frac{d}{dt}\frac{1}{N}\operatorname{Tr}[\varphi(a(t))] = \frac{1}{N}\operatorname{Tr}[\varphi'(a(t))\cdot a'(t)]$$

7.20 $1/n$ Expansion of the Green Functions

The most frequently used matrix models are the Gaussian orthogonal ensemble (GOE) of random $n \times n$ symmetric matrices and the GUE of random $n \times n$ Hermitian matrices. The density of the probability distribution in these ensembles has the form:

$$\mathbb{P}_n(\mathbf{X}) = Z_n^{-1}\exp\left(-n\operatorname{Tr}V(\mathbf{X})\right) \tag{7.89}$$

where $V(t) = t^2/4\sigma^2$ and Z_n is the normalization constant.

The probability distribution Equation 7.89 has two important features: (1) it is invariant w.r.t. either orthogonal or unitary transformations of $\mathbb{R}^n$ or $\mathbb{C}^n$, respectively; and (2) the matrix elements are independent random variables.

In this section, we consider the Wigner ensemble of $n \times n$ real symmetric matrices of the form

$$\mathbf{X} = [X_{ij}]_{i,j=1}^n, \quad X_{ij} = [1 + \delta_{ij}] W_{ij}/\sqrt{n} \tag{7.90}$$

where $W_{ij}, i \leq j$ are independent random variables such that

$$\mathbb{E}\{W_{ij}\} = 0, \quad \mathbb{E}\{W_{ij}^2\} = \sigma^2 \tag{7.91}$$

Here and thereafter $\mathbb{E}$ denotes averaging over all $W_{ij}, i \leq j$.

The distribution of W_{ij}'s may depend on (i, j), but we assume that they are independent of n. If W_{ij}'s are independent random variables, then the ensembles, Equations 7.90 and 7.91, coincide with the GOE. This justifies the presence of the term with δ_{ij} in Equation 7.90.

Let $\lambda_1, ..., \lambda_n$ be the eigenvalues of a real symmetric or complex Hermitian matrix of order n. Introduce the eigenvalue counting function (or the *counting measure*) $\mathcal{N}_n$ of its eigenvalues $\mathcal{N}_n(E) = \#\{\lambda_i : \lambda_i$ is an eigenvalue of $\mathbf{X}$ and $\lambda_i \leqslant E\}$. The *normalized counting measure is*

$$N_n(\lambda) = \frac{1}{n}\mathcal{N}_n(\lambda) \tag{7.92}$$

At the end of the 1950s, Wigner [64] proved that in the case of identically distributed W_{ij} having all moments, $N_n(E)$ converges in probability, as $n \to \infty$, to a non-decreasing function $N_{sc}(E)$ (the semicircle law) whose derivative is

$$\rho(E) = \begin{cases} \frac{1}{2\pi\sigma^2}\sqrt{4\sigma^2 - E^2}, & |E| \leqslant 2\sigma \\ 0, & |E| > 2\sigma \end{cases} \tag{7.93}$$

The modern formulation of Wigner's result is as follows. Let us consider random matrices, Equations 7.90 and 7.91, with mutually independent arbitrary distributed entries defined on a common probability space. Then the condition (the matrix analogue of the Lindeberg condition of probability theory)

$$\lim_{n\to\infty} \frac{1}{n^2} \sum_{i\leqslant j} \int_{|x>\nu n|} x^2 d\,\mathrm{Prob}\,[W_{ij} \leqslant x] = 0, \quad \text{for any } \nu > 0 \tag{7.94}$$

is sufficient [65] and necessary [66] for the limiting relation

$$\lim_{n\to\infty} N_n(E) = N_{sc}(E) \tag{7.95}$$

to hold for every E with probability 1. If we will not assume that W_{ij} are defined on the same probability space or if their probability distributions depend on n, then the same condition, Equation 7.94, will imply the convergence in probability in Equation 7.95.

As usual in spectral theory, this result admits a natural reformulation in terms of the resolvent (Green's function). Indeed, the normalized trace of the resolvent

$$g_n(z) = \frac{1}{n}\,\mathrm{Tr}\,(\mathbf{X} - z\mathbf{I})^{-1} \tag{7.96}$$

is simply the Stieltjes transform of $N_n(E)$:

$$g_n(z) = \frac{1}{n}\sum_{i=1}^n \frac{1}{\lambda_i - z} = \int \frac{1}{\lambda - z} dN_n(\lambda) \tag{7.97}$$

Denote the Stieltjes transform of the Wigner law, Equation 7.93, by $r(z)$,

$$r(z) = \int \frac{1}{\lambda - z} dN_{sc}(\lambda) = \frac{-z + \sqrt{z^2 - 4\sigma^2}}{2\sigma^2} \tag{7.98}$$

The obvious condition $\operatorname{Im} r(z) \operatorname{Im} z \geqslant 0$ determines the branch of the square root in Equation 7.98. Due to the *one-to-one correspondence* between nondecreasing functions and their Stieltjes transforms [67], Equation 7.95 is equivalent to the following limiting relation:

$$\lim_{n \to \infty} g_n(z) = r(z) \tag{7.99}$$

which holds with probability 1 for any nonreal z.

Equation 7.99 and the obvious bound $|g_n(z)| \leqslant \left|\frac{1}{\operatorname{Im} z}\right|$ imply that the variance of $g_n(z)$ vanishes as $n \to \infty$ and hence the moments

$$m_n^{(p)}(z_1, ..., z_p) = \mathbb{E}\left\{\prod_{l=1}^{p} g_n(z_l)\right\} \tag{7.100}$$

factorize:

$$m_n^{(p)}(z_1, ..., z_p) = \prod_{l=1}^{p} m_n^{(1)}(z_l) + o(1), \quad n \to \infty \tag{7.101}$$

This factorization, which follows already from the convergence in probability in Equations 7.95 or 7.99, is typical for the macroscopic regime and can be found hidden behind many calculations in this regime.

Since according to Equation 7.99

$$m_n^{(1)}(z) = \mathbb{E}\{g_n(z)\} = r(z) + o(1)$$

the leading term of $m_n^{(p)}(z_1, ..., z_p)$ is $\prod_{l=1}^{p} r(z_l)$.

7.21 Asymptotic Expansions for Gaussian Unitary Matrices

In this section, we study the asymptotic expansion of $\mathbb{E}\{g_n(z)\}$. Let $g : \mathbb{R} \to \mathbb{C}$ be a C^∞-function with all derivatives bounded and let tr_n be the normalized trace on the $n \times n$ matrices. Here C^∞ is the set of all smooth functions. Ercolani and McLaughlin [68] established asymptotic expansions of the mean value $\mathbb{E}\operatorname{tr}_n\{g(\mathbf{X}_n)\}$ for a rather general class of random matrices $\mathbf{X}_n$ including the GUE. Using an analytical approach, Haagerup and Thorbjornsen provide an alternative proof of this asymptotic expansion in the GUE case. Specifically we derive for a GUE random matrix $\mathbf{X}_n$ that

$$\mathbb{E}\operatorname{tr}_n\{g(\mathbf{X}_n)\} = \frac{1}{2\pi} \int_{-2}^{2} g(x)\sqrt{4 - x^2}dx + \sum_{j=1}^{k} \frac{\alpha_j(g)}{n^{2j}} + O\left(\frac{1}{n^{2k+2}}\right) \tag{7.102}$$

where k is an arbitrary positive integer. Considered as mappings of g, Haagerup and Thorbjornsen [69] determine the coefficients $\alpha_j(g), j \in \mathbb{N}$, as distributions (in the sense of L. Schwarts). They derive a similar asymptotic expansion for the covariance $\operatorname{Cov}\{\operatorname{Tr}_n[f(\mathbf{X}_n)], \operatorname{Tr}_n[g(\mathbf{X}_n)]\}$, where f is a function of the same kind as g, and $\operatorname{Tr}_n = n\operatorname{tr}_n$. For example,

$$g(x) = \frac{1}{\lambda - x} \quad \text{and} \quad f(x) = \frac{1}{\mu - x}$$

for nonreal values $\lambda, \mu \in \mathbb{C} \backslash \mathbb{R}$. In this case the mean and covariance considered peviously correspond to, respectively, the one- and two-dimensional Cauchy (or Stieltjes) transform of the $\mathrm{GUE}(n, 1/n)$.

7.22   Generally Correlated Gaussian Model for Distributed Streaming Data Sets

We are motivated for distributed streaming data sets that are spatially located across a number of servers, say $s = 100$. The data set in each server can be viewed as a zero-mean, Gaussian random vector $\boldsymbol{\xi}_i$ with a specific covariance matrix $\boldsymbol{\Sigma}_i, i = 1, ..., s$, that is, $\boldsymbol{\xi}_i \sim \mathcal{N}(\mathbf{0}, \boldsymbol{\Sigma}_i)$. In general, these covariance matrices are *different* from each other. We may estimate the true covariance matrix $\boldsymbol{\Sigma}_i$ of each random vector using its corresponding sample covariance matrix $\mathbf{S}_i$ for $i = 1, ..., s$. We are interested in the regime where the sizes of data matrices are very large so their asymptotic limits are relevant. The natural question is "What are the fundamental limits for this generally correlated Gaussian model?"

We study the spectrum of generally correlated Gaussian random matrices whose columns are zero-mean independent vectors but have different correlations, under the specific regime where the number of their columns and that of their rows grow at infinity with the same pace. Applications include multiuser multiple-input single-output (MISO) systems.

Further Comments

In many sections, we reproduce some contents from [1], which serves as our master guide. Section 7.2 collects some classical results taken from [6,7].

For the linear eigenvalue statistics defined in Section 7.10, historically, the first study was carried out by Jonsson [70] for the generalized variance $f(x) = \log x$, motivated by multivariate statistics. Section 7.4.2 is taken from [70].

Section 7.20 is taken from [8].

Section 7.21 is drawn from [69].

References

1. National Research Council. *Frontiers in Massive Data Analysis*. Washington, DC: The National Academies Press, 2013.

2. Theodore Wilbur Anderson. *An Introduction to Multivariate Statistical Analysis*, volume 2. Wiley New York, 1958.

3. Robert Caiming Qiu, Zhen Hu, Husheng Li, and Michael C Wicks. *Cognitive Radio Communication and Networking: Principles and Practice*. Hoboken, NJ: John Wiley & Sons, 2012.

4. Robert Qiu and Michael Wicks. *Cognitive Networked Sensing and Big Data.* Berlin, Germany: Springer, 2014.

5. Robert C Qiu and Paul Antonik. *Smart Grid and Big Data: Theory and Practice.* Hoboken, NJ: John Wiley & Sons, 2015.

6. Jean Jacod and Philip E Protter. *Probability Essentials.* Berlin, Germany: Springer Science & Business Media, 2003.

7. Albert N Shiryaev. *Probability.* Graduate Texts in Mathematics, 28(2):191-194, 1996.

8. Alexei M Khorunzhy, Boris A Khoruzhenko, and Leonid A Pastur. Asymptotic properties of large random matrices with independent entries. *Journal of Mathematical Physics*, 37(10):5033–5060, 1996.

9. Terence Tao and Van Vu. Random matrices: Universality of local eigenvalue statistics up to the edge. *Communications in Mathematical Physics*, 298(2):549–572, 2010.

10. Alice Guionnet, Manjunath Krishnapur, and Ofer Zeitouni. The single ring theorem. *arXiv preprint arXiv:0909.2214*, 2009.

11. LA Pastur. A simple approach to the global regime of Gaussian ensembles of random matrices. *Ukrainian Mathematical Journal*, 57(6):936–966, 2005.

12. Leonid Andreevich Pastur, Mariya Shcherbina, and Mariya Shcherbina. *Eigenvalue Distribution of Large Random Matrices*, Volume 171. American Mathematical Society Providence, RI, 2011.

13. L Pastur and V Vasilchuk. On the law of addition of random matrices. *Communications in Mathematical Physics*, 214(2):249–286, 2000.

14. Eugene P Wigner. Characteristic vectors of bordered matrices with infinite dimensions I. In *The Collected Works of Eugene Paul Wigner*, pages 524–540. Berlin, Germany: Springer, 1993.

15. Dan Voiculescu. Limit laws for random matrices and free products. *Inventiones Mathematicae*, 104(1):201–220, 1991.

16. Kurt Johansson et al. On fluctuations of eigenvalues of random Hermitian matrices. *Duke Mathematical Journal*, 91(1):151–201, 1998.

17. X He, Q Ai, C Qiu, W Huang, and L Piao. A Big Data architecture design for smart grids based on random matrix theory. *arXiv preprint arXiv:1501.07329*, 2015.

18. Zhidong Bai, Dandan Jiang, Jian-Feng Yao, and Shurong Zheng. Corrections to LRT on large-dimensional covariance matrix by RMT. *The Annals of Statistics*, 37:3822–3840, 2009.

19. Dandan Jiang, Tiefeng Jiang, and Fan Yang. Likelihood ratio tests for covariance matrices of high-dimensional normal distributions. *Journal of Statistical Planning and Inference*, 142(8):2241–2256, 2012.

20. Olivier Ledoit and Michael Wolf. A well-conditioned estimator for large-dimensional covariance matrices. *Journal of Multivariate Analysis*, 88(2):365–411, 2004.

21. Changchun Zhang and Robert Qiu. Massive MIMO as a Big Data system: Random matrix models and testbed. *Access IEEE*, 3: 837–851, 2015.

22. Rudolf Wegmann. The asymptotic eigenvalue-distribution for a certain class of random matrices. *Journal of Mathematical Analysis and Applications*, 56(1):113–132, 1976.

23. Yingshuang Cao, Long Cai, C Qiu, Jie Gu, Xing He, Qian Ai, and Zhijian Jin. A random matrix theoretical approach to early event detection using experimental data. *arXiv preprint arXiv:1503.08445*, 2015.

24. John K Hunter and Bruno Nachtergaele. *Applied Analysis*. River Edge, NJ: World Scientific Publishing Co., 2001.

25. László Erdős, Horng-Tzer Yau, and Jun Yin. Rigidity of eigenvalues of generalized wigner matrices. *Advances in Mathematics*, 229(3):1435–1515, 2012.

26. Jamal Najim and Jianfeng Yao. Gaussian fluctuations for linear spectral statistics of large random covariance matrices. *arXiv preprint arXiv:1309.3728*, 2013.

27. Zhidong D Ba, and Jack W Silverstein. CLT for linear spectral statistics of large-dimensional sample covariance matrices. *Annals of Probability*, 32(1): 553–605, 2004.

28. Jean Ginibre. Statistical ensembles of complex, quaternion, and real matrices. *Journal of Mathematical Physics*, 6(3):440–449, 1965.

29. PJ Forrester. Fluctuation formula for complex random matrices. *Journal of Physics A: Mathematical and General*, 32(13):L159, 1999.

30. Brian Rider. Deviations from the circular law. *Probability Theory and Related Fields*, 130(3):337–367, 2004.

31. B Rider and Jack W Silverstein. Gaussian fluctuations for non-Hermitian random matrix ensembles. *The Annals of Probability*, 34:2118–2143, 2006.

32. Ivan Nourdin and Giovanni Peccati. Universal Gaussian fluctuations of non-Hermitian matrix ensembles: from weak convergence to almost sure CLT. *arXiv preprint arXiv:1002.1212*, 2010.

33. Brian Rider and Bálint Virág. The noise in the circular law and the Gaussian free field. *International Mathematics Research Notices*, 2007:rnm006, 2007.

34. Bertrand Duplantier, Rémi Rhodes, Scott Sheffield, and Vincent Vargas. Log-correlated Gaussian fields: A overview. *arXiv preprint arXiv:1407.5605*, 2014.

35. Florent Benaych-Georges and Jean Rochet. Fluctuations for analytic test functions in the single ring theorem. *arXiv preprint arXiv:1504.05106*, 2015.

36. Vyacheslav L Girko. The elliptic law. *Teoriya Veroyatnostei i ee Primeneniya*, 30(4):640–651, 1985.

37. Charles Bordenave, Djalil Chafaï, et al. Around the circular law. *Probability Surveys*, 9:1–89, 2012.

38. ZD Bai and Jack W Silverstein. No eigenvalues outside the support of the limiting spectral distribution of large-dimensional sample covariance matrices. *Annals of Probability*, 26:316–345, 1998.

39. ZD Bai and Jack W Silverstein. Exact separation of eigenvalues of large dimensional sample covariance matrices. *Annals of Probability*, 27:1536–1555, 1999.

40. Alice Guionnet and Ofer Zeitouni. Support convergence in the single ring theorem. *Probability Theory and Related Fields*, 154(3–4):661–675, 2012.

41. Brian Rider. A limit theorem at the edge of a non-Hermitian random matrix ensemble. *Journal of Physics A: Mathematical and General*, 36(12):3401, 2003.

42. Friedrich Götze and Alexander Tikhomirov. The rate of convergence for spectra of GUE and LUE matrix ensembles. *Central European Journal of Mathematics*, 3(4):666–704, 2005.

43. Friedrich Götze and Aleksandr Nikolaevich Tikhomirov. The rate of convergence of spectra of sample covariance matrices. *Theory of Probability & Its Applications*, 54(1):129–140, 2010.

44. Friedrich Götze and Alexandre Tikhomirov. *On the Rate of Convergence to the Semi-circular Law*. Basel, Switzerland: Springer, 2013.

45. F Götze and A Tikhomirov. On the rate of convergence to the Marchenko–Pastur distribution. *arXiv preprint arXiv:1110.1284*, 2011.

46. EP Wigner. Distribution laws for the roots of a random Hermitian matrix. *Statistical theories of Spectra: Fluctuations*, 446–461, 1965.

47. Madan Lal Mehta. *Random Matrices*, volume 142. Amsterdam, the Netherlands: Academic Press, 2004.

48. John Wishart. The generalised product moment distribution in samples from a normal multivariate population. *Biometrika*, 29:32–52, 1928.

49. PL Hsu. On the distribution of roots of certain determinantal equations. *Annals of Eugenics*, 9(3):250–258, 1939.

50. Uffe Haagerup and Steen Thorbjornsen. Random matrices with complex Gaussian entries. *Expositiones Mathematicae*, 21(4):293–337, 2003.

51. A Erdélyi, Wilhelm Magnus, Fritz Oberhettinger, and Francesco G Tricomi. Higher transcendental functions, vol. I. *Bateman Manuscript Project, McGraw-Hill, New York*, 1953.

52. John Harer and Don Zagier. The euler characteristic of the moduli space of curves. *Inventiones Mathematicae*, 85(3):457–485, 1986.

53. Alexandre Stojanovic. Une majoration des cumulants de la statistique linéaire des valeurs propres d'une classe de matrices aléatoires. *Comptes Rendus de l'Académie des Sciences-Series I-Mathematics*, 326(1):99–104, 1998.

54. Vladimir A Marčenko and Leonid Andreevich Pastur. Distribution of eigenvalues for some sets of random matrices. *Sbornik: Mathematics*, 1(4):457–483, 1967.

55. Vladimir Igorevich Bogachev. *Gaussian Measures*. Number 62. Providence, RI: American Mathematical Soc., 1998.

56. N I Akhiezer and I M Glazman. *Theory of Linear Operators in Hilbert Space*. New York: Dover Publications, 1993.

57. Jakob Hoydis, Romain Couillet, and Pablo Piantanida. The second-order coding rate of the MIMO Rayleigh block-fading channel. *arXiv preprint arXiv:1303.3400*, 2013.

58. Walid Hachem, Oleksiy Khorunzhiy, Philippe Loubaton, Jamal Najim, and Leonid Pastur. A new approach for mutual information analysis of large dimensional multi-antenna channels. *Information Theory, IEEE Transactions on*, 54(9):3987–4004, 2008.

59. Philippe Loubaton. On the almost sure location of the singular values of certain Gaussian block-Hankel large random matrices. *Journal of Theoretical Probability*, 1–105, 2014.

60. Uffe Haagerup and Steen Thorbjornsen. A new application of random matrices: Ext($C_{\text{red}}^*(F_2)$) is not a group. *Annals of Mathematics*, 711–775, 2005.

61. Mireille Capitaine and Catherine Donati-Martin. Strong asymptotic freeness for Wigner and Wishart matrices. *Indiana University Mathematics Journal*, 56(2):767–804, 2005.

62. Mireille Capitaine, Catherine Donati-Martin, Delphine Féral, et al. Central limit theorems for eigenvalues of deformations of Wigner matrices. In *Annales de l'Institut Henri Poincaré, Probabilités et Statistiques*, 48(1): 107–133, 2009.

63. Pascal Vallet, Philippe Loubaton, and Xavier Mestre. Improved subspace estimation for multivariate observations of high dimension: The deterministic signals case. *Information Theory, IEEE Transactions on*, 58(2):1043–1068, 2012.

64. Eugene P Wigner. On the distribution of the roots of certain symmetric matrices. *Annals of Mathematics*, 67:325–327, 1958.

65. Leonid A Pastur. On the spectrum of random matrices. *Theoretical and Mathematical Physics*, 10(1):67–74, 1972.

66. Vyacheslav L Girko. Spectral theory of random matrices. *Russian Mathematical Surveys*, 40(1):77–120, 1985.

67. Michael Reed and Barry Simon. *Methods of Modern Mathematical Physics: Functional Analysis*, volume 1. New York: Academic Press, 1980.

68. Nicholas M Ercolani and KDT-R McLaughlin. Asymptotics of the partition function for random matrices via Riemann-Hilbert techniques and applications to graphical enumeration. *International Mathematics Research Notices*, 2003(14):755–820, 2003.

69. Uffe Haagerup and Steen Thorbjornsen. Asymptotic expansions for the Gaussian unitary ensemble. *Infinite Dimensional Analysis, Quantum Probability and Related Topics*, 15(01):1250003, 2012.

70. Dag Jonsson. Some limit theorems for the eigenvalues of a sample covariance matrix. *Journal of Multivariate Analysis*, 12(1):1–38, 1982.

Chapter 8

Big Data of Complex Networks and Data Protection Law
An Introduction to an Area of Mutual Conflict*

Florent Thouvenin

8.1 Introduction

From a legal perspective, at least at first sight, network analyses and law seem to be incompatible as law is associated with letters rather than numbers. To some extent, a lawyer's self-perception is built on the assumption that he or she cannot be replaced by any machine, understanding that computers can read legal texts but are nevertheless unable to understand the law, let alone to make the necessary value judgments. While this assumption is certainly true at the moment, approaches to quantify aspects of law may ultimately lead to a profound understanding of law by computers. (Certain approaches to encode statutes for further analysis have already been undertaken, see, e.g., [1].) At least in the long term, the legal profession will have to adapt to this evolving digital environment.

---

*This contribution has benefitted from valuable discussions in the newly incorporated Center for Information Technology, Society, and Law (ITSL) at the University of Zurich.

Until today, two different streams of legal network analysis can be identified. On the one hand, some research has been conducted on *contract networks*, that is, on individual contracts that are explicitly interlinked thereby building a network-like structure of contracts and the respective contractual terms (e.g., [2]; [3]; [4], p. 322). Examples for such networks are franchising systems, joint ventures, contracts on just-in-time production and the like ([5], pp.11 and 13; see also [6]). However, this research has mainly focused on the regulation of contractual networks and on the question of how to organize the interlinking of different contracts from a contract law perspective; both in theory and practice, these attempts have only had a limited impact. In addition, no formal work on contract networks has yet been carried out.

On the other hand, there have been some approaches to conduct network analyses on law and judicial decisions. For instance, Fowler et al. analysed the relevance of almost 27,000 majority opinions by the US Supreme Court issued between 1791 and 2005 [7]. The opinions served as network nodes. These nodes were then connected through citations, which were either classified as inbound (opinion is cited by another opinion) or outbound (opinion cites another opinion). The analysis resulted in two separate relevance measures: inward relevance identified influential cases while outward relevance indicated well-grounded cases. (Another follow-up study examined the use of precedent with regard to judgments of the European Court of Human Rights [8]; see also [9].) Besides case law, network analysis has also been applied to study structural connections within legal systems (e.g., [1,10,11]) or individual legislative texts (e.g., [12]; see also [13]). Such formal attempts seem to be quite promising as they may allow identification and analysis of yet unidentified structures within the law. In future research, the method of such analysis could be based on two steps: first, legal networks could be characterized by using quantitative graph measures [14] to classify the structure of such networks. Second, such networks could be compared structurally by using graph similarity measures to detect similarities between different legal networks.

The aim of this chapter, however, is considerably more modest. Given that Big Data of complex networks often include the processing of personal data, this chapter aims at pointing to the most important tensions between Big Data applications and the law, namely data protection law.

8.2   The Mutual Conflicts and the State of European Data Protection Law

Big Data and data protection law provide for a number of mutual conflicts: from the perspective of Big Data analytics, a strict application of data protection law as we know it today would set an immediate end to most Big Data applications. From the perspective of the law, Big Data is either a big threat that calls for a strict application of data protection law or a major challenge for international and national lawmakers to adopt today's data protection laws to the latest technological and economic developments.

With regard to Big Data applications carried out in Europe or processing personal data of European citizens, the current state of data protection law provides for an additional challenge: since data protection law has only been harmonized to a limited extent in the European Union (EU) by the *Data Protection Directive 95/46/EC* [15] (hereinafter "DPD"; with regard to the quite important national differences, see [16], p. 33; [17], p. 65; [18]), most Big Data applications will have to make sure that they are compliant with the national data protection laws of most, if not all, member states of the EU and other European states that are neither members of the EU nor of the European Economic Area, namely Switzerland. This national fragmentation of data protection law has led to significant obstacles for transnational businesses and research activities, including but not limited to Big Data applications, and has raised barriers for the enforcement of data protection law (for an in-depth discussion of the consequences of the current fragmentation, see [19], Section 3.2).

Against this background, the EU envisages a full harmonization of data protection law by adopting the *General Data Protection Regulation*. In January 2012, the EU Commission presented its draft of the regulation to the European Parliament and the Council of the EU [20]. An extensively amended version of this draft has been adopted by the Parliament in March 2014 [21] and after lengthy discussions, the Council presented its General Approach in June 2015 [22]. In December 2015, the Commission, the Parliament, and the Council finally reached an agreement on the new data protection rules. A consolidated version of the General Data Protection Regulation (hereinafter "GDPR", [23]) was adopted by the Council and shortly thereafter by the Parliament in April 2016. It will enter into force two years after its publication in the Official Journal of the European Union and is therefore expected to become effective in spring 2018 (see also GDPR Art. 99).

As opposed to today's state of law, this regulation establishes a single set of rules on data protection, valid across the EU, and abolishes the national fragmentation. The rules contained in this regulation will apply to all companies from across the world that provide their services in the EU, thereby enhancing fair competition between US and European players in the data market. In addition, and maybe more importantly, the regulation should be a major step forward with regard to the compliance of the big US players with the provisions of European data protection law and can therefore be interpreted as an attempt to limit the market power of data giants such as Google, Facebook, Amazon, Apple, and Microsoft. The twin aims of the regulation are to enhance the level of protection of personal data and to increase business opportunities in the digital single market by providing a single set of rules for all businesses throughout the EU ([22], p. 1). From this perspective, and opposed to the more important general tensions between Big Data applications and data protection law, the General Data Protection Regulation is advocated by the European legislator to be an enabler for Big Data services in the EU [24]. Whether and to what extent this promise holds true, however, will have to be analysed in the following.

Given that the GDPR has not yet become effective, the following analysis will deal both with the current state of the law as contained in the directive and the future rules contained in the regulation. The analysis will thereby focus on three different areas that are particularly important with regard to Big Data: the *notion of personal data*, some of the *general principles of data protection law*, namely the principle of transparency, the principle of purpose limitation, and the principle of data minimization, and the *criteria for making data processing legitimate*.

8.3   The Notion of Personal Data

The application of data protection law is triggered by any processing of personal data. Both notions, processing and personal data, are construed in a very broad sense. According to the definition of European data protection law, *processing* means "any operation or set of operations which is performed upon personal data or sets of personal data, whether or not by automated means, such as collection, recording, organization, structuring, storage, adaptation or alteration, retrieval, consultation, use, disclosure by transmission, dissemination or otherwise making available, alignment or combination, restriction, erasure or destruction" (GDPR Art. 4 (2); correspondingly: DPD Art. 2 (b)). The notion of personal data is equally broad and encompasses "any information relating to an identified or identifiable natural person" (GDPR Art. 4 (1); correspondingly: DPD Art. 2 (a)). An identifiable person thereby is any natural person who can be identified, directly or indirectly, by means reasonably likely to be used by any other person, "in particular by reference to an identifier such as a name, an identification number, location data, an online identifier or to one or more factors specific to the physical, physiological, genetic, mental, economic, cultural or social identity of that natural person" (GDPR Art. 4 (1); in comparison, location data, online identifiers, and the genetic identity are not explicitly mentioned in DPD Art. 2 (a)).

In an analogue environment, the notion of personal data allowed drawing a rather rigorous line between personal data and other data thereby separating data processing that had to comply with the provisions of data protection law and other processing that did not. In order to escape the application of data protection provisions, one could anonymize personal data by deleting all data that allowed for the identification of the individuals to which the data related. With the digitalization of all sorts of data, however, the line between personal and nonpersonal data became blurry since the combination of different data collections became easier thereby enhancing the risk of (re)identification. This trend has been taken one significant step further with Big Data analytics. Given the enormous potential of (re)combination of data, it is assumed that it is and will become increasingly difficult to exclude the possibility to reestablish the relation between any given data and the person the data relates to (see [25], p. 253; [26], p. 48; [27], p. 47).

As a result, Big Data analytics dramatically extends the field of application of data protection law to almost all processing of data. Against this background, it has been argued that the established differentiation between personal and nonpersonal data should be abandoned altogether (see [28], p. 1743; from a Swiss perspective [29], p. 55) and data protection law should be replaced by a new body of law regulating the processing of all sorts of data (see [28], p. 1742 f.; from a Swiss perspective [29], p. 57 f.).

The General Data Protection Regulation initiated another debate regarding the potential of recombining anonymized data. Under the current Directive, data is only personal (and therefore protected) when such data can be de-anonymized by the data controller. Contrary to this subjective point of view, the Regulation (GDPR, recital 26) suggests an objective approach. As a result, anonymized data would even have to be handled as personal data not only if the controller could identify a person but also in case any other person would have the ability to de-anonymize the data. Such perspective would result in an over-excessive understanding of personal data as—given the amount and variety of additional data—safe anonymization is becoming difficult (see [16], p. 41 ff.).

8.4 General Principles of Data Protection Law

8.4.1 Overview

The processing of personal data is only legitimate if such processing complies with a number of *general principles*. According to established European data protection law: (1) personal data must be processed fairly and lawfully (DPD Art. 6 (1) (a); GDPR Art. 5 (1) (a)); (2) personal data must be collected for specified, explicit, and legitimate purposes and must not be further processed in a way incompatible with those purposes (DPD Art. 6 (1) (b); GDPR Art. 5 (1) (b)); (3) personal data must be adequate, relevant, and not excessive in relation to these purposes (DPD Art. 6 (1) (c); GDPR Art. 5 (1) (c); compared with the DPD, the Regulation goes one step further and replaces the words "not excessive" by the more restricting notion that personal data must be "limited to what is necessary in relation to the purposes for which they are processed"); (4) personal data must be accurate and, where necessary, kept up to date (DPD Art. 6 (1) (d); GDPR Art. 5 (1) (d)); and (5) personal data must be kept in a form which permits identification of data subjects for no longer than is necessary for the purposes for which the data were collected or for which they are further processed (DPD Art. 6 (1) (e); GDPR Art. 5 (1) (e)).

In addition to these principles which are contained both in the Data Protection Directive and in the General Data Protection Regulation, the latter also provides for a general *transparency principle* and explicitly states the *data minimisation principle* ([30]; Chapter 3.4.2. For the clarification of the data minimization principle, see the following, Section 8.4.4). These principles, however, are not entirely new. Rather, the transparency principle is part of the directive's principle of fair and lawful processing ([31], p. 1137, para 32.27; see also DPD Art. 6 (1) (c). For this principle, see Section 8.4.2) and the data minimization principle is contained in the principle that personal data must be adequate, relevant, and not excessive in relation to the purposes for which it is collected ([32], p. 84. For this principle see Section 8.4.2).

8.4.2 Transparency principle

According to the transparency principle which is regulated together with the principle of fair and lawful processing, personal data must be processed in a transparent manner in relation to the data subject (GDPR Art. 5 (1) (a)). In particular, the *specific purposes* for which personal data are processed have to be *explicit and legitimate and must be determined at the time of the collection* of such data (GDPR, recital 39). This means that the data subject should be informed in particular of the existence of the processing operation and its purposes and should be provided with any further information necessary to guarantee fair and transparent processing with regard to the specific circumstances (GDPR, recital 60). This information should be given to the data subject at the time of collection of the personal data, or, if the data are not collected from the data subject, within a reasonable period, depending on the circumstances of the case (GDPR, recital 61). However, there is obviously no obligation to (re-)provide such information to the data subjects if they are already in possession of this information (GDPR, recital 62).

To further foster the transparency, the data controller—that is, the natural or legal person, public authority, agency, or any other body which alone or jointly with others determines the purposes, conditions, and means of the processing of personal data (DPD Art. 2 (d); GDPR Art. 4 (7))—must provide any information with regard to the processing of personal data and for the exercise of the data subjects' rights *in a concise, transparent, intelligible and easily accessible form* (GDPR Art. 12 (1)). This aspect of the principle of transparency requires that any information addressed to the public or the data subject should be easily accessible and easy to understand, which means that clear and plain language must be used, and that, where appropriate, visualizations should be included (GDPR, recital 39 and 58).

With regard to Big Data analytics, three issues are particularly challenging: first of all, the processing of personal data must be transparent, that is, the data subject must be *aware of the collecting and further processing* of the data. The awareness of data subjects is problematic in an increasing number of scenarios since today data is most often collected automatically, that is, without any specific activity of the data subject such as answering questions or filling out forms. This holds true for scenarios such as sensor generated data, connectivity data produced and collected while browsing the Internet, data gathered by

Wi-Fi providers about (mobile) devices that tried to connect or actually were connected to wireless networks, and/or data provided by mobile phones to mobile operators, to name just a few. Although some people will be aware of the permanent data collection taking place when moving in the physical world or navigating the Internet, probably most people still are not. Given the lack of transparency with regard to the collection of data and given that Big Data analytics are most often based on or at least include such data, most Big Data applications are not compliant with the requirements of the transparency principle.

Second, and in addition to the collecting of personal data, the processing of such data, and namely its purpose, often remains unclear. Even if data subjects are aware of the collecting of their data, they may not know what the data is collected and used for. Although this is a general issue that relates to all sort of data processing, the lack of transparency with regard to the purpose of the processing seems to be particularly problematic in a Big Data context since Big Data applications are often based on combining different sets of data and using the data to gain insights that had not been envisaged when the data were collected. Although to be compliant with the transparency principle, it would be sufficient to provide data subjects with information about the processing of their data, using the data for purposes that are different from the purposes the data was collected for is very problematic with regard to the principle of purpose limitation. Therefore, this issue will be further elaborated in the following (see: Section 8.4.3).

Third, the principle of transparency requires data controllers to provide transparent and easily accessible *privacy policies*. However, even if such policies are provided, as is the case on an increasing number of websites, privacy policies most often remain very unclear with regard to the kind of data collected, the purpose of the collection, and the further processing. Language such as "for marketing purposes" or "to enhance the quality of our services" will have to be qualified as clearly insufficient to meet the requirements of the transparency principle.

8.4.3 Purpose limitation

The principle of purpose limitation contains two aspects that are key with regard to the (non)compliance of Big Data applications with the requirements of data protection law: first, personal data must be collected for specified, explicit, and legitimate purposes (DPD Art. 6 (1) (b); GDPR Art. 5 (1) (b)). Second, personal data may not be further processed in a way incompatible with the purposes for which the data were collected (DPD Art. 6 (1) (b); GDPR Art. 5 (1) (b)). However, the processing of personal data for historical, statistical, or scientific purposes must not be considered as incompatible with the purpose the data was collected for (DPD Art. 6 (1) (b); GDPR Art. 5 (1) (b) and Art. 89, which state additional conditions for processing for historical, statistical, and research purposes).

8.4.3.1 Specified, explicit, and legitimate purposes

The requirement that personal data must be collected for specified, explicit, and legitimate purposes means, first of all, that the purpose of the data processing must be determined prior to or at the time of the collection of such data. As a consequence, personal data may not be collected without determining its future use and without making sure that such use is intelligible for the data subject. This condition can be problematic with regard to Big Data applications since it prohibits the collection of data without any specific purpose, for example, with regard to a potential but yet undefined and unclear future use. However, this aspect of the principle of purpose limitation does not compromise compliance with data protection law if data collected for one specific purpose is used for a different purpose at a later stage.

Whereas the condition of a *legitimate purpose*, that is, a purpose which is not prohibited by law and does not infringe individual rights, is rather straightforward, the notion of a *specified and explicit purpose* is rather vague. Although the European Legislator has not clearly stated how this notion has to be construed and the European Court of Justice has not yet had a chance to provide substantial guidance, it seems to be safe to assume that the condition of an *explicit purpose* does not mean that the data subject must be actively informed about the processing of its data in oral or written form. Rather, it must be sufficient if the data subject is able to understand what purpose the data is collected for by drawing from the circumstances, for example, from the fact that a video camera is installed at a train station or the like. The condition of a *specified purpose* is at the same time the most important and the most unspecific notion within the principle of purpose limitation. Although it seems to be clear that this notion excludes all sorts of ambiguous definitions and requests some degree of specification as to what purpose the data will be processed for, the level of specification remains unclear and can only be fully defined on a case-by-case basis. By way of example, however, it can be assumed that very broad notions such as "for marketing purposes" would be too unspecific, whereas "for the purpose of providing the data subject with advertisements that fit its interests" would probably be regarded as sufficiently specific.

8.4.3.2   No processing for incompatible purposes

The most important conflict of Big Data applications and data protection law is triggered by the second aspect of the purpose limitation, that is, the requirement that personal data may not be processed in a way incompatible with the purposes for which the data were collected. Since Big Data applications often combine different sets of data and use the data for applications that had not been envisaged when the data were collected, most often, personal data will be used for purposes that are different from the purposes the data were collected for.

Whether such applications are compliant with the requirements of the principle of purpose limitation depends on the understanding of the "specified purpose" and the "incompatible processing." As with the "specified purpose" (see earlier, Section 8.4.3.1), the notion of "incompatible processing" has not been defined by the European Legislator or the European Court of Justice. As a consequence, as with the former, the latter could be construed in a rather narrow or in a much broader sense. The only aspect that seems to be clear is that the purpose of a later processing does not have to be identical with the initial purpose. It remains very unclear, however, what kind of nonidentical processing still qualifies as being compatible with the purpose of the initial processing.

The uncertainty is enhanced by the fact that the notions of the "specified purpose" and the "incompatible processing" are closely interlinked: if the condition of the "specified purpose" could be met by defining the purpose in a rather unspecific way, a vast number of rather diverse Big Data applications would be compatible with the requirement of the purpose limitation, as the circle of compliant applications would encompass all applications covered by the unspecific definition of the purpose and all other applications carried out for similar purposes. Should the condition of a "specified purpose" demand a rather high level of specification, however, the circle of compliant Big Data applications would be much smaller since both the number of applications covered by the specific definition of the purpose and the number of applications that could be regarded as compatible with this purpose would decline. From this perspective, the "specified purpose" and the "incompatible processing" must be understood as two regulating screws that can be tightened or loosened either independently or jointly thereby defining the scope of Big Data applications that are compliant with the principle of purpose limitation. Given that these regulating screws have not yet been put in their final position, the risk of a lack of compliance must be regarded as very relevant if personal data that has been collected for one purpose is used for a different purpose at a later stage.

8.4.4 Principle of data minimization

According to the principle of data minimization, personal data must be adequate, relevant, and *limited to what is necessary* in relation to the purposes for which they are processed (GDPR Art. 5 (1) (c). The wording of DPD Art. 6 (c) is both less explicit and less specific than the wording of the Regulation and only states "that personal data must be adequate, relevant and not excessive in relation to the purposes for which they are collected and/or further processed."). In addition, personal data should only be processed if, and as long as, the purpose of the data processing could not be fulfilled by processing information that does not involve personal data. This means that the period for which the data are stored must be limited to a strict minimum and that time limits should be established by the data controller for erasure of the data or for a periodic review (GDPR, recital 39).

As with the principle of purpose limitation, the conflict of the data minimization principle and Big Data applications is obvious. While Big Data applications most often rely on as much data as can be gathered and processed, thereby aiming at using the maximum of data available, the data minimization principle demands the opposite, that is, the reduction of personal data to a minimum. Although it seems to be possible for Big Data applications to comply with the data minimization principle by demonstrating that the amount of personal data processed actually is the minimum of data needed to achieve the purpose of the processing, the fundamental tension between Big Data applications and the data minimization principle remains.

8.5 Criteria for Making Data Processing Legitimate

8.5.1 Overview

Irrespective of being compliant with the data protection principles (see earlier, Section 8.4), the processing of personal data is only legitimate if at least one out of a set of six criteria for making such processing legitimate is met. According to this catalogue, the processing of personal data is legitimate: if the data subject has given his or her consent (DPD Art. 7 (a); GDPR Art. 6 (1) (a)); if the processing is necessary for the performance of a contract to which the data subject is party (DPD Art. 2 (b); GDPR Art. 6 (1) (b)); if the processing is necessary for compliance with a legal obligation to which the data controller is subject (DPD Art. 2 (c); GDPR Art. 6 (1) (c)); if the processing is necessary to protect the vital interests of the data subject (DPD Art. 2 (d); GDPR Art. 6 (1) (d)); if the processing is necessary for the performance of a task carried out in the public interest (DPD Art. 2 (e); GDPR Art. 6 (1) (e)); or if the processing is necessary for the purposes of the legitimate interests pursued by the controller, except where such interests are overridden by the interests or fundamental rights and freedoms of the data subject (DPD Art. 2 (f); GDPR Art. 6 (1) (f)). In addition, according to the Regulation, the processing of personal data which is necessary for the purposes of *historical, statistical, or scientific research* shall be lawful if these purposes cannot be otherwise fulfilled by processing data which does not or no longer permit the identification of the data subject and if data that enables the attribution of information to an identified or identifiable data subject is kept separately from the other information (GDPR recital 156 and Art. 89). Both in general and with regard to Big Data applications, the most important criteria for making data processing legitimate are consent and the legitimate interests of the data controller. These two criteria will therefore be looked at more closely.

8.5.2 Consent of data subject

The notion of consent is very broad and encompasses any freely given, specific, informed and unambiguous indication of the data subject's wishes by which he or she signifies his or her agreement to the processing of personal data that relates to him or her (DPD Art. 2 (h); GDPR Art. 4 (11)). According to the Data Protection Directive the data subject must have given his or her consent unambiguously. With a view to the enormous practical importance of consent, the European Legislator has clarified several issues regarding consent in the General Data Protection Regulation.

First of all, according to the Regulation, consent must be *explicit*. Such explicit consent can be given either by a statement or by a clear affirmative action by the data subject (GDPR Art. 4 (11)). The condition of explicit consent should ensure that data subjects are aware of giving their consent to the processing of their personal data. A possible statement would be the ticking of a box when visiting a website on the Internet, a clear affirmative action the using of a car park which is under video surveillance if the data subject is informed about the video surveillance and its purpose. The mere silence or inactivity of a data subject, on the other hand, would not constitute consent (GDPR, recital 32).

Second, the Regulation clearly states that the data subject has the *right to withdraw* his or her consent at any time (GDPR Art. 7 (3) sentence 1). Although such withdrawal does not affect the lawfulness of processing based on consent before its withdrawal (GDPR Art. 7 (3) sentence 2), any future processing would be illegitimate if no other criteria for making such processing legitimate is fulfilled.

Third, consent always relates to *one or more specific purposes* (GDPR Art. 6 (1) (a)). The processing of personal data for other purposes which are not compatible with the initial purpose is therefore only legitimate if the data subject gives his or her explicit consent for this other purpose or if the data controller can rely on another criterion for making the data processing legitimate (cf. GDPR Art. 6 (1)).

In addition, the Regulation states a number of rather *formal conditions* for consent: namely, if consent is to be given in the context of a written declaration which also concerns another matter, for example, in general terms and conditions made available on a website, the requirement to give consent must be presented distinguishable in its appearance from such other matter (GDPR Art. 7 (2)). Furthermore, the burden of proof for the data subject's consent always lies with the data controller (GDPR Art. 7 (1)), which means that, in case of doubt about consent, the data controller will have to prove that the data subject has given his or her consent for the processing of his or her personal data for the specific purpose.

For Big Data applications, the requirement of consent provides for a fundamental conflict with data protection law which is mainly triggered by the tying of *consent to one or several purposes*. Since Big Data applications are often based on the combination and analysis of massive amounts of personal data and aim at gaining new insights, the data is most often processed for a different purpose than the one it was collected for. If the notions of "specific purpose" and "incompatible purposes" are interpreted in a rather strict sense (see earlier, Section 8.4.3), most Big Data applications would only be legitimate if the data subjects would actually give their consent to the processing of their personal data for the respective application. Receiving consent from all data subjects involved, however, is often very burdensome if not impossible if consent must be sought for after the collection of the data. A strict interpretation of the requirement of consent for one or several specific purposes would therefore put a quick end to numerous potentially useful Big Data applications.

This conflict is exacerbated by the fact that data subjects may always withdraw their consent. As a consequence, even in cases where the purpose of the Big Data application would be compatible with the purpose the data was initially collected for, the data controller must make sure that the data subjects involved did not withdraw their consent since the time the data was collected. As with the case of a different purpose, running through this process may be very burdensome and may prove to be almost impossible if the data controller did not design specific technical mechanisms that allow for the tracking of the withdrawal of consent.

8.5.3   Legitimate interests of data controller

In the case of lack of consent of the data subject, the legitimate interests of the data controller may, amongst others (see earlier, Section 8.5.1), provide a basis for making the processing of personal data legitimate. This criterion requires a *balancing of the interests of the data subject(s) and the data controller* which must include the weighing of the fundamental rights and freedoms of the data subject (GDPR Art. 6 (1) (f)), namely the right to the protection of personal data as laid down in the Charter of Fundamental Rights of the EU and the Treaty on the Functioning of the European Union (Charter of Fundamental Rights of the European Union Art. 8; Treaty on the Functioning of the European Union Art. 16). Given that such balancing of interests must be made on a case-by-case basis, general statements with regard to the application of this criterion for making data processing legitimate are difficult. What can be said, however, is that the balancing of interests always requires a *careful assessment* of the interests involved (GDPR, recital 47; [31], p. 1140, para 32.24). In addition, the data controller is obliged to explicitly inform the data subject on the legitimate interests pursued and on the *right to object* to the data processing (GDPR Art. 13 (1) (d) and Art. 13 (2) (b)). The right to object thereby vests with the data subject if the data controller relies on its overriding interests for making the data processing legitimate; such right is, however, subject to the data controller's demonstration of compelling legitimate grounds for the data processing which override the interests and the fundamental rights and freedoms of the data subject (GDPR Art. 6 (1) (f) and recital 69).

8.6   Conclusion

Big Data applications and data protection law provide for a number of mutual conflicts. The most important conflicts are triggered by the general data protection principles of purpose limitation (see earlier, Section 8.4.3) and data minimization (see earlier, Section 8.4.4). In addition, the requirement of consent of the data subject for making the processing of his or her personal data legitimate is particularly burdensome for most Big Data applications (see earlier, Section 8.5.2). With a view to the potential of Big Data, European and national legislators should thoroughly consider whether the application of today's data protection law is really beneficial both for data subjects and the society at large. Although the European legislator advocates its General Data Protection Regulation as being beneficial for Big Data applications (see earlier, Section 8.2), this promise seems to be only (very) partly reflected in the Regulation.

The mutual conflicts of Big Data and data protection law points to the even more fundamental question of how a body of law that aims at protecting the privacy of data subjects should be construed. As opposed to the (very) precautionary, protective, and restrictive approach of the Data Protection Directive and the General Data Protection Regulation, a different approach could be envisaged:

First, instead of providing a rigid set of norms that should ensure an *ex ante protection of privacy* and which are enforced mainly by national data protection authorities, data protection law could be based on a set of remedies available in case of a privacy infringement and rely on private enforcement only. Although such an approach would considerably weaken the position of data subjects, it would be a—admittedly quite radical—means to ensure that protection of privacy is only granted where data subjects really feel a need for it and are ready to act in order to ensure it.

Second, with regard to data subject's need for protection, it might be worthwhile considering to establish a *fundamental distinction between the processing of personal data within an organization ("inside") and the making available of such data to the public ("outside")*, for example, on the Internet. Such a distinction between "inside" and "outside" could either serve as a basis for a differentiated set of legal sanctions for privacy infringements and might even allow to refrain from regulating the processing of personal data "inside" an organization altogether.

Third, instead of building up important restrictions to the processing of personal data, for example, by the requirement of the purpose limitation (see earlier, Section 8.4.3), data protection law could be limited to the *general principle of transparency* (see earlier, Section 8.4.2) and the *granting of a number of rights for data subjects*, namely the already established right to withdraw consent (see earlier, Section 8.5.2), to object to the processing of personal data (see earlier, Section 8.5.3), to have access to such data (DPD Art. 12 (a); GDPR Art. 15), and to obtain the rectification of such data from the data controller (DPD Art. 12 (b); GDPR Art. 16).

These three approaches could be cumulated or integrated into a blended mix and might serve as building blocks for a different, more nuanced and more "Big Data friendly" approach to grant the—undoubtedly very much needed—protection of the privacy of data subjects.

References

1. Patricia M. Sweeney et al. Network analysis of manually-encoded state laws and prospects for automation. *Paper presented at the XIV International Conference on AI and Law, Workshop "Network Analysis in Law,"* 14 June 2013. http://www.leibnizcenter.org/~winkels/NAiL2013Proceedings.pdf.

2. Gunther Teubner. *Networks as Connected Contracts.* Oxford, UK: Hart Publishing, 2011.

3. Richard M. Buxbaum. Is "network" a legal concept? *Journal of Institutional and Theoretical Economics* 1993;149(4):698–705.

4. Walter W. Powell. Neither market nor hierarchy: Network forms of organization. *Research in Organizational Behavior* 1990;12:295–336.

5. Gunther Teubner. Coincidentia oppositorum: Hybrid networks beyond contract and organisation. In: Marc Amstutz and Gunther Teubner (eds.). *Networks: Legal Issues of Multilateral Co-operation.* Oxford, UK: Hart Publishing, 2009, pp. 3–30.

6. Erich Schanze. Symbiotic Arrangements. In: Peter Newman (ed.). *The New Palgrave Dictionary of Economics and the Law.* Vol. 3. London, UK: Palgrave Macmillan, 1998, pp. 554–559.

7. James H. Fowler et al. Network analysis and the law: Measuring the legal importance of precedents at the U.S. Supreme Court. In: *Political Analysis.* Vol. 15, Oxford, UK: Oxford University Press, 2007, pp. 324–346.

8. Yonatan Lupu and Erik Voeten. Precedent in international courts: A network analysis of case citations by the European Court of Human Rights. *British Journal of Political Science* 2012;42:413–439.

9. Krzysztof J. Pelc. The politics of precedent in international law: A social network application. *American Political Science Review* 2014;108:547–564.

10. Romain Boulet et al. A network approach to the French system of legal codes—Part I: Analysis of a dense network. *Artificial Intelligence and Law* 2011;19:333–355.

11. Romain Boulet et al. Network analysis of the French environmental code. In: Pompeu Casanovas et al. (eds.). *AI Approaches to the Complexity of Legal Systems.* Berlin, Germany, 2012, pp. 39–53.

12. Michael J. Bommarito II and Daniel M. Katz. A mathematical approach to the study of the United States code. *Physica A* 2010;389:4195–4200.

13. Nicola Lettieri and Sebastiano Faro. Computational social science and its potential impact upon law. *European Journal of Law and Technology* 2012;3(3).

14. Matthias Dehmer and Frank Emmert-Streib (eds.). *Quantitative Graph Theory: Mathematical Foundations and Applications.* Chapman and Hall/CRC Press: Boca Raton, FL, 2015.

15. Directive 95/46/EC of the European Parliament and of the Council of 24 October 1995 on the protection of individuals with regard to the processing of personal data and on the free movement of such data, Official Journal L 281, 23 November 1995, 31–50.

16. Juergen Hartung. Neue Regulierungsaspekte in der EU-Datenschutzreform. In: Rolf H. Weber and Florent Thouvenin (eds.). *Neuer Regulierungsschub im Datenschutzrecht.* Zurich, Switzerland: Schulthess Juristische Medien AG, 2012, pp. 31–54.

17. Peter Hustinx. The reform of EU data protection. In: Normann Witzleb et al. (eds.). *Emerging Challenges in Privacy Law.* Cambridge, UK: Cambridge University Press, 2014, pp. 62–71.

18. Monika Kuschewsky (ed.). *Data Protection & Privacy, Jurisdictional Comparisons.* London, UK: Thomson Reuters, 2012.

19. European Commission. Commission Staff Working Paper, Impact Assessment, 25 January 2012, SEC (2012) 72 final.

20. Proposal for a Regulation of the European Parliament and of the Council on the protection of individuals with regard to the processing of personal data and on the free movement of such data (General Data Protection Regulation), 25 January 2012, COM/2012/011 final.

21. European Parliament legislative resolution of 12 March 2014 on the proposal for a regulation of the European Parliament and of the Council on the protection of individuals with regard to the processing of personal data and on the free movement of such data (General Data Protection Regulation). Adopted on 12 March 2014, 1st reading/single reading. http://www.europarl.europa.eu/sides/getDoc.do?type=TA&language=EN&reference=P7-TA-2014-0212.

22. General Approach of the Council of the European Union. Agreed on 15 June 2015. http://data.consilium.europa.eu/doc/document/ST-9565-2015-INIT/en/pdf.

23. General Data Protection Regulation. Adopted on 8 April 2016 by the Council and on 14 April 2016 by the Parliament. http://eur-lex.europa.eu/legal-content/EN/TXT/PDF/?uri=CONSIL:ST_5419_2016_REV_1&from=EN.

24. EU Commission. What will the EU data protection reform bring for startup companies and Big Data? 2015. http://ec.europa.eu/justice/newsroom/data-protection/news/150415_en.htm.

25. Eric Horvitz and Deirdre Mulligan. Data, privacy, and the greater good. *Science* 2015;349:253–255.

26. Rongxing Lu et al. Toward efficient and privacy-preserving computing in Big Data era. *IEEE Network* 2014;28(4):46–50.

27. James R. Kalyvas and Michael R. Overly. *Big Data, A Business and Legal Guide*. Boca Raton, FL: CRC Press, 2015.

28. Paul Ohm. Broken promises of privacy: Responding to the surprising failure of anonymization. *UCLA Law Review* 2010;57:1701–1777.

29. Bruno Baeriswyl. Big Data zwischen Anonymisierung und Re-Individualisierung. In: Rolf H. Weber and Florent Thouvenin (eds.). *Big Data und Datenschutz—Gegenseitige Herausforderungen*. Zurich, Switzerland: Schulthess Juristische Medien AG, 2014, pp. 45–59.

30. European Commission. Draft General Data Protection Regulation, Explanatory Memorandum, 25 January 2012, COM(2012) 11 final.

31. Juergen Hartung. Datenschutzrecht der Europäischen Union. In: Nicolas Passadelis, David Rosenthal, and Hanspeter Thuer (eds.). *Datenschutzrecht, Beraten in Privatwirtschaft und öffentlicher Verwaltung*. Basel, Switzerland: Helbing Lichtenhahn Verlag, 2015, pp. 1123–1156.

32. Rosa Barcelo and Peter Traung. The emerging European Union security breach legal framework: The 2002/58 e-privacy directive and beyond. In: Serge Gutwirth et al. (eds.). *Data Protection in a Profiled World*. Dordrecht, the Netherlands: Springer, 2010, pp. 77–104.

Chapter 9

Structure, Function, and Development of Large-Scale Complex Neural Networks

Joaquín J. Torres

9.1 Introduction

In recent years, some initiatives within neuroscience research have started to study in detail, at a large scale, the behavior, function, and structure of the brain (see for instance the Human Brain Project [1], the US Brain Initiative [2], the Human Connectome Project [3], and the Openworm project [4], to name the most important). The improvement of the recording techniques for the acquisition of data in neuroscience, for example, the developing of large multiarray electrodes, which can simultaneously record more precisely the activity of a large population of neurons, or the use of noninvasive techniques in the study of neural systems, such as functional magnetic resonance imaging, computed tomography, electro and magneto encephalography, positron emission tomography, and so on, has made it so that presently the neuroscience community has a huge amount of data that is sometimes difficult or impossible to analyze in detail in a reasonable time with current data analysis techniques. This makes the study of the brain as a whole or, in general, any large-scale neural system with a very large number of degrees of freedom, very suitable systems to be studied within the "Big Data" computational analytical techniques and approaches.

Large-scale neural systems and, in particular, the brain of mammals are traditionally considered as paradigms of what is called a *complex system*. Among other causes, this complexity arises from the fact that neural systems are, in general, constituted by a very large number of simple excitable units, the *neurons* (about 10^{11} in the human brain [5]), that generate electrical signals which are transmitted through complex inter-neuron structures or *synapses*. Normally, a typical neuron can have also a large number—about 10^3—of synaptic contacts with other neurons [5]. Neurons and the connections or synapses among them constitute a complicated network whose structure, commonly, is assumed to be the result of evolution and some biophysical processes that occur during development and regeneration. These factors can induce both anatomical and functional changes in such a network [5] that can determine its computational properties.

For more than a century, scientists have been trying to understand the structure and functioning of different neural systems, including some brain areas. This effort has resulted in an almost complete understanding of the basic biophysical mechanisms for the generation of the electrical signals at the neurons, the so-called *action potentials,* and how neurons transmit these signals among them through the synapses. From this knowledge some *relatively* simple *in vitro* or/and *in vivo* experiments have been reproduced with different mathematical models, but a full understanding of how large-scale neural systems work as a whole still is lacking. This fact has been assumed to be important for the understanding of complex and high-level brain functions, for instance, memory consolidation and retrieval, spatio-temporal pattern formation [6] and, in last term, it could shed some light on the nature of consciousness.

In recent decades, there has been an increasing interest in the study of the role that the network topology could have in the emergent behavior observed in actual neural systems and, in particular, in the brain. This is particularly interesting since many large-scale networked natural systems exhibit intriguing emergent features that are the result of their network topology. One example is the so-called *small-world property,* that is, the average path length between two nodes in these networks is very small compared with the network size and there is a high level of *clustering* in the network (i.e., there is a large number of links among the neighbors of a given node). This property is a main characteristic of some social interacting networks [7], or the topology of Internet [8], but also can emerge in different biological systems [9] and, in particular, in a neural medium, as has been recently reported, for instance, in a set of *in vitro* experiments of growing cultured neurons [10,11]. Moreover, it has been revealed that small-world features in such neural systems result from self-organizing processes involving large local and global network efficiency and network degree–degree correlations [11]. The possible computational implications that these network characteristics could have, for instance, in complex brain functions are far from being understood yet in depth.

Other large-scale complex networks with interest in science, and very recently in biology, are the so-called *scale-free* (SF) network topologies. SF networked structures occur in many different situations, including the Internet [8,12], e-mail [13] and scientific-citation networks [14], ecological, and protein [15] and gene interaction networks [9], to name a few (see also [16] and references therein). These SF networks, in addition to small-world features, present *scale invariance,* that is, the network has the same aspect at any scale of observation. The SF property is manifested in the fact that some statistical properties of the network, as for instance, the node degree k_i distribution, are power-law distributed, that is, $p(k) \sim k^{-\gamma}$. The last implies that, after a growing process, the network includes a relatively large number of nodes with small connectivity—defining the so-called network boundary—and a relatively low number of nodes with a high connectivity or *hubs.* As a consequence, SF networks also exhibit the small-world property that, as we already mentioned, implies a very small average path length between two nodes compared to the network size and a large level of clustering in the network. Barabási and Albert (BA) first demonstrated that complex networks with power-law degree distributions can emerge after a process of network growing by adding new nodes which are linked to the old ones by means of a *preferential attachment* rule [12].

Despite the large number of studies reported in the last decade concerning the existence of SF networks in nature and the study of their statistical properties (see for instance [17]), only recently have the specific consequences that such architecture has on the emergent behavior of simple neural networks models that try to mimic the functioning of actual neural systems been reported. For instance, in [18] and [19] the effect of a BA SF topology on the performance of neural networks that present the so-called *associative memory* property has been analyzed. The authors in [18] show that such networks are able to store and retrieve a set of patterns of neural activity with a lower computer-memory cost compared with the case of fully connected networks. On the other hand, in [19] a zero temperature study of the behavior of different BA SF, small-world, and diluted network topologies is presented. The authors found that better network performance is achieved for the random diluted case than for the other network topologies. However, the BA SF network allows for a non-uniformly distributed information retrieval along the network, with a more robust and efficient information retrieval associated to the subset of highly connected nodes or *hubs*. Nevertheless, the relative large mean connectivity used in this study (around $\langle k \rangle = 50$) impedes in practice the SF property of the BA network and, therefore, this result lacks interest.

The previously mentioned examples illustrate that networked systems constituted by a large number of simple excitable units, as traditional attractor neural networks (ANNs), can be a very suitable framework to investigate the role of different evolving large-scale network topologies in their emergent behavior. In many cases, these systems can be analytically treated using, for instance, statistical physics techniques, since one can use the fact that the number of elements in the system is very large (the so-called thermodynamic limit), and easily simulated in a computer due to the *relatively* simple dynamics characterizing the basic elements (e.g., neuron and synapses) of such systems (see for instance [20]).

9.1.1   Structure of the chapter

This chapter deals with the use of computational network methods for the study of large-scale neural systems using statistical physics techniques, which is a very suitable theoretical framework for the study of systems constituted by a huge number of degrees of freedom. The chapter also describes the most relevant and intriguing computational properties of their emergent collective behavior. More precisely, in this chapter, I review some of our recent research concerning the role of different aspects of the underlying network topology in the emergent behavior of biologically inspired large-scale neural networks.

The chapter is structured as follows: first, in Section 9.2, I briefly report on the role that the neuron degree distribution has on the ability of the network to store and retrieve relevant information. In particular, the performance of different SF and homogenous topologies are compared. This study reveals that fully connected neural networks exhibit, in general, a better performance than networks with sparse connectivity. However, it is more reasonable to think that fully connected topologies are biologically unrealistic. It is unlikely, for instance, that natural evolution leads to such a large connectivity mainly due to energy consumption limitations. This study demonstrated as well that, in the presence of noise or nonzero temperature conditions, heterogeneous networks with SF topology have a large capacity to store and retrieve patterns of neural activity than more homogeneous random highly diluted (HD) networks with the same number of synapses. On the other hand, at zero temperature, the study concludes that the performance of SF networks decays more slowly with the number of stored patterns than in the case of more homogeneous random HD networks, and that the relative performance of the SF networks with respect to the HD networks increases with increasing values of the distribution power-law exponent.

In Section 9.3, I explain a recent work that extends the previous study to another particularly interesting issue, that is, the influence that different types of node *degree–degree* correlations in the network topology can have on the performance of auto-associative neural networks during recall processes. Using a recent method proposed to include node degree–degree correlations according to some predefined functional form [21], I describe in detail the case of SF correlated networks with degree–degree correlations that are also power-law distributed. Depending on the value and sing of the exponent of such power-laws, one can generate networks with neutral (zero exponent), positively correlated or assortative (positive exponent), and negatively correlated or disassortative degree–degree correlations (negative exponent). The study presented here shows that assortative networks perform better than neutral and disassortative ones during associative memory tasks in the presence of noise, and that memory recall enhances as the degree of assortativity increases.

Finally, in Section 9.4, I report on a general theoretical framework for stochastic network topology evolution. I describe the particular case of some evolving large-scale complex networks with simple nonlinear link attachment and detachment stochastic rules. For some values of the relevant parameters characterizing these *transition rates* and starting from a random homogeneous network topology, the system can reach a stationary state which is characterized by a power-law node degree distribution with given correlations. Moreover, the general framework at this critical stationary state can be used to investigate the effect that different nonlinear mechanisms defining such stochastic rules can have on the development of large-scale neural systems. In particular, I illustrate that for certain choices of these rules, the model is able to well reproduce the synaptic-pruning curves observed in the human brain.

The models and results I present here illustrate the usefulness that different computational and analytical network methods can have on the understanding of the emergent behavior of large-scale neural systems, or the understanding of brain development and function.

9.2 Influence of the Network Degree Distribution on Associative Memory Property

In a large number of studies concerning networked biological systems, it is common to assume a trivial fully connected network topology, which in some cases permits the obtaining of exact solutions, which at the time seemed more important than attempting to introduce biological realism. The topology of neural systems and in particular that of the brain—whether at the level of neurons and synapses, cortical areas, or functional connections—is obviously far from being so trivial [22–25]. In this section, I explore the role that nontrivial network topologies could have in the basic functioning of large-scale neural systems constituted by a large number of relatively simple units. As a starting point of this analysis, which will be extended to other aspects in the next section (for instance, considering a network with node degree–degree correlations), I present here recent results concerning how the overall network degree distribution of nonfully connected networks can affect the computational properties of simple ANN models [26]. It is worth mentioning that the influence of the degree distribution on dynamical behavior is found in many other settings, such as the more general situation of systems of coupled oscillators [27]. The results presented in this section refer to the ability of a neural network to store and retrieve relevant information in the form of particular patterns of neural activity, that is, the so-called *associative memory* property, and how this property is affected for varying topology in the presence of thermal fluctuations and noise originated by the interference among stored patterns.

Let us first consider a BA SF evolving network with N nodes and $\eta(N - N_0)$ links, where N_0 is the initial number of nodes to generate the network. On the other hand, $\eta \leq N_0$ is the number of links that are added during the evolution at each time step, with N being the final number of nodes in the network reached after evolution. In order to have a neural system with such topology, one can place at each node i a binary neuron with state $s_i = 1$ or 0 depending if it is firing or silent. Also, to allow the associative memory property in the system, we assume it *stores* P binary random fixed patterns $\xi^\mu \equiv \{\xi_i^\mu = 1, \text{ or } 0\}, \mu = 1, \ldots, P$ of neural activity with mean activity level $\langle \xi_i^\mu \rangle \equiv \frac{1}{N} \sum_i \xi_i^\mu = 1/2$. In practice, this can be done by defining a *synapse* at each link connecting two given nodes, with an intensity ω_{ij} chosen according to the so-called *Hebbian learning rule* [28], that is

$$\omega_{ij} = \frac{1}{\mathcal{K}} \sum_\mu (2\xi_i^\mu - 1)(2\xi_j^\mu - 1) \qquad (9.1)$$

where $\mathcal{K}$ is a constant usually taken equal to the network size N. The associative memory mechanism is nowadays used routinely for tasks such as image identification, memory recall, and pattern recognition. What is more curious is that it has been established that something similar to Equation 9.1 is implemented in nature via the processes of long-term potentiation and depression at the synapses [29], and these phenomena are indeed required for learning in actual neural systems [30].

To understand in detail the effect that heterogeneous networks, with the SF property, can have on complex brain tasks as associative memory, it is convenient to compare their performance with more homogeneous topologies with the same number of synapses, such as random HD networks. These HD topologies can be generated from a fully connected network by randomly suppressing synapses until, for instance, only $\eta(N - N_0)$ of them remain, which allows for a direct comparison between random HD and BA SF networks. It is worth mentioning that to maintain the SF behavior in the BA network η must be very small compared with the network size, that is, $\eta \ll N$ [18]. In the inset of Figure 9.1a the connectivity or degree distribution of a random HD network (circles) is illustrated and compared with a BA SF network (squares). The main differences are that the latter has no typical connectivity value, and nodes with large connectivity (hubs) have a nonzero probability to appear. More specifically, the SF network distribution is a power-law while the random HD distribution has a maximum and a Gaussian decay and, consequently, may be characterized by a (typical) mean connectivity. A relevant magnitude to monitor in order to compare the performance of different network topologies in associative memory tasks is the so-called *overlap* of the current network activity with a given stored activity pattern, defined as

$$m^\mu = \frac{2}{N} \sum_i (2\xi_i^\mu - 1)s_i \qquad (9.2)$$

The main graph of Figure 9.1a depicts a direct comparison of the performance of BA and random HD networks through Equation 9.2. The inset of Figure 9.1a also illustrates the corresponding degree distributions for both type of networks. The comparison shown in the figure corresponds to the case of a single stored pattern ($P = 1$) and $\eta = 3$, although similar behavior can be found for large number of patterns and other values of η. The main graph clearly shows that, excluding very low temperatures, the retrieval of information, as characterized by nonzero m^μ, is better for SF than for random HD networks. For both network topologies, the retrieval of information deteriorates as P is increased, which is a well known phenomenon widely described in the ANN literature (see for instance [20]). This fact is mainly due to the interference among stored patterns during retrieval process, which induces the appearance of mixture and spin-glass states [31]. The figure clearly depicts that, at finite temperature, the performance of the SF networks increases significantly compared with random HD networks. The reason is the positive effect that hubs have in retaining the information encoded in the patterns in the presence of noise, as has been reported in [26]. This work also reported that, even at high temperature, the mean local overlap—defined

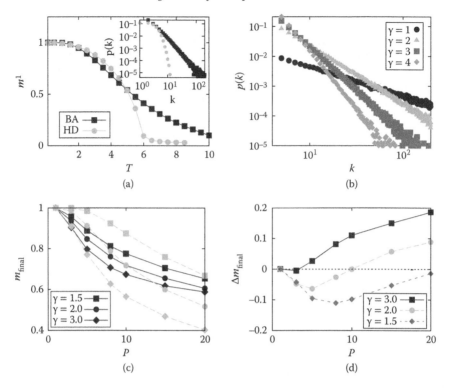

FIGURE 9.1: Performance of ANN with different complex topologies during diverse associative memory tasks. (a) The main graph depicts a comparison of the performance of a BA SF network (squares) with a random HD network (circles) during the retrieval of the information stored in a single pattern, as a function of the noise or stochasticity in the network. The inset shows the main differences between both topologies used in this study, that is, the lack of a typical mean connectivity and the existence of a relatively small number of hubs in the case of SF networks. These histograms have been obtained after an average over 100 random realizations of a network with $N = 1600$ neurons with $\eta = N_0 = 3$. (b) SF connectivity degree distributions generated with different power-law decays, namely $p(k) \sim k^{-\gamma}$, to study the effect that the parameter γ has on the performance of ANN during pattern retrieval tasks (see also panels (c) and (d)). Each data point corresponds to an average of over 100 network realizations with $N = 4000$ neurons each. (c) Zero temperature performance of random SF networks with varying topology as a function of the exponent γ (solid lines) compared with the performance of different random HD networks with the same number of synapses (dashed lines), and as a function of the number of stored patterns P. Each data point has been obtained after averaging over 50 network and pattern realizations. (d) Relative difference in the performance (see main text for an explanation) for the cases depicted in panel (c).

as the average, over neurons with the same connectivity degree k, of the local overlaps $m_i^\mu \equiv (2\xi_i^\mu - 1)s_i$—for hubs can be very close to one whereas it is very small for neurons with low connectivity degree. These findings are in agreement with the zero temperature behavior reported in [19] and can be well reproduced within simple and standard mean field theories [32].

Another important issue to investigate concerning ANN with SF topologies refers to the possible influence of the exponent of the node degree distribution on the performance of the network, for instance, during associative memory tasks. To analyze this, one can study networks characterized by power-law distributions with different exponents and compare their pattern retrieval capacity. With this end, one can generate, using standard techniques, nonevolving random networks with SF network topologies [33] with $p(k) \sim k^{-\gamma}$, where γ is a tunable parameter, as is illustrated in Figure 9.1b. The figure shows that the number of neurons in the network with different connectivity degrees increases as γ is decreased. In other words, the heterogeneity of the network increases for lower values of γ. Even more interesting is when one compares the behavior of such random SF networks with that of random HD networks with the same number of synapses, in the limit $T = 0$ (cf. Figure 9.1c). As thermal fluctuations are then suppressed, the network performance is only perturbed by the interference among the stored patterns. To consider the limit of interest, one can start with the activity of the network clamped in one of the stored patterns, and compute the state of each neuron according to the deterministic rule $s_i(t + 1) = \Theta[h_i(t)]$, where $\Theta(x)$ is the Heaviside step function, and $h_i \equiv \sum_j \omega_{ij} s_j$ is the *local field* or *synaptic current* associated with neuron i, where the sum extends to all neurons j connected to it. At the end of the system evolution, the overlap m_{final} with the starting pattern is recorded. Figure 9.1c shows the behavior of m_{final} as a function of the total number of stored patterns P at $T = 0$ for both random SF and random HD networks. Each data point corresponds to an average of over 50 networks and pattern realizations. The figure clearly illustrates that for low P, random HD networks perform memory retrieval tasks better than SF networks. As much as P increases, however, the noise induced by the interference among patterns affects the SF network performance less than that of the HD network, mainly due to the presence of hubs in the case of SF networks, which helps to maintain the activity of the network around the starting pattern. To visualize more clearly this difference in performance between SF and HD random networks it is convenient to define $\Delta m_{final} \equiv m_{final}^{SF} - m_{final}^{HD}$, and see how this magnitude varies as a function of P for different values of γ. This is illustrated in Figure 9.1d where one can see that SF networks have better and better performance as compared with the random HD networks as the number of stored patterns is increased. This effect also is enlarged as γ is increased. Both findings can be understood by considering the different decays of $p(k)$ for large k in both SF and HD networks and the fact that, for increasing values of γ, the mean connectivity degree $\langle k \rangle$ decreases for both topologies. Then, for a given number of stored patterns and due to the power-law decay of the degree distribution in the case of SF networks, there exists a nonzero number of hubs which have a high capacity to retain memory information. However, the exponential decay of the degree distribution of HD networks impedes the presence of such hubs and, therefore, its positive effect in maintaining the activity around the starting pattern, in particular when pattern interference increases.

9.3 Effect of Network Degree–Degree Correlations on Neural Network Performance

Most of the studies focused on the role of network topology in natural systems have paid much attention to investigate the effect of the degree distribution on the emergent properties of such systems. These degree distributions tend to be highly heterogeneous for most real networks and, in many cases, are approximately SF (i.e., described by power-laws) [16,17,34]. By including this topological feature in simple ANN models in the previous section we have seen that that degree heterogeneity increases the performance of the system at high levels of thermal noise and in the presence of many interfering patterns. This is due to the presence of hubs (high degree nodes) in heterogeneous SF networks which are able to retain information at levels well above the usual critical noise. Standard mean-field theories can be developed to analytically reproduce these findings well by considering the configurational ensemble of networks (the set of random networks with a given degree distribution), despite the approximation inherent to the mean-field techniques when the network is not fully connected.

Another relevant property of empirical networks to take into account also in the study of complex networks is the existence of correlations between the degrees of nodes and those of their neighbors [35–37]. If the average degree–degree correlation is positive the network is said to be *assortative*, while it is called *disassortative* if negatively correlated. Most heterogeneous networks are disassortative [17], a fact that recently has been demonstrated to be due to their equilibrium (maximum entropy) state given the constraints imposed by the degree distribution [21]. However, there are probably often mechanisms at work which drive systems from equilibrium by inducing different correlations. This feature, known as assortativity or mixing by degree, is also relevant for processes taking place on networks. For instance, assortative networks have lower percolation thresholds and are more robust to targeted attack [36,37], while disassortative ones (as in ecosystems networks) are more synchronizable [38].

Normally, the procedure used in a computational study of correlated networks is to generate a network from the configuration ensemble and then introduce correlations (positive or negative) by some stochastic rewiring process [39]. This method, however, produces results that may depend on the details of this mechanism, since there is no guarantee that one is correctly sampling the phase space of networks with given correlations. Other methods generate correlated networks using a kind of *hidden variables* from which the correlations are originated [40]. In this last case, the method may not be appropriate if one wishes to consider all possible networks with given degree–degree correlations, independently of how these may have arisen. A recent method suggested in [21] permits one to computationally generate random networks with given degree–degree correlations, without recurring to hidden variables to generate such correlations, and that are representative of the ensemble of interest (i.e., they are model independent).

Using this last method, in this section I report a recent study [41] where we studied the effect of node degree–degree correlations on the performance of auto-associative neural networks like those described in the previous section. The system under study consists of an ANN of N binary neurons, each with an activity state given by a variable $s_i = 1, 0$, representing, respectively, a firing or silent neuron. Every time step, the state of each neuron is updated according to the stochastic transition probability $P(s_i \rightarrow \sigma) = \frac{1}{2} + (\sigma - \frac{1}{2}) \tanh [2T^{-1}(h_i - \theta_i)]$, with $\sigma = 1, 0$, where, as in the previous section, the local field h_i is the combined effect on neuron i of all its neighbors, $h_i = \sum_j \omega_{ij} a_{ij} s_j$, the parameter $\theta_i \equiv \frac{1}{2} \sum_j \omega_{ij}$ is firing voltage threshold, and T is a *temperature* which controls the level of random fluctuations in the environment. Also, to take into account the topology of the network one has to introduce in h_i the element a_{ij} of the adjacency matrix which represents the number of directed edges (usually interpreted as synapses in a neural network) from node j to node i. For simplicity, we shall assume the adjacency matrix to be symmetric, so each node is characterized by a single degree $k_i = \sum_i a_{ij}$. The mean degree of the network then is $\langle k \rangle \equiv \frac{1}{N} \sum_i k_i$. Moreover and as before, this system can store P given network activity configurations (memory patterns) $\xi^\nu \equiv \{\xi_i^\nu = 1, 0, i = 1, \ldots, N\}$ by setting the interaction strengths ω_{ij} (synaptic weights) according to the Hebb rule [28] introduced previously in Equation 9.1 with $\mathcal{K} = \langle k \rangle$. In this way, each pattern becomes an attractor of the dynamics, and the system will evolve toward whichever one is closest to the initial state it is placed in.

To study node degree–degree correlations in the previously mentioned system, it is convenient to define the expected value of the mean degree of the neighbors of a given node in the network as $\bar{k}_{nn,i} = k_i^{-1} \sum_j [a_{ij}] k_j$. Here, $[\cdot]$ means an average over the ensemble of all networks that are randomly wired according to some constraints. For instance, to study networks with a given degree sequence $(k_1, \ldots, k_N)$, it is common to assume the *configuration ensemble*, where one has $[a_{ij}]^{conf} = \frac{k_i k_j}{\langle k \rangle N}$, and which is useful to study networks with a given degree distribution $p(k)$, as in the case of the BA SF evolving network [42]. It is straightforward to see that $\bar{k}_{nn,i}$ is independent of k_i for the configuration ensemble. However, as mentioned previously, real networks often display degree–degree correlations which imply $\bar{k}_{nn,i} = \bar{k}_{nn}(k_i)$. Then, if $\bar{k}_{nn}(k)$ increases with k, the network is said to be assortative, whereas it is disassortative if it decreases with k (see the inset of Figure 9.2a). Positive assortativity means that edges are more than randomly likely to occur between nodes of a similar degree.

The ensemble of all networks with a given degree sequence $(k_1, \ldots, k_N)$ contains a subset for all members of which $\bar{k}_{nn}(k)$ is constant (the configuration ensemble) but also subsets displaying other functions $\bar{k}_{nn}(k)$. One can identify each one of these subsets (regions of phase space) with an expected adjacency matrix $[a_{ij}]$ that simultaneously satisfies $\sum_j k_j [a_{ij}] = k_i \bar{k}_{nn}(k_i)$ and $\sum_j [a_{ij}] = k_i$ by choosing the *ansatz*

$$[a_{ij}] = [a_{ij}]^{conf} + \int d\nu \frac{f(\nu)}{N} \left[\frac{(k_i k_j)^\nu}{\langle k^\nu \rangle} - k_i^\nu - k_j^\nu + \langle k^\nu \rangle \right] \tag{9.3}$$

where $\nu \in \mathbb{N}$ and the function $f(\nu)$ is in general arbitrary [21]. This choice yields

$$\bar{k}_{nn}(k) = \frac{\langle k^2 \rangle}{\langle k \rangle} + \int d\nu f(\nu) \kappa_{\nu+1} \left[\frac{k^{\nu-1}}{\langle k^\nu \rangle} - \frac{1}{k} \right] \tag{9.4}$$

with $\kappa_{b+1} \equiv \langle k^{b+1} \rangle - \langle k \rangle \langle k^b \rangle$. In practice, one could find an appropriate function $f(\nu)$ to adjust Equation 9.4 to fit any given $\bar{k}_{nn}(k)$, and then build a network with the desired degree–degree correlations just generating random numbers according to Equation 9.3. This method guarantees that every network generated with this particular function $\bar{k}_{nn}(k)$ and no other ones is contained in the ensemble defined by $[a_{ij}]$ [41]. Therefore, if we are able to consider random networks drawn according to this matrix, we can be confident that we are correctly taking account of the whole ensemble of interest. In other words, one can study the effects of degree–degree correlations using the information contained on $p(k)$ and $\bar{k}_{nn}(k)$ by obtaining the associated matrix $[a_{ij}]$. However, all topological properties cannot be captured in this way, as for instance, higher-order correlations and modularity, which would require concentrating on a subpartition of those with the same $p(k)$ and $\bar{k}_{nn}(k)$.

Many empirical networks have $\bar{k}_{nn}(k) = A + Bk^\beta$, with $A, B > 0$ [11,35], the mixing being assortative if β is positive and disassortative when negative. Using Equation 9.4, such a case can be obtained, within the theoretical framework exposed previously, by choosing the bimodal function $f(\nu) = C[\frac{\kappa_2}{\kappa_{\beta+2}} \delta(\nu - \beta - 1) - \delta(\nu - 1)]$ with C a positive constant, since this choice yields

$$\bar{k}_{nn}(k) = \frac{\langle k^2 \rangle}{\langle k \rangle} + C\kappa_2 \left[\frac{k^\beta}{\langle k^{\beta+1} \rangle} - \frac{1}{\langle k \rangle} \right] \tag{9.5}$$

In the following, we shall concentrate on ensembles of networks with SF degree distributions $p(k) \sim k^{-\gamma}$ and correlations given by Equation 9.5 with $C \approx 1$, which generally implies a maximum in their entropy [21]. Then, the only parameter we need to take into account is β that then can be used as an assortativity parameter defining the type of degree–degree correlations in the ensemble. After plunging the choice for previously mentioned $f(\nu)$

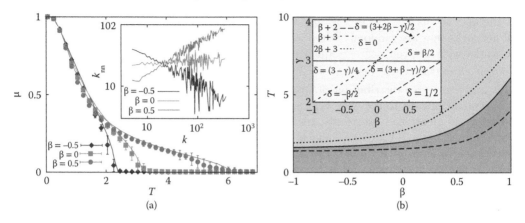

FIGURE 9.2: (a) Stable stationary value of the weighted overlap μ_1 against temperature T for large scale-free networks with degree distribution $p(k) \sim k^{-\gamma}$ with $\gamma = 2.5$ and $\langle k \rangle = 12.5$, and with correlations according to $\overline{k}_{nn} \sim k^\beta$ (see inset) for $\beta = -0.5$ (disassortative), 0.0 (neutral), and 0.5 (assortative). Symbols correspond to MC simulations of the system with $N = 10^4$ neurons, with error bars representing standard deviations, and solid lines corresponding to steady-state solutions computed from the map, Equation 9.7. (b) The main graph depicts the theoretically computed phase diagram (T, β) of a network with scale-free topology and power-law distributed degree–degree correlations. The critical temperature for associative memory T_C has been plotted as a function of the assortativity parameter β for $\gamma = 2.5$ (dotted line), 3 (solid line), and 3.5 (dashed line), respectively. Each critical line separates a memory phase at relatively low temperature (light gray area here shown for $\gamma = 3$) from a nonmemory phase at relatively high temperature (here the dark gray region depicted for $\gamma = 3$). The inset shows the regions in the parameter space (γ, β) where the critical temperature has different asymptotic behavior for very large N in the form of $T_C \sim N^\delta$.

into Equation 9.3 one finds that the ensemble of networks exhibiting correlations given by Equation 9.5 with $C = 1$ is defined by the mean adjacency matrix

$$\begin{aligned}[a_{ij}] &= \tfrac{1}{N}\left[k_i + k_j - \langle k \rangle\right] \\ &+ \frac{\kappa_2}{\kappa_{\beta+2}}\frac{1}{N}\left[\frac{(k_i k_j)^{\beta+1}}{\langle k^{\beta+1}\rangle} - k_i^{\beta+1} - k_j^{\beta+1} + \langle k^{\beta+1}\rangle\right]\end{aligned} \tag{9.6}$$

Equation 9.6 allows us to investigate the effect of degree–degree correlations in an auto-associative neural network under a mean-field approach by substituting the adjacency matrix a_{ij} appearing in the local field h_i for its expected value $[a_{ij}]$, and the neuron states for their expected values according to the transition probabilities $P(s_i \to \sigma)$, that is, $s_i \to P(s_i \to 1) = \tfrac{1}{2} + \tfrac{1}{2}\tanh[2T^{-1}(h_i - \theta_i)]$. The result for a single stored pattern ($P = 1$, $\xi_i \equiv \xi_i^1$, $i = 1, \ldots, N$) (see details in [41]) is a 3D map of closed coupled equations for the macroscopic overlap observables μ_0, μ_1, and $\mu_{\beta+1}$, with $\mu_\alpha \equiv \langle k_i^\alpha (2\xi_i - 1)(2s_i - 1)/\langle k^\alpha \rangle$, which describes—within this mean-field approximation—the dynamics of the system as follows:

$$\mu_0(t+1) = \int p(k)\tanh[G(t)/(\langle k \rangle T)]dk$$

$$\mu_1(t+1) = \frac{1}{\langle k \rangle}\int p(k)k\tanh[G(t)/(\langle k \rangle T)]dk \tag{9.7}$$

$$\mu_{\beta+1}(t+1) = \frac{1}{\langle k^{\beta+1}\rangle}\int p(k)k^{\beta+1}\tanh[G(t)/(\langle k \rangle T)]dk$$

with $G[t] = k\mu_0(t) + \langle k \rangle [\mu_1(t) - \mu_0(t)] + (\kappa_2/\kappa_{\beta+2})(k^{\beta+1} - \langle k^{\beta+1} \rangle)[\mu_{\beta+1}(t) - \mu_0(t)]$. These *order parameters* measure the extent to which the system is able to recall information in the presence of thermal fluctuations. For the first order one has $\mu_0 = \langle (2\xi_i - 1)(2s_i - 1) \rangle$ which is similar to the overlap function introduced previously in Equation 9.2 for random unbiased patterns, and is a standard measure in neural networks to visualize network performance during memory recall tasks. On the other hand, μ_1 weighs the local overlap in each neuron with its degree, and therefore it measures information per synapse instead of per neuron. This fact implies that μ_1 is more closely related to the empirical measure of neural activity in neurological experiments, since the total electric potential in an area of tissue is likely to depend on the number of synapses transmitting action potentials. In any case, a comparison between the two order parameters is a good way of assessing to what extent the performance of neurons depends on their degree since neurons with larger degree can, in general, store more efficiently information at higher temperatures [26].

The map, Equation 9.7, can be easily computed for any degree distribution $p(k)$. For $\beta = 0$ (the neutral or uncorrelated case) the system collapses to the 2D map obtained in [26], while it becomes the typical 1D case for a homogeneous $p(k)$, say, a fully connected network [43]. In the limit $t \rightarrow \infty$ one can easily study the steady-state solutions of the map and their stability as a function of the degree distribution $p(k)$, the assortativity parameter β, and the temperature T. Figure 9.2a illustrates, for $P = 1$, the behavior of the stationary weighted overlap μ_1 as a function of the temperature for disassortative (with $\beta = -0.5$), neutral ($\beta = 0$), and assortative (with $\beta = 0.5$) SF networks. In all cases, there is a critical temperature T_C, at which the system will undergo the characteristic second-order phase transition from a phase in which it exhibits memory to one in which it does not. This critical temperature can be analytically derived from the map, Equation 9.7, in the stationary state (see details in [41]) which is the solution of a third-order polynomial equation with coefficients being functions of different moments of $p(k)$ and, therefore, will depend on both γ and β. For the neutral (uncorrelated) case one has $T_C = \langle k^2 \rangle / \langle k \rangle^2$ as expected [44]. The figure also shows that these theoretical findings agree with weighted overlap obtained from Monte Carlo (MC) simulations of SF neural networks (data points in the main figure) generated from the ensemble of networks exhibiting correlations given by Equation 9.5 and characterized by the mean adjacency matrix, Equation 9.6. The figure shows that the system behaves very similar at low temperatures for different types of degree–degree correlations, but as T increases its behavior begins to differ substantially for different values of β. Thus, the figure depicts that the disassortative network is the least robust to noise. However, the assortative one is capable of retaining some information at temperatures considerably higher than the critical value, T_C, of neural networks. This effect is mainly due to the ability of the high-degree nodes to retain information in the presence of noise. Since in an assortative network a subgraph of hubs will have more edges than in a disassortative one, it has a higher effective critical temperature. Therefore, even when most of the nodes are acting randomly, the set of nodes of sufficiently high degree nevertheless displays associative memory.

The typical phase diagram (T, β) of the previously mentioned system is depicted in the main graph of Figure 9.2b. This shows the behavior of the analytically computed critical temperature T_C (see [41] for details), as a function of the assortativity parameter β and for three increasing values of the exponent γ characterizing the SF features of the network degree distribution (dotted, solid, and dashed curves depicted, respectively from top to bottom). Each line separates a phase in which the system is able to recall the stored pattern (light gray area here drawn for $\gamma = 3$) from a nonmemory phase (dark gray region also drawn for $\gamma = 3$). This shows that T_C for associative memory in SF networks grows with degree heterogeneity (decreasing γ) and increases very significantly with positive degree–degree correlations (increasing β).

As already seen, T_C for associative memory depends on γ and β. Moreover, for large values of N, it can be shown that the critical temperature scales as $T_C \sim N^\delta$ with $\delta = \delta(\gamma, \beta)$ [41]. In fact, the theoretical computed T_C depends on several moments of $p(k)$ which can have different asymptotic behavior for large N depending on the values of γ and β. This implies that the scaling exponent δ differs from one region to another in the space of these parameters as is depicted in inset of Figure 9.2b. For more heterogeneous networks with $\gamma < 3$, from lowest to highest assortativity, one has scaling exponents which are dependent on only γ, only β, both γ and β, and perhaps most interestingly, neither of the two parameters with T_C scaling, in the latter case, as $N^{1/2}$. For more homogeneous networks, that is, $\gamma > 3$, divergence of the critical temperature with N appears despite the fact that the second moment of $p(k)$ is finite, simply as a result of assortativity.

When one stores a large number of patterns in the synaptic weights, the behavior and performance of the correlated networks do not qualitatively differ from that observed for the single-pattern situation so that it can be confirmed that the main conclusions explained previously concerning the influence of assortativity in auto-associative neural networks for $P = 1$ are robust and generalizable to a larger number of stored patterns.

9.4   Evolving Critical Networks

In the previous sections, we have explored the emergent properties of large neural networks with complex topologies, for example, SF regardless of the dynamic processes involved in the generation and evolution of such networks. This approach is based on the fact that, sometimes, the mere consideration of a network consisting of nodes and links between them is enough to capture the essence of cooperation in complex systems. As we have seen, the study of such networks is usually made simpler by considering statistical properties, for example, the degree distribution $p(k)$ or some correlation functions as the node degree–degree correlations. However, very few studies have focused on the study of the potential consequences that different types of network evolution could have on the features of the final structure of the network. Another important related aspect to consider here is that different types of network evolution could determine the computational properties of networked dynamical systems built on such networks, not only in their stationary states but also during evolution or development of the network.

Prominent examples of such systems are neural media. In the brain, for instance, the underlying network topology can change at different time scales, as for instance, during brain development, but also at shorter time scales as a consequence of learning, which induces certain synapse strengthening and weakening of other connections. Other topology changes on the brain structure can occur as a consequence of some diseases, as for instance Alzheimer and Parkinson diseases. In general, all these brain changes could explain the existence of different cognitive abilities at different stages of brain development during life.

In this section, I illustrate a general theoretical framework useful to investigate all these aspects of network evolution and their possible computational implications. As we will see, it has a direct application to studying the development of neural systems and synaptic pruning.

9.4.1   Nonlinear rewiring

As we already mentioned, one of the pioneering works that presented a simple procedure for network evolution was due to BA [12]. This method was based on a preferential attachment probabilistic rule which links a new node to the existing ones with a probability which is linear in their degree, and which leads asymptotically to a network with a SF degree distribution. Preferential attachment seems to be behind the emergence of many real-world, continuously growing networks. Although many different works based on this simple model have been reported in the past (for a review see [16]) even considering nonlinear probabilities [45], one has to assume at the end a linear preferential attachment in order to obtain a final power-law degree distribution.

Despite the importance of the preferential attachment rule to explain the structure of many natural networks, not all networks in which some nodes at times gain (or lose) new edges have a continuously growing number of nodes. The brain network may be a relevant example. Once formed, the number of neurons does not seem to continually augment, and yet its structural topology is dynamic [46]. Synaptic growth and dendritic arborization have been also shown to increase with electric stimulation [47,48]—and, in general, the more connected a neuron is, the more current it receives from the sum of its neighbors. It is reasonable to think that this dependence on the neuron connectivity degree must be nonlinear since both neuron activity and synaptic transmission are affected by different nonlinear processes including, for instance, a firing threshold mechanism and activity dependent synaptic modifications at short and long time scales which can have strong computational implications during brain development. Within this context, it has been recently shown how topological phase transitions and SF solutions could emerge in the case of nonlinear rewiring in fixed-size networks [49]. In this work, an evolving network model with preferential rewiring according to nonlinear (power-law) probabilities has been studied. In addition, it has been assumed that the number of nodes and edges is conserved but the topology evolves, arriving eventually at a macroscopically (nonequilibrium) stationary state—as described by global properties such as the degree distribution. Depending on the exponents chosen for the rewiring probabilities, the final state can be either fairly homogeneous, with a typical mean connectivity degree, or highly heterogeneous, with the emergence of star-like structures. In the critical case marking the transition between these two regimes, the degree distribution is shown to follow a nonlinear diffusion equation. This describes a tendency toward stationary states that are characterized either by SF or by mixed SF distributions, depending on parameters.

The starting point of the proposed model is a random network with N nodes with degrees k_i, $i = 1, \ldots N$, and $N\langle k\rangle/2$ links. At each time step a node i is chosen with a probability $\rho(k_i)$ and at random one of its edges is removed, which is again reconnected to another node j chosen according to a probability $\pi(k_j)$. That is, an edge is broken and another one is created, and the total number of edges, as well as the total number of nodes, are conserved. The functions $\pi(k)$ and $\rho(k)$ are arbitrary, but for simplicity we assume a power-law dependency, namely $\pi(k) \sim k^\alpha$ and $\rho(k) \sim k^\beta$, that capture the essence of a wide class of nonlinear monotonous response functions and are easy to handle analytically. Probabilities π and ρ can be interpreted then as transition probabilities between node connectivity states. The probability to have a given connectivity degree at a time t then evolves following the master equation:

$$
\begin{aligned}
\frac{dp(k,t)}{dt} &= \frac{(k-1)^\alpha}{\overline{k}_\alpha} p(k-1,t) + \frac{(k+1)^\beta}{\overline{k}_\beta} p(k+1,t) \\
&\quad - \left[\frac{k^\alpha}{\overline{k}_\alpha} + \frac{k^\beta}{\overline{k}_\beta} \right] p(k,t)
\end{aligned}
\tag{9.8}
$$

where $\bar{k}_a = \bar{k}_a(t) \equiv \sum_k k^a p(k,t)$. One can easily verify that any stationary solution of Equation 9.8 must satisfy the condition $p_{st}(k+1)(k+1)^\beta \bar{k}_\alpha^{st} = p_{st}(k)k^\alpha \bar{k}_\beta^{st}$, which, for $k \gg 1$, implies

$$\frac{\partial p_{st}(k)}{\partial k} = \left(\frac{\bar{k}_\beta^{st}}{\bar{k}_\alpha^{st}} \frac{k^\alpha}{(k+1)^\beta} - 1 \right) p_{st}(k) \tag{9.9}$$

Therefore, the stationary degree distribution $p_{st}(k)$ will have an extremum at $k_e = \left(\bar{k}_\alpha^{st} / \bar{k}_\beta^{st} \right)^{1/(\alpha-\beta)}$ (where we have approached $k_e \approx k_e+1$). If $\alpha < \beta$, this will be a maximum, signaling the peak of the distribution. On the other hand, if $\alpha > \beta$, k_e will correspond to a minimum. Therefore, most of the distribution will be broken in two parts, one for $k < k_e$ and another for $k > k_e$. The critical case for $\alpha = \beta$ will correspond to a monotonously decreasing stationary distribution, but such that $\lim_{k\to\infty} \partial p_{st}(k)/\partial k = 0$, as can be easily derived from Equation 9.9. In fact, Equation 9.8 for $\alpha = \beta$ is the discrete version of a nonlinear diffusion equation,

$$\frac{\partial p(k,\tau)}{\partial \tau} = \frac{\partial^2}{\partial k^2} [k^\alpha p(k,\tau)]$$

where the time variable has been rescaled to $\tau = t/\bar{k}_\alpha(t)$. If no border effects are considered, the solutions of this equation are of the form $p_{st}(k) \sim Ak^{-\alpha} + Bk^{-\alpha+1}$ with A and B constants. For $\alpha > 2$ and given A, one can always find B such that $p_{st}(k)$ is normalized in the thermodynamic limit. For instance, if one assumes the lower possible degree value $k \geq 1$, one has $B = (\alpha-2)[1 - A/(\alpha-1)]$. However, if $1 < \alpha \leq 2$, then only A can remain nonzero, and $p_{st}(k)$ will be a pure power-law. For $\alpha \leq 1$ both constants must tend to zero as $N \to \infty$, so no normalized stationary solution can be found.

Starting from an initial degree probability distribution $p(k,0) = \delta(k - \langle k \rangle)$, Figure 9.3a illustrates the final distribution obtained after simulation of the master Equation 9.8 (symbols) during $t = 10^5$ Monte Carlo steps (MCS), for $k \geq 1$, $\beta = 1$, and three different values of α. The figure also depicts the resulting degree distributions obtained by direct numerical integration of such equation (solid lines). For $\alpha < \beta$, or subcritical regime, the network will evolve to have a characteristic connectivity degree, centered around $\langle k \rangle$. At the critical case $\alpha = \beta$, all connectivity degrees appear, according either to a pure or a composite power-law, as detailed previously. Finally, for $\alpha > \beta$—the supercritical situation—a quasi-bimodal degree distribution starts to develop characterized by star-like structures with a great many nodes connected to just a few hubs.

The previously mentioned results show that the grade of heterogeneity of the stationary distribution obtained is seen to depend crucially on the relation between the exponents modeling the probabilities a node has of obtaining or losing a new edge. Moreover, the study presented here illustrates how SF distributions, with a range of exponents, may emerge for nonlinear rewiring, although only in the critical situation in which the probabilities of gaining or loosing edges are the same.

9.4.2   Synaptic pruning

The formalism for network evolution, a consequence of nonlinear rewiring, that we have described previously can be extended to a more general scenario [51] making only minimal assumptions about the attachment/detachment probabilistic rules. For instance, taking into account some biological motivation related with neuroscience, one can assume that these probabilities factorize into two parts: a local term which depends on node degree, and a global term, which is a function of the mean degree of the network. The assumption of

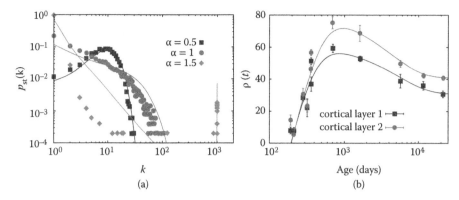

FIGURE 9.3: Evolving critical networks and emergence of synaptic pruning. (a) Degree distributions $p(k, t)$ obtained after $t = 10^5$ MCS of evolution starting from an initial degree distribution with all nodes with the same degree $k_i = \langle k \rangle = 10$, for subcritical ($\alpha = 0.5$), critical ($\alpha = 1$), and supercritical ($\alpha = 1.5$) rewiring exponents. Symbols from MC simulations and corresponding solid lines from numerical integration of Equation 9.8. Other parameters were $\beta = 1$ and $N = 1000$ in all cases. (b) Synaptic densities in layers 1 (squares) and 2 (circles) of the human auditory cortex against time from conception. Data from [50], obtained by directly counting synapses in tissues from autopsies. Lines follow the best fits to Equation 9.13 where the parameters were: for layer 1, $\tau_g = 151.0$ and $\tau_P = 5221$; and for layer 2, $\tau_g = 191.1$ and $\tau_P = 4184$, all in days (we have approximated t_0 to the time of the first data points, 192 days).

a local term depending on the node degree is based (as we already mentioned) on the experimental evidence that enhanced neuron electric activity induces synaptic growth, and such activity depends on the net current received from its neighbors, which is higher the more neighbors the neuron has. On the other hand, synaptic growth and death also depend on concentrations of various molecules, which can diffuse through large areas of tissue and therefore cannot in general be considered local. A feature of brain development in many animals is *synaptic pruning*—the large reduction in synaptic density undergone throughout infancy. Moreover, it is well known that an elimination of less needed synapses can reduce the energy consumed by the brain while maintaining near optimal memory performance [52,53]. It is reasonable to assume, therefore, attachment/detachment rules depending on the mean degree of the network—or mean synaptic density —to control energy consumption in the system.

In [51], network evolution and its application to synaptic pruning within this general framework has been studied using only topological information for the attachment/detachment rules. The proposed formalism, however, can be extended to any situation in which the dynamics of the elements at the nodes can be functionally related to degrees. The study illustrates how appropriate choices of functions for local and global attachment/detachment rules induce the system to evolve toward heterogeneous (sometimes SF) networks while undergoing synaptic pruning in quantitative agreement with experiments (cf. Figure 9.3b).

As before, let us consider a network of N nodes where initially their degrees follow some distribution $p(k, 0)$ with mean $\kappa \equiv \langle k \rangle$. We wish to study the evolution of networks in which nodes can gain or lose edges according to stochastic rules which only take into account local and global information on degrees. It is reasonable to assume then that at every time step, each node i has a probability of gaining a new edge, P_i^{gain}, to a random node; and a probability of losing a randomly chosen edge, P_i^{lose}. Under the general assumptions made previously, one can consider that these probabilities factorize as $P_i^{gain} = u(\kappa)\pi(k_i)$ and

$P_i^{lose} = d(\kappa)\sigma(k_i)$, where u, d, π, and σ can be arbitrary functions, but impose nothing else other than normalization. For each edge that is withdrawn from the network, two nodes decrease in degree: i, chosen according to $\sigma(k_i)$, and j, a random neighbor of i's; so there is an added effective probability of loss $k_i/(\kappa N)$. Similarly, for each edge placed in the network, not only l chosen according to $\pi(k_l)$ increases its degree; a random node m will also gain, with the consequent effective probability N^{-1}. Then, introducing the notation $\tilde{\pi}(k) \equiv \pi(k) + N^{-1}$ and $\tilde{\sigma}(k) = \sigma(k) + k/(\kappa N)$ the network evolution can be seen (as in the case of nonlinear rewiring described previously) as a stochastic *one step process* [54] with transition rates $u(\kappa)\tilde{\pi}(k)$ and $d(\kappa)\tilde{\sigma}(k)$ defined by a master equation for the degree distribution:

$$\frac{dp(k,t)}{dt} = u(\kappa)\tilde{\pi}(k-1)p(k-1,t) + d(\kappa)\tilde{\sigma}(k+1)p(k+1,t) \\ -[u(\kappa)\tilde{\pi}(k) + d(\kappa)\tilde{\sigma}(k)]p(k,t) \tag{9.10}$$

Assuming now that $p(k,t)$ evolves toward a stationary distribution, namely $p_{st}(k)$, it is straightforward to see that $p_{st}(k)$ then satisfies

$$\frac{\partial p_{st}(k)}{\partial k} = \left[\frac{u(\kappa_{st})\tilde{\pi}(k)}{d(\kappa_{st})\tilde{\sigma}(k+1)} - 1\right] p_{st}(k) \tag{9.11}$$

where $\kappa_{st} = \langle k \rangle_{st} = \sum_k k p_{st}(k)$. Note that Equation 9.11 is similar to Equation 9.9 and, as in this case, $p_{st}(k)$ has an extremum at the k_e solution of the equation $\tilde{\pi}(k_e) = \tilde{\sigma}(k_e+1)$. To obtain this last condition, one has to use $u(\kappa_{st}) = d(\kappa_{st})$ since the normalization of $u(\kappa)$ and $d(\kappa)$ implies that at each time step exactly $u(\kappa)$ nodes are chosen to gain edges and $d(\kappa)$ to lose them, and in the stationary state the total number of edges is conserved. A maximum in $p_{st}(k)$ implies a relatively homogeneous distribution, while a minimum means $p_{st}(k)$ will be split in two, and therefore highly heterogeneous. More intuitively, if for nodes with large enough k there is a higher probability of gaining edges than of losing them, the degrees of these nodes will grow indefinitely, leading to heterogeneity. If, on the other hand, highly connected nodes always lose more edges than they gain, one will obtain quite homogeneous networks. From this reasoning, a particularly interesting case, which turns out to be *critical*, is that in which $\pi(k)$ and $\sigma(k)$ are such that $\tilde{\pi}(k) = \tilde{\sigma}(k) \equiv v(k) \forall k$. Using this last condition in Equation 9.11 means that for large k, $\partial p_{st}(k)/\partial k \to 0$, and $p_{st}(k)$ flattens out—as for example a power-law does. Also at this critical condition, the corresponding Fokker–Planck equation (see for instance [54]) for the stochastic process defined by Equation 9.10 becomes

$$\frac{\partial p(k,t)}{\partial t} = \frac{1}{2}\left[u(\kappa_{st}) + d(\kappa_{st})\right]\frac{\partial^2}{\partial k^2}\left[v(k)p(k,t)\right] \\ - \left[u(\kappa_{st}) - d(\kappa_{st})\right]\frac{\partial}{\partial k}\left[v(k)p(k,t)\right]$$

Then, the stationary solution must satisfy, on the one hand, $v(k)p_{st}(k) = Ak + B$, so that the diffusion is stationary and, on the other, $u(\kappa_{st}) = d(\kappa_{st})$, to cancel out the drift. For this situation to be reachable from any initial condition, $u(\kappa)$ and $d(\kappa)$ must be monotonous functions, decreasing and increasing, respectively.

A straightforward application of the theory developed previously is synaptic pruning. As a simple example, we will first consider global probabilities which have the linear forms

$$u(\kappa) = \frac{n}{N}\left(1 - \frac{\kappa}{\kappa_{max}}\right) \quad \text{and} \quad d(\kappa) = \frac{n}{N}\frac{\kappa}{\kappa_{max}}$$

where n is the expected value of the number of additions and deletions of edges per time step, and κ_{max} is the maximum value the mean degree can have. This choice describes a situation in which the higher the density of synapses, the less likely new ones are to sprout and the more likely existing ones are to atrophy—a situation that might arise, for instance, in the presence of a finite quantity of nutrients. From the definition of $\kappa(t)$, using the master Equation 9.10 and taking into account that π and σ are normalized to one, one easily finds that

$$\frac{d\kappa(t)}{dt} = u[\kappa(t)] - d[\kappa(t)] = \frac{n}{N}\left[1 - 2\frac{\kappa}{\kappa_{max}}\right]$$

which is independent of the local probabilities. Therefore, the mean degree will increase or decrease exponentially with time, from $\kappa(0)$ to $\frac{1}{2}\kappa_{max}$. Assuming that the initial condition is, say, $\kappa(0) = \kappa_{max}$, and expressing the solution in terms of the mean synaptic density—that is,, $\rho(t) \equiv \kappa(t)N/(2V)$, with V the total volume considered—one has

$$\rho(t) = \rho_f \left(1 + e^{-t/\tau_P}\right) \tag{9.12}$$

where $\rho_f \equiv \rho(t \to \infty)$ and the time constant for pruning is $\tau_P = \rho_f V/n$. Equation 9.12 gives a decay in time of density of synapses starting from some given initial density $\rho(0) = \kappa_{max}N/(2V)$, phenomenon which can be related with synaptic pruning. In addition, one can assume a previous initial overgrowth of synapses due, for instance, to the transient existence of some kind of growth factors affecting axon development. To account for this fact, it is enough to include a nonlinear, time-dependent term $g(t) \equiv a\exp(-t/\tau_g)$ in the global probability of growth, that is, $u(\kappa) = (n/N)[1 - \kappa(t)/\kappa_{max} + g(t)]$, leaving $d[\kappa(t)]$ as before. With these choices one has

$$\rho(t) = \rho_f \left[1 + e^{-t/\tau_P} - (1 + e^{-t_0})e^{-\frac{t-t_0}{\tau_g}}\right] \tag{9.13}$$

where t_0 is the time at which synapses begin to form ($t = 0$ corresponds to the moment of conception) and τ_g is the time constant related to growth. Equation 9.13 was fitted in Figure 9.3b to experimental data on layers 1 and 2 of the human auditory cortex obtained during autopsies [50]. Since the contour conditions ρ_f and t_0 are simply taken as the value of the last experimental data point and the time of the first one, respectively, the time constants τ_P and τ_g are the only parameters needed for the fit.

9.5   Conclusions

In this chapter, I presented recent results concerning the relevance that several features of the underlying topology of simple neural networks can have in its emergent behavior, and that can determine their computational properties. Also, I described how these topological features, including scale invariance, can emerge naturally using a simple stochastic model for network growth and development. This theoretical framework allows us to investigate the basic mechanisms that induce the development of networks with given topological structure and correlations.

I reported first how the topology of a neural network characterized in terms of the degree distribution has a key role in the processes of memorization and retrieval of patterns. In particular, neural networks with SF topology may exhibit better performance than Hopfield-like networks with the same number of synapses distributed randomly over the network. These results can be useful to understand the role of regions with different connectivity degrees in actual neural systems during memorization and retrieval of information. In particular, they may improve our understanding of how fluctuations or perturbations in the typical number of synapses of some brain areas can affect the processing of information and memorization in these regions. Furthermore, these findings suggest the convenience of developing new methods to store the more relevant information into the hubs, increasing in this way the effective network performance and efficiency.

Second, I explained a recent work that has investigated the effect that node degree–degree correlations have on the performance of auto-associative neural networks. In this chapter, it has been shown that assortative networks of simple model neurons are able to exhibit associative memory in the presence of levels of noise that uncorrelated and disassortative networks cannot match. This finding seems to be in contradiction with a recent result that shows that disassortative networks increase synchronization of a set of coupled oscillators [38]. The contradiction is only apparent since the latter is expected to occur at a low level of noise—since noise tries to desynchronize oscillators or destabilize memories. In fact, at a high level of noise, the recall of memories is only possible with the ability of the high-degree nodes to retain information. This is more likely to occur in an assortative network where a subgraph of hubs will have more edges than in a disassortative one and, therefore, it will retain the memory information better in the presence of noise.

Finally, I reported a study that investigated the consequences that the dynamics that characterizes the evolution of networks by means of simple stochastic rules has on the final stationary states of such evolution. I first introduced a network model that evolves while leaving the total numbers of nodes and edges roughly constant by means of a nonlinear rewiring. The grade of heterogeneity of the stationary degree distribution obtained is seen to depend crucially on the relation between the exponents modeling the probabilities a node has of obtaining or losing a new edge. The previously mentioned model shows how SF distributions, with a range of exponents, may emerge for nonlinear rewiring, although only in the critical situation in which the probabilities of gaining or losing edges are the same. It could be that this nontrivial relation between the microscopic rewiring actions and the emergent macroscopic degree distributions could shed light on a class of biological, social, and communications networks.

Also, I described a more general scenario of stochastic network evolution in which any choice of transition probabilities which depend on local and/or global degrees can be treated analytically, thereby obtaining some model-independent results. I also reported how such a framework can be applied to realistic biological scenarios, as for instance, synaptic pruning in humans, for which the use of nonlinear global probabilities reproduces the initial increase and subsequent depletion in synaptic density in good agreement with experiments. Furthermore, the model can be applied to understand the topological structure of the neural network of simple living organisms such as the *Caenorhabditis elegans* [51]. These examples indicate that it is not far-fetched to contemplate how many structural features of the brain or other networks—and not just the degree distributions—could arise by simple stochastic rules like the ones considered here.

The results presented in this chapter can be very useful for computational network theory, since they can be applied to design new paradigms of biologically-inspired neural networks which are optimal for certain desirable computational effects. For instance, the general theory introduced in Section 9.4 allows testing of which are the most appropriate microscopic dynamics controlling network evolution to obtain topologies with given computational characteristics. In addition, the analytical methods considered in this chapter are regularly used to study systems with a large number of degrees of freedom, whose macroscopic behavior can only be understood when one assumes the thermodynamic limit ($N \to \infty$). This fact implies that those methods are also perfectly useful for analysis of "Big Data" problems.

References

1. The Human Brain Project. See for details: https://www.humanbrainproject.eu/.

2. The Brain Initiative. See for more details: http://braininitiative.nih.gov/.

3. The Human Connectome Project. See for details: http://www.humanconnectomeproject.org/.

4. The Open Worm Project. See also for details: http://www.openworm.org/.

5. E. R. Kandel, J. H. Schwartz, and T. M. Hessel. *Principles of Neural Science.* NY: McGraw-Hill, 2000.

6. H-J. Park and K. Friston. Structural and functional brain networks: From connections to cognition. *Science*, 342(6158):1238411, 2013.

7. D. J. Watts and S. H. Strogatz. Collective dynamics of "small-world" networks. *Nature*, 393:440–442, 1998.

8. J. Shudong and A. Bestavros. Small-world characteristics of internet topologies and implications on multicast scaling. *Computer Networks*, 50(5):648–666, 2006.

9. V. van Noort, B. Snel, and M. A. Huynen. The yeast coexpression network has a small-world, scale-free architecture and can be explained by a simple model. *EMBO Rep.*, 5(3):280–284, 2004.

10. O. Shefi, I. Golding, R. Segev, E. Ben-Jacob, and A. Ayali. Morphological characterization of *in vitro* neuronal networks. *Phys. Rev. E*, 66:021905, 2002.

11. D. de Santos-Sierra, I. Sendiña-Nadal, I. Leyva, J. A. Almendral, S. Anava, A. Ayali, D. Papo, and S. Boccaletti. Emergence of small-world anatomical networks in self-organizing clustered neuronal cultures. *PLoS ONE*, 9(1):e85828, 2014.

12. A. L. Barabási and R. Albert. Emergence of scaling in random networks. *Science*, 286(5439):509–512, 1999.

13. H. Ebel, L.I. Mielsch, and S. Bornholdt. Scale-free topology of e-mail networks. *Phys. Rev. E*, 66:035103, 2002.

14. S. Redner. How popular is your paper? An empirical study of the citation distribution. *Eur. Phys. J. B*, 4(2):131–134, 1998.

15. S. Maslov and K. Sneppen. Specificity and stability in topology of protein networks. *Science*, 296(5569):910–913, 2002.

16. S. Boccaletti, V. Latora, Y. Moreno, M. Chavez, and D.-U. Hwang. Complex networks: Structure and dynamics. *Phys. Rep.*, 424:175–308, February 2006.

17. M. E. J. Newman. The structure and function of complex networks. *SIAM REVIEW*, 45:167–256, 2003.

18. D. Stauffer, A. Aharony, and J. Adler L. da Fontoura-Costa. Efficient Hopfield pattern recognition on a scale-free neural network. *Eur. Phys. J. B*, 32(3):395–399, 2003.

19. P.N. McGraw and M. Menzinger. Topology and computational performance of attractor neural networks. *Phys. Rev. E*, 68:047102, 2003.

20. P. Peretto. *An Introduction to the Modeling of Neural Networks*. Cambridge, UK: Cambridge University Press, 1992.

21. S. Johnson, J. J. Torres, J. Marro, and M. A. Muñoz. Entropic origin of disassortativity in complex networks. *Phys. Rev. Lett.*, 104:108702, 2010.

22. L. A. N. Amaral, A. Scala, M. Barthélémy, and H. E. Stanley. Classes of small-world networks. *Proc. Natl. Acad. Sci. USA*, 97(21):11149, 2000.

23. O. Sporns, D. R. Chialvo, M. Kaiser, and C. C. Hilgetag. Organization, development and function of complex brain networks. *Trends Cogn. Sci.*, 8:418–425, 2004.

24. V. M. Eguíluz, D. R. Chialvo, G. A. Cecchi, M. Baliki, and A. V. Apkarian. Scale-free brain functional networks. *Phys. Rev. Lett.*, 94:018102, 2005.

25. E. Bullmore and O. Sporns. Complex brain networks: Graph theoretical analysis of structural and functional systems. *Nat. Rev. Neurosci.*, 10:186–198, 2009.

26. J. J. Torres, M. A. Muñoz, J. Marro, and P. L. Garrido. Influence of topology on the performance of a neural network. *Neurocomputing*, 58–60:229–234, 2004.

27. M. Barahona and L. M. Pecora. Synchronization in small-world systems. *Phys. Rev. Lett.*, 89:054101, 2002.

28. D. O. Hebb. *The Organization of Behavior: A Neuropsychological Theory*. New York: Wiley, 1949.

29. O. Paulsen and T. J. Sejnowski. Natural patterns of activity and long-term synaptic plasticity. *Curr. Opin. Neurobiol.*, 10:172–179, 2000.

30. A. Gruart, M. D.Muñoz, and J. M. Delgado García. Involvement of the ca3-ca1 synapse in the acquisition of associative learning in behaving mice. *J. Neurosci.*, 26:1077–1087, 2006.

31. D. J. Amit. *Modeling Brain Function: The World of Attractor Neural Network*. Cambridge, UK: Cambridge University Press, 1989.

32. J. J. Torres, M. Muñoz, J. Marro, and P. L. Garrido. *Mean-field theory of attractor neural networks on a scale-free topology*. Internal report, Granada, Spain: University of Granada, 2004.

33. M. Molloy and B. Reed. A critical-point for random graphs with a given degree sequence. *Random Struct. Algorithms*, 6(2-3):161–179, 1995.

34. A. L. Barabási and Z. N. Oltvai. Network biology: Understanding the cell's functional organization. *Nat. Rev. Genet.*, 5:101–113, 2004.

35. R. Pastor-Satorras, A. Vázquez, and A. Vespignani. Dynamical and correlation properties of the Internet. *Phys. Rev. Lett.*, 87:258701, 2001.

36. M. E. J. Newman. Assortative mixing in networks. *Phys. Rev. Lett.*, 89:208701, 2002.

37. M.E. J. Newman. Mixing patterns in networks. *Phys. Rev. E*, 67:026126, 2003.

38. M. Brede and S. Sinha. Assortative mixing by degree makes a network more unstable. *arXiv:cond-mat/0507710*, 2005.

39. S. Maslov, K. Sneppen, and A. Zaliznyak. Detection of topological patterns in complex networks: Correlation profile of the Internet. *Physica A*, 333:529–540, 2004.

40. M. Boguñá and R. Pastor-Satorras. Class of correlated random networks with hidden variables. *Phys. Rev. E*, 68(3):036112, 2003.

41. S. de Franciscis, S. Johnson, and J. J. Torres. Enhancing neural-network performance via assortativity. *Phys. Rev. E*, 83(3):036114, 2011.

42. G. Bianconi. Mean field solution of the Ising model on a Barabási–Albert network. *Phys. Lett. A*, 303(2):166–168, 2002.

43. J. J. Hopfield. Neural networks and physical systems with emergent collective computational abilities. *Proc. Natl. Acad. Sci. USA*, 79:2554–2558, 1982.

44. S. Johnson, J. Marro, and J. J. Torres. Functional optimization in complex excitable networks. *EPL*, 83(4):46006, 2008.

45. P. L. Krapivsky, S. Redner, and F. Leyvraz. Connectivity of growing random networks. *Phys. Rev. Lett.*, 85:4629–4632, 2000.

46. A. Y. Klintsova and W. T. Greenough. Synaptic plasticity in cortical systems. *Curr. Opin. Neurobiol.*, 9(2):203–208, 1999.

47. K.S. Lee, F. Schottler, M. Oliver, and G. Lynch. Brief bursts of high-frequency stimulation produce two types of structural change in rat hippocampus. *J. Neurophysiol.*, 44:247–258, 1980.

48. M. De Roo, P. Klauser, P. Mendez, L. Poglia, and D. Muller. Activity-dependent PSD formation and stabilization of newly formed spines in hippocampal slice cultures. *Cereb. Cortex*, 18:151–161, 2008.

49. S. Johnson, J.J. Torres, and J. Marro. Nonlinear preferential rewiring in fixed-size networks as a diffusion process. *Phys. Rev. E*, 79:050104(R), 2009.

50. P. R. Huttenlocher and A. S. Dabholkar. Regional differences in synaptogenesis in human cerebral cortex. *J. Comp. Neurol.*, 387:167–178, 1997.

51. S. Johnson, J. Marro, and J. J. Torres. Evolving networks and the development of neural systems. *J. Stat. Mech.*, P03003, 2010.

52. G. Chechik, I. Meilijson, and E. Ruppin. Synaptic pruning in development: A computational account. *Neural Comp.*, 10:1759–1777, 1998.

53. G. Chechik, I. Meilijson, and E. Ruppin. Neuronal regulation: A mechanism for synaptic pruning during brain maturation. *Neural Comp.*, 11:2061–2080, 1999.

54. N. G. van Kampen. *Stochastic processes in physics and chemistry*. Amsterdam, the Netherlands: North-Holland, 1990.

Chapter 10

ScaleGraph
A Billion-Scale Graph Analytics Library

Toyotaro Suzumura

10.1 Introduction

Graph mining has been widely used in a variety of domains to gain insight into graph properties, relationships among vertices, and complex dynamics underlying a graph. Recently, with the advance of information technology, a number of gigantic graphs have emerged especially in the form of social networks, web link graphs, Internet topology graphs, and so on, whose sizes range from millions up to billions of edges [1]. A single SMP machine cannot deal with such graphs owing to the limitations of physical memory and CPU resourcs.

Many libraries and frameworks, such as the parallel boost graph library (PBGL) [2], Apache Giraph [3], Google Pregel [4], Pagasus [5], and GraphLab [6], have been introduced to address the constraints of a single machine by using distributed systems. The libraries and frameworks concerned about performance usually use the C/C++ programming language and message passing interface (MPI) following the distributed memory paradigm, which requires users with extensive programming skill. The others that are concerned about productivity and portability usually use the Java programming language, which is easier and more widely used in software programs but has less performance than C/C++.

Achieving high-performance and reducing software complexity are the utmost challenges of developing a library for high-performance computing systems, when we scale up from peta-scale to exa-scale systems. The partition global address space (PGAS) programming model is a programming model that merges the merits of a single program, multiple data (SPMD) programming style for distributed memory systems and the data referencing semantics of shared memory systems. The PGAS programming model provides built-in notations for manipulating remote data. Therefore, the mechanism for sending and receiving data across machines is transparent to users, which makes users' code less complicated. Using PGAS languages such as X10, Chapel, and UPC is considered a promising approach for achieving highly productive computing systems. There have been prevailing libraries developed in other programming languages for large-scale graph analytics, although such support is rare in PGAS implementations.

X10 [7] is designed for multicore and cluster systems, with the awareness of developer productivity as well as software performance, following the philosophy of performance and productivity at scale. By considering the aforementioned issues and the features of X10, developing an X10-based graph analytics library that harnesses the performance and productivity of X10 is of interest to us. We would like to address the following problems. How can we implement graph kernels and frameworks that harness the features of X10 such as referential semantics and Cilk-style fork-join parallel model? What is needed to improve the performance of X10-based libraries?

We propose the ScaleGraph library, which is implemented using the X10 programming language. We implemented a number of algorithms and evaluated them using both synthetics and real graphs. The contributions are the following:

- Carefully designed an X10-based graph analytics library that utilizes the features of X10 such as referential semantics, collective communication, the Cilk-style fork-join parallel model, and native code integration to achieve high productivity as well as high performance

- Optimizations for X10 runtime, optimized X10 Team, and explicitly managed memory (EMM), which leads to significant improvements on performance and memory utilization

- Performance and scalability analysis of an X10-based graph analytics library and comparison with other prevailing libraries

The rest of the chapter is organized as follows. Section 10.2 describes related work. Section 10.3 describes how we optimized the X10 runtime. Section 10.4 describes ScaleGraph architecture and graph kernels. In Section 10.5, we present the results of the performance and scalability evaluation of each algorithm and optimization techniques as well as the performance and scalability evaluation against other same-class libraries. In Section 10.6 and 10.7 we conclude and discuss future work, respectively.

10.2 Related Work

Recently, large-scale graph analytics has become a popular topic because of the emergence of gigantic graphs, especially in the form of social networks, web link graphs, and Internet topology graphs. Many graph analytics frameworks and libraries have been introduced with different focuses, but they share the same goal to deal with gigantic graphs. [13–19]

PBGL [2], written in C++ on top of MPI, is a parallel version of the boost graph library. The library provides basic graph abstractions and graph visitor abstractions in the form of C++ generic templates that users can override to implement their own graph kernels without the need to know the underlying graph implementation. PBGL also encompasses a number of graph kernels and graph generators. To use PBGL efficiently, users are required to be well versed in C++ generic templates, which is very difficult for inexperienced programmers and graph analysts.

Google's Pregel [4] is a graph-processing framework that aims at scalability and fault-tolerance. Pregel introduces a new computation model, a vertex-centric message passing model inspired by the bulk synchronization processing model. The model is sufficiently flexible to express arbitrary graph algorithms. As for fault tolerance, Pregel uses regular "ping" messages that a master sends to check the health of workers and vice versa. ScaleGraph adapts Pregel's computation model as a building block for implementing algorithms.

Apache Giraph [3] is an open-source graph-processing framework based on Hadoop, a widely used implementation of the MapReduce computation model. Apache Giraph employs Pregel's computation model as its building block for implementing graph kernels. Apache Giraph provides rich third-party interfaces of data storage such as Hive, Gora, and Rexster. Its applications are managed executables (Java), while those of the ScaleGraph library are native executable (C/C++), so ScaleGraph has a performance advantage over Apache Giraph.

GraphLab [6] is a distributed graph-processing framework that originated from the machine learning area. The execution model of GraphLab consists of three main steps: remove, execute, and add vertices. The model allows execution to perform asynchronously, which is different from Pregel's computation model; this is of benefit to algorithms that do not require step synchronization. Graph Lab uses MPI for communication, which is the same as ScaleGraph, but GraphLab mainly focuses on algorithms for machine learning, while ScaleGraph covers a wide range of graph analytics algorithms.

10.3 X10 Optimization

In this section, we first present an overview of the X10 programming language and then propose our optimization method techniques that help improve the performance of the ScaleGraph library.

10.3.1 X10 overview

X10 is a type-safe, imperative, concurrent, object-oriented programming language being developed as open-source software by IBM Research in collaboration with many academic institutions around the world. The programming model of X10 is the asynchronous partitioned global address space (APGAS) model which merges the advantages of an SPMD programming style for distributed memory systems and the data referencing semantics of shared memory systems as well as permits asynchronous task creation between a local place and a remote place. The syntax of X10 is reminiscent of a combination between Java and Scala, which are well-known, modern high-level programming languages. The developers who are familiar with those languages will adapt to X10 effortlessly.

X10 introduces the concepts of place and activity. A place, which is analogous to a process, has its own local data and activities. An activity, which is similar to a lightweight thread, is a statement that is being executed. The activity has its own local data and it shares Place's local data with one another. The activity can access remote data (i.e., the data located on another place) via a keyword or collective communication routines of the X10.util.Team class. The activity can also migrate with all referenced objects (i.e., deep copying) from its current place to another place and continue its execution seamlessly on the new place. X10 supports Cilk-style fork-join programming for managing the parallelism of activities. The async keyword creates an activity and puts it into the job queue of the worker. The finish keyword stalls the execution of the main work until all spawned activities are executed. The X10 programming model is shown in Figure 10.1.

10.3.2 MPI collective communication optimization

Collective communication is communication in which all machines involved build up peer-to-peer communication for manipulating a shared piece of information that is distributed among processes, that is, places in X10. X10 provides collective communication

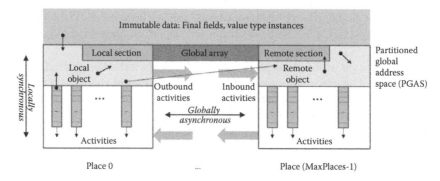

FIGURE 10.1: X10 programming model.

routines through the Team class in the x10.util package. The class encompasses the allreduce, alltoall, barrier, broadcast, reduce, and scatter collectives.

MPI, which is one of the X10 supported communication APIs, is an open standard message-passing protocol. MPI has the functions for collective communication and they are often optimized with the awareness of underlying communication hardware and topology. X10 Team also has the same functions of synchronization and collective computation as MPI, but it supports only a part of MPI's collective communication functions.

X10RT (X10 runtime transport) is a communication layer of the X10 runtime library. X10RT is responsible for sending and receiving messages and data between places. X10RT makes use of various transportation frameworks through X10RT interfaces, which include point-to-point communication interfaces, and collective communication interfaces. The restriction of the interfaces is that they must be able to be called in a nonblocking manner. Because nonblocking collective communication routines were not introduced until MPI 3.0, the collective communication routines of X10 Team for MPI transport are emulated using MPI point-to-point communication at the X10RT level. Using the emulator instead of MPI collective communication routines, which is highly optimized, incurs overhead that leads to degrading the performance of collective communication.

We propose a new X10 Team class for dealing with the aforementioned issues. The key features of the new Team class are that the new Team class encompasses all collective communication routines that are available in MPI and these routines directly use MPI collective communication routines without emulating at the X10RT level. Our new Team class supports the implementations of both blocking collective communication routines (MPI 2.0) and nonblocking collective communication routines (MPI 3.0). For blocking collective communication routines, when they are called from the caller, a new background thread is created to process the routines and then the routine returns to the caller immediately. For nonblocking collective communication routines, when they are called from the caller, the routine calls the MPI collective communication routines directly without creating a new thread. By avoiding the emulation, we achieved a significant performance gain over the original X10 Team as presented in Section 10.5 for performance evaluation.

The new Team class also supports parallel serialization to transfer the data that needs to be serialized. The data that has references to other objects has to be serialized to transfer to a remote place. The serialization for the collective communication is performed in parallel to improve performance. Any complex data structure is supported for transfer with collective communication.

10.3.3 New memory management system

Next, we propose a method for optimizing memory management in X10. The X10 native back-end uses the Boehm–Demers–Weiser conservative garbage collector (BDWGC) for automatic memory management. BDWGC is the most popular garbage collector (GC) that can be used with C/C++. But there is an issue of inefficient memory utilization. When it is used with large memory allocation, the GC heap grows dramatically. The GC increases its heap to fit the size of the new requested memory when the available fragment memory in its heap is not large enough. The GC does not separate heaps for small memory requests and large memory requests, which tends to cause memory fragmentation accelerating the growth of its heap, which leads to inefficient memory utilization. Furthermore, the growth of the GC heap increases the number of false references since the size of the GC heap is the same as the size of address space range that the GC assumes a word is a pointer. This issue is very critical since it leads to memory leaks.

According to the aforementioned issues, we propose a new memory management system named EMM. In ScaleGraph, EMM can be used through an array, MemoryChunk.

MemoryChunk can be used in the same manner as X10's native array. It is designed to deal with a large number of items. The memory allocation in MemoryChunk consists of two modes for small memory requests and large memory requests, respectively. The appropriate mode is determined internally from the size of requested memory and a certain memory threshold. For small memory requests, MemoryChunk uses BDWGC's allocation scheme, while for large memory requests, MemoryChunk explicitly uses malloc and free system calls, passing the memory management to the operating system, which manages memory more efficiently than BDWGC. Since, in the X10 language, the internal region of an array cannot be pointed to by other object, the address space for the array can be ignored by GC; malloc and free system calls can be used to allocate the memory of an array. By using EMM, we achieved significant improvement of memory utilization as shown in Section 10.5.4.

10.4   ScaleGraph

10.4.1   ScaleGraph overview

ScaleGraph is an open-source library for large-scale graph analytics built on top of the productive X10 programming language. ScaleGraph 2 is redesigned from scratch using the experiences from developing ScaleGraph 1.0 [8] in which many issues regarding performance, user usability, and data I/O were raised because of the design choices and unexpected issues of X10 when working with very large data. We designed ScaleGraph 2 with an awareness of the aforementioned issues. We developed new readers that are able to read and write data in parallel. We optimized X10 Team, which is considered the most important feature for working with multiple machine nodes. We also developed an XPregel framework, which is an implementation of the Pregel-based computation model, to help users develop their graph kernels easier.

The current release (ScaleGraph 2.2) of the ScaleGraph library is composed of three main components: XPregel framework, basic linear algebra subprograms (BLAS) for sparsematrix, and File IO (Figure 10.2). ScaleGraph provides a number of graph algorithms including degree distribution, betweenness centrality, breadth first search (BFS), HyperANF, maximum flow, PageRank, spectral clustering (SC), single-source shortest path

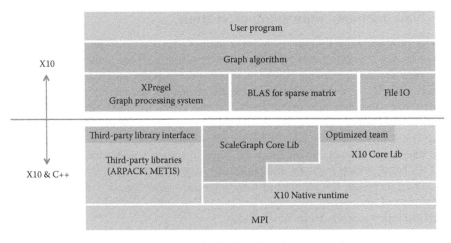

FIGURE 10.2: ScaleGraph software stack.

(SSSP), and strongly connected components, targeting massive-multicore distributed systems. According to the APGAS model and off-the-shelf notions for manipulating intercommunication and activity parallelism, we achieved the high productivity of implementing the ScaleGraph library in just 1800 lines of X10 code for the XPregel framework implementation and 60 lines of X10 code for the degree distribution algorithm while Giraph [3] has more than 11,000 lines of Java code only for its communication component (comm package of Giraph 1.0.0). Furthermore, X10 allows us to make use of the prevailing third-party libraries through native code integration. This helps us not to recreate duplicated code from scratch that might not be as efficient as the long renowned libraries.

10.4.2 Graph representation

Graph data in the ScaleGraph library can be represented as either a distributed graph object or a distributed sparse matrix. A distributed graph stores the distributed mutable lists of edges, vertex attributes, and edge attributes, the distribution of which is the same as the data given by a user or a file reader. A distributed graph supports graph manipulation, such as modification of edges and attributes; it does not support any query operations such as querying vertex neighbors. A distributed sparse matrix, which is supposed to be converted from a distributed graph object, is an efficient approach for storing an immutable, sparse, adjacency matrix of graph data; it supports various query operations such as querying vertex neighbors, vertex attributes, and edge attributes. The reason that ScaleGraph has two representations for graph data is to provide modification functionalities as well as fast graph querying.

A distributed sparse matrix in the ScaleGraph library is stored in compressed sparse row format, the distribution of which can be two-dimensional block distribution, one-dimensional column-wise distribution, or one-dimensional row-wise distribution, where the last two are special cases of the first one. All distributions are based on the method proposed in [9]. Figure 10.3. shows a directed weighted graph represented as an adjacency matrix, where row indices represent source vertices and column indices represent target vertices. The matrix can be transposed to change from out-edge notation to in-edge notation. For two-dimensional block distribution, the sparse matrix will be partitioned into blocks (Figure 10.4). The number of the blocks is given by RC and must match the number of given places, where R is the number of rows and C is the number of columns to partition. The local id and the place of a vertex can be determined from the vertex id itself by using only bit-wise operations, which help graph algorithms, which usually frequently check which place is the owner of given vertices, reduce computation overhead.

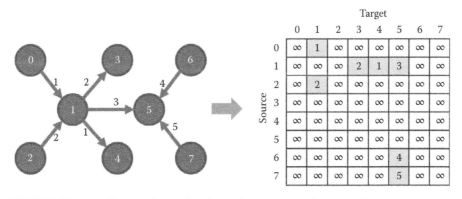

FIGURE 10.3: Directed weighted graph represented as an adjacency matrix.

(a) Target, Source

	0	1	2	3	4	5	6	7
0	∞	1	∞	∞	∞	∞	∞	∞
1	∞	∞	∞	2	1	3	∞	∞
2	∞	2	∞	∞	∞	∞	∞	∞
3	∞	∞	∞	∞	∞	∞	∞	∞
4	∞	∞	∞	∞	∞	∞	∞	∞
5	∞	∞	∞	∞	∞	∞	∞	∞
6	∞	∞	∞	∞	∞	4	∞	∞
7	∞	∞	∞	∞	∞	5	∞	∞

(b) Target, Source

	0	1	2	3	4	5	6	7
0	∞	1	∞	∞	∞	∞	∞	∞
1	∞	∞	∞	2	1	3	∞	∞
2	∞	2	∞	∞	∞	∞	∞	∞
3	∞	∞	∞	∞	∞	∞	∞	∞
4	∞	∞	∞	∞	∞	∞	∞	∞
5	∞	∞	∞	∞	∞	∞	∞	∞
6	∞	∞	∞	∞	∞	4	∞	∞
7	∞	∞	∞	∞	∞	5	∞	∞

P0   P1   P2   P3

(c) Target, Source

	0	1	2	3	4	5	6	7
0	∞	1	∞	∞	∞	∞	∞	∞
1	∞	∞	∞	2	1	3	∞	∞
2	∞	2	∞	∞	∞	∞	∞	∞
3	∞	∞	∞	∞	∞	∞	∞	∞
4	∞	∞	∞	∞	∞	∞	∞	∞
5	∞	∞	∞	∞	∞	∞	∞	∞
6	∞	∞	∞	∞	∞	4	∞	∞
7	∞	∞	∞	∞	∞	5	∞	∞

FIGURE 10.4: Various distributions of distributed sparse matrix on four places: (a) 2D block (R=2, C=2), (b) 1D column-wise (R=1, C=4), and (c) 1D row-wise (R=4, C=1).

10.4.3   Parallel loading and writing data

Loading and writing data from IO storage are considered equally important to executing graph kernels. When loading a large graph, if the graph loader is not well designed, the time to load a graph will be significantly longer than that of executing a graph kernel because of network communication overhead and the large latency of IO storage. ScaleGraph provides a parallel text file reader/writer. At the beginning, an input file will be separated into even chunks, the number of which is equal to the number of places available. Each place will load only its respective chunk, and it then separates the chunk into smaller, even chunks equal to the number of worker threads and assigns these smaller chunks to respective threads. All readers and writers are designed with an awareness of cache alignment and lock free, which helps the ScaleGraph library utilize CPU power efficiently.

10.4.4   XPregel framework

The XPregel framework is a framework that is inspired by the Pregel vertex-centric message-passing model, implemented using the X10 programming language. The framework serves as a building block for implementing graph kernels in the ScaleGraph library. In this section, we describe an overview of the XPregel framework, its implementation, and the optimization techniques that enhance the performance of the framework.

10.4.4.1   Pregel computational model

Google's Pregel [4] model is inspired by Valiant's bulk synchronous parallel model. Pregel computations consist of a sequence of iterations called supersteps. During each superstep, a user-defined function is invoked for each vertex, conceptually in parallel. The function specifies behaviors of a given single vertex V at superstep S. It can (i) receive the messages sent to V in superstep S+1, (ii) send messages to the other vertices that will receive the messages at superstep S+1, and (iii) modify the state and the outgoing edges of V. Messages are typically sent along outgoing edges, and could be also sent to any vertex whose identifier is known. The Pregel model defines three main interfaces as follows:

Message passing. Each vertex class implements a compute method. The framework guarantees that all the messages sent to the vertex will be passed to the compute method of the vertex. The framework will call the compute method of each vertex to process the state of the vertex. The compute method is also a place where a vertex sends messages to other vertices typically via their vertex id.

Combiner. The Pregel computation model also defines a Combiner concept like the MapReduce model. The Combiner is used to reduce the number of messages that are sent to the same destination vertex by applying a given operation on the messages and then produces a single output message that will be sent to the remote place.

Aggregator. Aggregators are mechanisms for global communication and monitoring. Each vertex provides a value to an aggregator in superstep S, then the master combines those values using a reduction operator, and the resulting value is sent to all vertices in workers in superstep S + 1.

10.4.4.2 XPregel overview

The architecture of X-Pregel is shown in Figure 10.5. The graph for XPregel is represented as a distributed sparse adjacency matrix by one-dimensional row-wise distribution. Each place processes only its local partition of a graph in parallel using multiple X10 activities. Each place maintains the buffer for each activity and each activity runs in a lock-free manner. After every activity finishes processing its local graph partition, each merges the buffer of its activities, exchanges messages with other places, and computes the aggregation using all-to-all collective communication and all-reduce collective communication. XPregel exposes both interplace (machine node) and intraplace parallelism.

10.4.4.3 Interfaces

To use the XPregel framework, users are required to create an XPregelGraph through factory methods and specify the types of a vertex (V) and an edge (E), and a distributed sparse matrix. Unlike Giraph in which a graph kernel class has to override the provided interfaces, the XPregrel framework uses closures, which provides more flexibility for implementing graph kernels. The core algorithm of a graph kernel can be implemented by calling the iterate method of XPregelGraph as shown in the following. Users are also required to specify the type of messages (M) as well as the type of aggregated value (A). The method accepts three closures: a compute closure, an aggregator closure, and an end closure. A compute closure is invoked for each vertex; it passes vertex information and messages to each vertex. Within a compute closure, a vertex may vote to halt and add data for aggregation. An aggregator closure is invoked when all vertices have been processed; the closure accepts

Distributed sparse adjacency matrix (ID row-wise distribution)

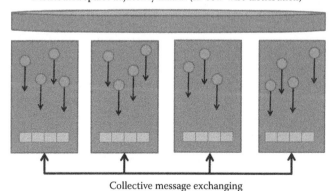

Collective message exchanging

FIGURE 10.5: XPregel framework architecture.

the values to be aggregated and users may use mathematical operations such as sum, min, max, and average to perform aggregation. An end closure accepts two parameters, the number of current iterations, and the aggregated value of the current iteration. The closure is required to return true or false to indicate whether the execution should continue.

Figure 10.6 shows a PageRank example using XPregel. PageRank is a link analysis algorithm to measure the relative importance of a vertex within a graph. At the beginning, each vertex sets its initial value. In each superstep, a vertex contributes its value, which depends on the number of links to its neighbors. Each vertex summarizes the score from its neighbors and then sets its value to the sum. The computation continues until the aggregated value of change in a vertex's values is less than a given criteria or the number of iterations is more than a given iteration limit.

10.4.4.4 SendAll optimization technique

Network communication is the critical bottleneck of distributed systems. The SendAll technique is aimed at reducing messages when a vertex happens to send the same messages to all of its neighbors. In a normal situation, sending the same message to all neighbors creates many identical messages that might be sent to the same place. If SendAll is enabled, the source place will send only one message to the destination places for each vertex and then each destination place will duplicate the message passing to respective destination vertices. Users can use this feature by calling the SendMessageToAllNeighbors() method. An applicable algorithm for the SendAll technique is an algorithm where every vertex sends an identical message to all of its neighbors, for example, PageRank and BFS.

10.4.5 BLAS for sparse matrix

Graph problems that rely on the computation of each vertex, such as BFS, SSSP, PageRank, maximum flow, and minimum spanning forest, can be intuitively expressed by the Pregel vertex-centric message passing model using the XPregel framework. There are also graph problems, such as PageRank, graph diameter estimation, SC, and connected component, that can be expressed using repeated matrix-vector multiplication. The Scale-Graph library provides BLAS. Users can use BLAS for implementing a graph kernel for the aforementioned graph problems.

10.4.6 Native library integration

X10 provides the notions for integrating other native libraries, which allows X10 users to use existing libraries rather than write new source code from scratch (Figure 10.7). This is considered one of important features of X10 that greatly increases a programmer's productivity.

The current release of the ScaleGraph library provides the interfaces for Parallel ARPACK (PARPACK) and ParMetis. PARPACK, an extension of the ARPACK library targeting distributed systems, is a library for solving large-scale eigenvalue problems on

```
public def iterate[M,A](
    compute :(ctx:VertexContext[V,E,M,A],
              messages:MemoryChunk[M]) => void,
    aggregator :(MemoryChunk[A])=>A,
    end :(Int,A)=>Boolean)
```

FIGURE 10.6: XPregel interface.

```
xpgraph.iterate[Double,Double](
                (ctx :VertexContext[Double, Double, Double, Double], messages :MemoryChunk[Double]) => {
                        val value :Double;
                        if(ctx.superstep() == 0) {
                                        // calculate initial page rank score of each vertex
                                        value = 1.0 / ctx.numberOfVertices();
                        } else {
                                        // for step onward,
                                        value = (1.0-damping) / ctx.numberOfVertices() +
                                                        damping * MathAppend.sum(messages);
                        }
                        ctx.aggregate(Math.abs(value - ctx.value()));
                        ctx.setValue(value);
                        ctx.sendMessageToAllNeighbors(value / ctx.outEdgesId().size());
                },
                // calculate aggregate value
                (values :MemoryChunk[Double]) => MathAppend.sum(values),
                (superstep :Int, aggVal :Double) => {
                        return (superstep >= maxIter || aggVal < eps);
                }
);
```

FIGURE 10.7: Source code for PageRank algorithm in ScaleGraph.

top of MPI. PARPACK is used in implementing SC of the ScaleGraph library. ParMetis, an extension of the Metis library for distributed systems, is an MPI-based parallel library for partitioning unstructured graphs and meshes, and computing fill-reducing orderings of sparse matrices.

10.5 Evaluation

10.5.1 Experimental environment

We conducted the performance and scalability evaluation on a TSUBAME 2.5 supercomputer. TSUBAME 2.5 is a production supercomputer ranked as the 13th most powerful in June 2014 with peak performance of 5736 teraFLOPS. All compute nodes, that is, machine nodes, are equipped with two Intel Xeon X5760 2.93 GHz CPUs with each CPU having 6 cores and 12 hardware threads (12 cores and 24 hardware threads in total for each compute node), and 54GB of memory. All compute nodes are connected with InifinitBand. We used the ScaleGraph 2.2 release, optimized X10 2.3.1 release, and MVAPICH2 1.9 release. We measured only wall-clock time of core computation regardless of the time for loading a graph and writing the result.

10.5.2 Evaluation data set

We used both synthetic graphs and real-world graphs. We used the RMAT [10] graph and the Erdős-Rényi random graph for synthetic graphs and the Twitter follower network graph for a real-world graph. For the RMAT graph generation parameter, we used $A = 0.45$, $B = 0.15$, $C = 0.15$, and $D = 0.30$. We use the term scale to specify the size of synthetic graphs. When the size of a graph is scale N, the graph has 2^N vertices and $ef \times 2^N$ edges, where ef is an edge factor, which is a ratio of the number of edges to the number of vertices. We used 16 for the edge factor throughout the evaluation of this chapter (as is recommended by the Graph500 benchmark). A detailed description of the Twitter network graph is in Section 10.5.8.

10.5.3   Team optimization evaluation

We evaluated our optimized X10 Team against the original X10 Team and X10-level emulator with various collective communication routines. The evaluation is performed with 128 nodes by exchanging 8 MB data for each place. Figure 10.8 shows the speedup of the optimized Team implementation compared to the faster of the two existing X10's methods, the original X10 Team, which emulates collective communication routines using MPI point-to-point routines, and the X10-level emulator, which emulates collective communication routines using X10 at statement. The optimized X10 outperforms the existing X10's methods significantly, except for the barrier routine, which is just synchronization and has no data transfer.

10.5.4   Evaluation on explicitly managed memory

We evaluated the performance of Explicit Memory Management by comparing the wall-clock time of computing 30 iterations of PageRank and the memory consumption after the computation. The graph kernel runs with two places on a single machine node. The number of threads per place for X10 is set to 12. The graph used in this experiment is a synthetic RMAT graph with 16 million vertices and 268 million edges.

With EMM enabled, the execution time reduced from 163 to 135 and the memory consumption after the computation reduced from 28.3 to 7.2 GB, thus running 17% faster and consuming 21.1 GB less memory than that without EMM enabled (Table 10.1). This feature helps the ScaleGraph library deal with very large graphs efficiently.

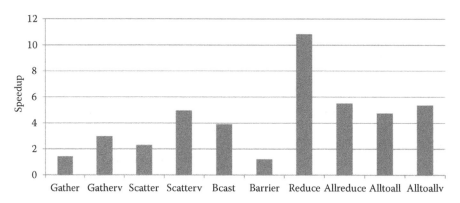

FIGURE 10.8: Speedup of optimized Team against the existing X10 communication methods.

TABLE 10.1:   Execution time and memory consumption for computing PageRank on RMAT scale 24 graph

Comparison	Without EMM	With EMM
Execution time (30 iterations)	163.1 s	135.2 s
Memory consumption (after 30 iterations)	28.3 GB	7.2 GB

10.5.5   XPregel optimization

We evaluated XPregel using PageRank with the following scenarios: normal configuration and configurations that enable the SendAll and Combine functions. In this experiment, we used a billion-scale RMAT graph of scale 26. The graph is composed of 67 million vertices and 1.07 billion edges.

For PageRank computation (see Figure 10.9), the SendAll-enabled scenario achieved two to three times speedup over the normal configuration and the Combine-enabled scenarios, while the Combine-enabled and the normal configuration scenarios did not have much difference in performance. The Combine technique can reduce the number of transferred messages but combining messages requires additional high computation cost. Duplicating messages of the SendAll technique also has additional computation cost but it is much smaller than that of the Combine technique. This is the reason why the SendAll technique is faster than Combine technique.

According to the characteristic of PageRank in which a vertex sends messages to all its neighbors in every superstep, by applying the SendAll and Combine techniques the number of sent messages substantially reduces as shown in Figure 10.9. The Combine technique combines multiple messages into one message that consists of a destination vertex id and payload, while the SendAll technique sends an array of payload along with the bitmap of all vertices. Roughly speaking, a message sent in the SendAll technique is composed of 1 bit and payload. Figures 10.10 and 10.11 show that the SendAll technique reduces the size of messages sent among machine nodes better than the Combine technique.

10.5.6   Evaluated graph kernels of ScaleGraph

We evaluated the performance of the following seven graph kernels implemented in the ScaleGraph library.

- PageRank

- BFS

- SSSP

- Weakly connected component (WCC)

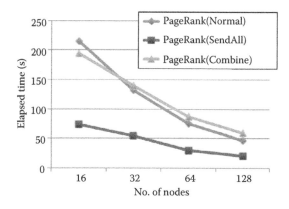

FIGURE 10.9: The wall-clock time for computing PageRank with normal configuration, SendAll enabled, and Combine enabled.

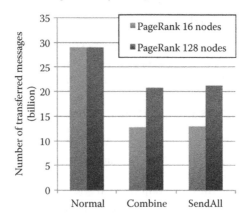

FIGURE 10.10: The number of messages sent during computing PageRank with normal configuration, SendAll enabled, and Combine enabled on 16 and 128 machine nodes.

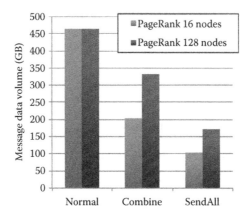

FIGURE 10.11: The size of messages sent during computing PageRank with normal configuration, SendAll enabled, and Combine enabled on 16 and 128 machine nodes.

- SC

- HyperANF [11]

- Degree distribution

All kernels are implemented using the XPregel framework, except for SC, which is implemented using the sparse BLAS of ScaleGraph and the PARPACK library. For PageRank, we measured the time of 30 iterations. HyperANF is a graph kernel to approximate the average distance of all vertex pairs by using Hyper LogLog Counter. We used $B = 5$ as the parameter for the size of the counter and we measured the time of two iterations of HyperANF in the following experiments.

10.5.7   Graph kernel performance evaluation on synthetic graphs

To understand the performance and scalability of ScaleGraph graph kernels, we performed weak-scaling and strong-scaling analysis on various scales of RMAT graphs and random graphs.

Table 10.2 shows the result of weak-scaling analysis. The execution time increases gradually as the number of machine nodes increases because the graph kernels require heavy communication among machine nodes. Table 10.3 shows the result of strong-scaling analysis. We used an RMAT graph with scale 28 (268 million vertices and 4.3 billion edges) for the strong-scaling analysis. For all the graph kernels, the execution time is reduced as the number of machine nodes increases. The result of these analyses shows that the ScaleGraph library has good scalability.

10.5.8 Graph kernel performance evaluation on real Twitter graph

To reflect the true performance for real-graph scenarios, we evaluated the graph kernels using the real Twitter graph that was crawled using the top 1000 followed users as seed nodes [1]. The graph is composed of 496 million users and 28.5 billion following relationships. We performed the evaluation with this billion-scale graph on 128 machine nodes. The result is shown in Table 10.4. The comparison with other graph libraries is one of our future works, but it would be estimated using an RMAT graph in next section.

10.5.9 Performance evaluation against PBGL and Giraph

To better understand the performance of ScaleGraph, we compared ScaleGraph performance relatively to same-class, well-known libraries, that is, PBGL and Giraph in both a strong-scaling and weak-scaling manner. The number of threads per place for ScaleGraph is set to 12. Giraph was configured to run one worker per node with each having 12 threads and the number of splits is 12 times the number of workers. PBGL was configured to run four processes per node, the number of which is the configuration that gives the best elapsed

TABLE 10.2: Weak-scaling performance of each algorithm (wall-clock time in seconds)

Problem Size and Number of Nodes	PageRank	BFS	SSSP	WCC	SC	HyperANF	Degree
RMAT, Scale 22, 1 node	13.7	1.9	8.9	5.6	351.1	50.3	33.1
RMAT, Scale 26, 16 nodes	28.3	4.0	13.5	12.0	701.4	88.9	36.3
RMAT, Scale 28, 64 nodes	37.9	7.5	18.8	17.0	1166.0	103.5	39.4
RMAT, Scale 29, 128 nodes	45.3	11.2	24.5	22.1	1438.8	142.3	41.1
Random, Scale 29, 128 nodes	46.5	8.8	20.6	21.4	1106.6	162.3	42.7

TABLE 10.3: Strong scaling performance of each algorithm for RMAT scale 28 graph (wall-clock time in seconds)

Number of Nodes	PageRank	BFS	SSSP	WCC	SC	HyperANF	Degree
16 nodes	124.1	21.9	65.8	55.9	2969.9	38.0	16.1
32 nodes	91.7	18.7	36.9	30.2	1639.0	27.0	11.6
64 nodes	38.1	7.5	20.1	17.2	1169.9	10.6	4.9
128 nodes	26.5	5.8	14.7	10.5	706.4	6.8	3.1

TABLE 10.4:   Elapsed running time on 128
machine nodes

Algorithms	Elapsed time
In-degree and out-degree calculation	42 s
PageRank (30 iteration steps)	278 s
Spectral clustering (2 clusters)	29 min

TABLE 10.5:   Strong-scaling performance on RMAT
scale 25

Nodes	ScaleGraph (s)	Giraph (s)	PBGL (s)
1	158.9	-	-
2	85.0	-	966.8
4	44.9	2885.1	470.3
8	23.4	443.1	309.5
16	13.3	125.3	290.9

time result. The result is shown in Table 10.5. The absent elapsed time in the result for PBGL and Giraph is due to an out-of-memory problem. This shows that ScaleGraph is more efficient than PBGL and Giraph in the sense of memory utilization.

For strong-scaling performance analysis, ScaleGraph outperforms the two libraries significantly. Giraph exposes CPU bottlenecks, as more nodes added, the elapse greatly reduces. PBGL has performance saturation from eight nodes. The evaluation result shows that ScaleGraph is more efficient than PBGL and Giraph in terms of performance and memory utilization.

10.6   Conclusion

In this chapter, we presented the ScaleGraph library, which is an X10-based distributed graph-processing library. According to the PGAS model and the Cilk-style fork-join parallel model in X10, ScaleGraph's XPregel framework exposes internode and intranode parallelism, which makes it utilize computation resources of distributed multicore systems efficiently.

We optimized X10 Team, which is a critical component for exchanging data among machine nodes in distributed processing, to directly use MPI collective communications resulting in significant performance improvement over the original X10 Team. Furthermore, we introduced EMM that is able to manage the allocation of a very large chunk of memory. XPregel's SendAll optimization also shows an impressive result. According to X10 primitive and the many optimization techniques applied in the ScaleGraph library, we achieved high productivity, scalability, and performance as presented in the evaluation results.

Comparing with the same-class libraries and frameworks emphasizes the superiority of the ScaleGraph library; we achieved significant speedup over PBGL and Apache Giraph, respectively. The ScaleGraph library is available from [12].

10.7 Future Work

As for future work, we plan to improve the ScaleGraph library in many aspects such as performance, functionality, and usability, to make it better support various types of users that have different backgrounds, various scales of problems, and various configurations of system environments.

Regarding graph analysts who are typically not familiar with programming, we are considering integrating the ScaleGraph library with R, which is a programming language and software environment that is widely used among statisticians and data miners. Furthermore, we are investigating deploying the ScaleGraph library onto cloud environments such as Amazon EC2 to provide online graph analytics services to everyone around the world.

Acknowledgments

This research has been partly supported by the Japan Science and Technology Agency. We appreciate the efforts of all developers who were involved in developing the ScaleGraph library.

References

1. M. Watanabe and T. Suzumura. How social network is evolving?: A preliminary study on billion-scale Twitter network. In *Proceedings of the 22nd International Conference on World Wide Web Companion*, WWW '13 Companion, pages 531–534, Republic and Canton of Geneva, Switzerland, 2013. International World Wide Web Conferences Steering Committee.

2. D. Gregor and A. Lumsdaine. Lifting sequential graph algorithms for distributed-memory parallel computation. In *Proceedings of the 20th Annual ACM SIGPLAN Conference on Object-Oriented Programming, Systems, Languages, and Applications*, OOPSLA '05, pages 423–437, New York, NY, 2005. ACM.

3. The Apache Software Foundation. *Giraph—welcome to Apache Giraph!* http://giraph. apache.org, 2014.

4. G. Malewicz, M.H. Austern, A.J.C. Bik, J.C. Dehnert, I. Horn, N. Leiser, and G. Czajkowski. Pregel: A system for large-scale graph processing. In *Proceedings of the 2010 ACM SIGMOD International Conference on Management of Data*, SIGMOD '10, pages 135–146, New York, NY, 2010. ACM.

5. U. Kang, C.E. Tsourakakis, and C. Faloutsos. Pegasus: A peta-scale graph mining system implementation and observations. In *Data Mining, 2009. ICDM '09. Ninth IEEE International Conference on*, pages 229–238, Dec 2009.

6. Y. Low, D. Bickson, J. Gonzalez, C. Guestrin, A. Kyrola, and J.M. Hellerstein. Distributed GraphLab: A framework for machine learning and data mining in the cloud. *Proc. VLDB Endow.*, 5(8):716–727, April 2012.

7. P. Charles, C. Grothoff, V. Saraswat, C. Donawa, A. Kielstra, K. Ebcioglu, C. von Praun, and V. Sarkar. X10: An object-oriented approach to non-uniform cluster computing. *SIGPLAN Not.*, 40(10):519–538, October 2005.

8. M. Dayarathna, C. Houngkaew, and T. Suzumura. Introducing ScaleGraph: An X10 library for billion scale graph analytics. In *Proceedings of the 2012 ACM SIGPLAN X10 Workshop*, X10 '12, pages 6:1–6:9, New York, NY, 2012. ACM.

9. A. Yoo, E. Chow, K. Henderson, W. McLendon, B. Hendrickson, and U. Catalyurek. A scalable distributed parallel breadth-first search algorithm on BlueGene/L. In *Supercomputing, 2005. Proceedings of the ACM/IEEE SC 2005 Conference*, pages 25–25, Nov 2005.

10. D. Chakrabarti, Y. Zhan, and C. Faloutsos. R-MAT: A recursive model for graph mining. In *SDM*, volume 4, pages 442–446. SIAM, 2004.

11. P. Boldi, M. Rosa, and S. Vigna. Hyper ANF: Approximating the neighbourhood function of very large graphs on a budget. In *Proceedings of the 20th International Conference on World Wide Web*, WWW '11, pages 625–634, New York, NY, 2011. ACM.

12. T. Suzumura and K. Ueno. ScaleGraph: A high-performance library for billion-scale graph analytics. In *Big Data (Big Data), 2015 IEEE International Conference on*, Santa Clara, CA, 2015, pp. 76–84.

13. T. Suzumura and K. Ueno. Performance characteristics of Graph500 on large-scale distributed environment. In *IEEE IISWC 2011* (IEEE International Symposium on Workload Characterization), Austin, TX, Nov 2011.

14. K. Ueno and T. Suzumura. Highly scalable graph search for the Graph500 benchmark. In *HPDC 2012* (The 21st International ACM Symposium on High-Performance Parallel and Distributed Computing), Delft, the Netherlands, June 2012.

15. K. Ueno and T. Suzumura. Parallel distributed breadth first search on GPU. In *HiPC 2013* (IEEE International Conference on High Performance Computing), India, Dec 2013.

16. Toyotaro Suzumura, M. Ganse, and S. Nishii. Towards graph stream processing platform. Large-Scale Network Analysis Workshop in conjunction with World Wide Web 2014 (WWW '14), Seoul, Korea, April 2014.

17. C. Houngkaew and T. Suzumura. X10-based distributed and parallel betweenness centrality and its application to social analytics. *HiPC 2013* (IEEE International Conference on High Performance Computing), India, Dec 2013.

18. H. Ogata and T. Suzumura. Towards X10-based link prediction for large-scale network. Large-Scale Network Analysis Workshop in conjunction with World Wide Web 2014 (WWW '14), Seoul, Korea, April 2014.

19. K. Shirahata, H. Sato, T. Suzumura, and S. Matsuoka. A scalable implementation of a MapReduce-based graph processing algorithm for large-scale heterogeneous supercomputers. 13th IEEE/ACM International Symposium on Cluster, Cloud and Grid Computing (CCGrid), Delft, the Netherlands, May 2013.

Chapter 11

Challenges of Computational Network Analysis with R

Shiwen Sun, Shuai Ding, Chengyi Xia, and Zengqiang Chen

11.1   Introduction

Big Data has become the driver for innovation in recent years, from strategy decision-making to scientific computing [1]. Several leading high-tech transnational enterprises, including Google, International Business Machines Corporation (IBM), Alibaba, McKinsey & Company, Facebook, and Amazon, have started to apply Big Data to better serve their customers around the world. Examples of Big Data include the Google search index, Alibaba.com's product list, and the database of Tencent user profiles. Big Data emerged due to the incentive of two key trends in information technology (IT) development [2]. First, a wide variety of emerging data infrastructures (social network, Internet of things, cloud computing, open source software, etc.) make it easier and cheaper to generate, acquire, store, process, and analyze data. Second, in a trend called the commercialization of the data service, not just leading IT enterprises, but many types of individual users, enterprises, governments, schools, hospitals, and other institutions, have become intimately involved in the processing and consuming data.

Big Data has the broadest development and application in the computer domain, especially in the aspects of data storage, data analysis, and data sharing [3]. Because of the huge quantity of data, the methods and models of data processing that differ from traditional ones require achievement of scientific storage in computers. Being different from simple copy and backups, the method of scientific storage realizes effective preservation, management, and utilization of mass data through scientific methods of data management, appropriate software and hardware platforms, and advanced storage technology. Scientific storage mainly adopts the methods of centralized storage and backups, construction of metadata documents, and a document search index. At the same time, analysis of Big Data brings enormous challenge to traditional and conventional techniques of data analysis. The effective and real-time management of huge heterogeneous data is an urgent problem in the technology for analysis of Big Data. Experts in the Big Data field conduct data analysis under the environment of Big Data by improving the traditional genetic algorithm, neural network calculation, data-mining technology, regression analysis methods, and so on. Data sharing is the restricted visit and application of data under the premise of assuring data security. Data sharing under the environment of Big Data should pay attention to privacy protection of the following aspects: the collection of Big Data, the sharing and publishing of Big Data, the analysis of Big Data, the life cycle of Big Data, and so on.

Big Data has been widely used and fully developed in physics, chemistry, biology, and so on [4]. In physics, especially in geophysics, the successful storage and analysis of massive data for crustal movement and geomorphology can provide reliable data and a theoretical basis for earthquake prediction, resource exploration, weather forecasts, and so on. To some extent, it also masters the law of earth operation. In chemistry, Big Data has been widely used in inorganic chemistry, organic chemistry, and analytical chemistry and helps in further development and application in the calculation of chemical output. In biology, owing to the advances for storage and calculation ability of Big Data, great breakthroughs have been made in acquiring and mapping sequences of human genome, the analysis of disease gene sequences, and the targeted disease treatment and finally reached a new development platform. Thus, as can be seen, Big Data technology has attracted intensive attention and full development in basic disciplines.

Meanwhile, Big Data also brings new opportunities and challenges for development of management science. The application of Big Data plays an important and irreplaceable role, especially in government management and government decision-making [5]. Aiming at government decision-making for emergencies, Big Data technology, through the integrating, analyzing, and application of data from geographic systems, regional transportation, health care, community information, and other relevant information, enables the government to process a large amount of data concerning city spatial information and properties relating to emergencies in the shortest time. On this basis, government can establish a linkage mechanism and handle grid management, an important task. As an example, in emergencies of traffic safety, Big Data technology can make sensors and other mobile terminals that render drivers and transport equipment a data source of traffic condition emergencies. Meanwhile, it can also send the position, the velocity, and other information back to a Big Data center to form a complete road traffic conditions map for predicting traffic conditions and possible risks to anticipate emergencies.

In practice, we also face enormous challenges in the new era of Big Data [6]. Flexible data infrastructure needs to be constructed for the acquisition and preparation of Big Data, because big/fast data cleaning and noise modeling causing energy consumption and high costs are becoming two essential constraints in Big Data applications. Considering the fact that Big Data are commonly dynamically aggregated on heterogeneous data sources, intelligent data conversion and database design for various types of data (specifically semistructured and unstructured data) is a crucial issue to data scientists. Moreover, most existing samples-oriented information retrieval, statistical analysis, and data-mining technologies have become powerless in the process of analysis and modeling of Big Data. There is an urgent need to address non-SQL Big Data analytics and distributed data mining. These thriving directions require continued investigation. Today, worldwide Big Data is entering a time of unprecedented application-driven research.

Big graph data, for example, large biological networks or airline networks, is a kind of complex unstructured Big Data [7]. Strongly promoted by the leading biotechnology companies and Internet companies, high-dimensional graph data analysis and mining is quickly becoming popular in recent years. Data relationships and potential knowledge in big graph data are typically large-scale and complicated. However, utilizing traditional graph analytics technology in Big Data for domain-driven knowledge discovery is unrealistic in reality, since performing a large number of high-dimensional matrix calculations is time consuming and lacks accuracy. To attack this critical challenge, many emerging technologies have been proposed to enhance the accuracy of big graph structure modeling and evolutionary analysis, such as deep learning and hypergraph-based clustering. Deep learning is a new field of machine learning; the concept was proposed by Hinton et al. [8] in 2006, whose motivation was to build and simulate the neural network of the human brain to analyze and interpret complex data. In a Big Data environment, the huge amount of data will help in improving the accuracy of deep learning model training.

Social network analysis of Big Data aims to analyze the social network relationship of individuals and mine out the hidden potential relationship in flow of information and value exchange, allowing enterprises to establish visual and measurable models to get commerce opportunities. For instance, the 175 million Twitter users create 95 million microblogs every day, and Facebook processes 25 TB of data on 300,000 servers per day. Facing large-scale data like this, MapReduce can process and analyze these data more efficiently. The distributed computing model utilizes the core operation mechanisms of MapReduce. It divides Big Data into smaller pieces for distributed computers on the network to calculate, and then summarize the results to form final conclusions. In addition to the MapReduce model, social network analysis of Big Data also includes topological property analysis of a network, node impact analysis, frequent subgraph mining, community structure analysis, community evolution analysis, propagation mechanisms and dynamics analysis, and so on.

The main contribution of the chapter deals with some related concepts and frameworks of Big Data analytics, and the statistical analyses of complex network properties regarding several typical networking systems, such as degree distribution, communication efficiency, characteristic path length, clustering coefficient, assortative mixing, and so on. So, we explore the attack robustness and optimization of real-world complex networks. Another contribution focuses on the trust-enhanced cloud service selection method under online social networks. By combining various types of quality of service (QoS) properties, we find that the proposed method has better QoS properties than previous ones and can obtain the optimum values of parameters based on extensive experiments.

The chapter is structured as follows. Section 11.2 describes some basic concepts of Big Data processing and analytics; In Section 11.3, we present some case studies regarding the application of Big Data analytics in complex networks; then, based on Big Data technology, we introduce a trust-enhanced cloud service selection model ensuring the QoS of Internet applications. Finally, we end this chapter with some concluding remarks.

11.2   Basic Concepts on Big Data Processing and Analytics

In recent years, Big Data has drawn many scientific institutions and scholars into the new era of Big Data research. In particular, the Obama administration announced in March 2012 that the United States would invest $200 million to perform a Big Data research plan which will cover efforts from multiple federal agencies, such as the Defense Advanced Research Projects Agency (DARPA), National Institutes of Health (NIH), and National Science Foundation (NSF). There has been a great effort in these institutions to explore the scientific landscape of Big Data. Here we introduce the definitions, architectures, and related analytics of Big Data in brief.

11.2.1   Definitions

Many definitions and discussions of Big Data from different perspectives were developed. Wikipedia proposed, "Big Data is an all-encompassing term for any collection of data sets so large and complex that it becomes difficult to process using traditional data processing applications." However, the definition of Big Data is very diverse and industry and academia is far from reaching a consensus. Big Data especially signifies some important features differentiating it from the concepts of "massive data" and "very large data" as well as a large plethora of data. Several typical classes of definitions which are often used in [10] are listed as follows:

- *Attributive definition.* As early as in 2011, International Data Corporation (IDC), which is a pioneer in the field of Big Data, defined Big Data and deemed that "Big Data technologies describe a new generation of technologies and architectures, designed to economically extract value from very large volumes of a wide variety of data, by enabling high-velocity capturing, discovery, and/or analysis" [10]. This definition pointed out the four salient features of Big Data for the first time, that is, *volume, variety, velocity, and value.* After that, the "4Vs" definition has been widely used to illustrate some Big Data–related concepts. As an example, IBM also stated that *Big*

Data is best understood in terms of four dimensions: volume, velocity, variety, and veracity, which is slightly different from IDC's definition [11].

- *Comparative definition.* Mckinsey's report in 2011 considered Big Data as "data sets whose size is beyond the ability of typical database software tools to capture, store, manage and analyze." This definition is rather subjective and does not provide any specific description about Big Data in terms of any particular metric [2].

- *Architectural definition.* The National Institute of Standards and Technology (NIST) [12] pointed out that "Big Data is where the data volume, acquisition velocity or data representation limits the ability to perform effective analysis using traditional relational approaches or requires the use of significant horizontal scaling for efficient processing." Meanwhile, they further classified Big Data into Big Data science and Big Data frameworks, in which Big Data science is "the study of techniques covering the data acquisition, conditioning, and evaluation of Big Data" and Big Data frameworks are "software libraries along with their associated algorithms that enable distributed processing and analysis of Big Data problems across clusters of computer units."

In addition, from the perspective of management science, Big Data can also be considered as an important strategic resource which is always characterized by several features: decision usefulness, privacy risk, massiveness, heterogeneity, growth, complexity, and repeatable mining, and the potential value can be found through the given processing techniques and means. In addition, several important issues that were raised repeatedly in the scientific community include data usability and security, data service pricing, social media data and spatio-temporal data, personalization, green computing, and energy consumption.

11.2.2   Big Data's system architecture

In this section, we briefly introduce Big Data's system architecture from two different viewpoints, named the value-chain view and layered view, respectively, which are proposed first in [9].

11.2.2.1   A value-chain view

In fact, a value-chain view adopts the tradition systems-engineering approach to understand Big Data architecture. According to the life cycle of Big Data during its evolution, from its birth to death, the value-chain view divides the Big-Data system into four continuous stages or modules which include data generation, data acquisition, data storage, and data analytics as follows.

- *Data generation.* Within this stage, the raw data will be generated from the physical sensors or other digital sources which include audio, video, click streams, web pages or information systems, and so on. Since available data sources are rather diverse and different, raw data is very complex and distributed, and also correlated with different levels of domain-specific values. Thus, it poses great challenges to the real-world application of Big Data.

- *Data acquisition.* This stage describes the specific process of obtaining information from the raw data, which covers data collection, data transmission, and data preprocessing. First, data collection refers to transforming the raw data, which comes from various sources and even distributed environments, into a unified data format. Second, after collecting the given raw data, data transmission will send the data into some suitable and available storage center to deal with as soon as possible. Finally, noise and redundant information may be included inside the transmitted data, which need to be removed so as to reduce the volume of storage, and even to enhance the efficiency of data analysis. Thus, some necessary data preprocessing operations must be performed before effective storage and analytics.

- *Data Storage.* At this stage, data will be stored in a particular physical media (e.g., magnetic disks and disk array, storage class memory) and need to be organized effectively with the help of data management software. Thus, a Big Data storage system usually consists of two subsystems: hardware infrastructure and data management systems. On one hand, the hardware system provides enough shared information and communications technology (ICT) resources in a flexible way to diverse tasks in response to their instantaneous requests; at the same time it should be able to scale up and down and dynamically reconfigure according to various application requirements. On the other hand, the data management software is installed on top of the configurable hardware infrastructure so as to efficiently and reliably manage large-scale data sets. Furthermore, the data storage system also needs to provide multiple scalable access interfaces to query, analyze, and program the vast quantity of data.

- *Data analytics.* This is the most important aspect differentiating Big Data from traditional data analysis since Big Data analytical methods and tools are leveraged to inspect, transform, visualize, and model data to extract the potential values hidden behind the real Big Data. Regarding many application fields, such as web search engines, scientific research, log files, and so on, a great variety of types of data and filed-specific analytical methods can be utilized to derive the intended value, performance, and impact. Although diverse fields may cope with different application requirements and data characteristics, several common underlying technologies can be applied into various fields and six emerging analytical researches are structured data, text, multimedia, web, network, and mobile analytics.

The previously mentioned main properties and techniques within each stage are summarized in Table 11.1.

11.2.2.2 A layered view

Also, the Big Data system can be understood as a layered structure and is illustrated in Table 11.2, similar to the TCP/IP protocol for computer networks. The whole system will be divided into three layers including the application layer, computing layer, and infrastructure layer, from top to bottom.

TABLE 11.1: Big Data system architecture from the value-chain view

Stage	Main Functions, Modules, Characteristics, and Properties
Generation	Universe observation, web page, business data, social networks, large-scale scientific experiments, UGC, e-commerce, health care
Acquisition	Crawler, WDM, data integration, radio telescope, data compression, sensors, optic interconnects, RFID, OFDM, 3-tier tree, 2-tier tree
Storage	Shared-nothing parallel databases, NoSQL, Google file system, MapReduce, PNUTS, MongoDB, Dynamo, Dryad, SimpleDB, Voldemort
Analytics	Data mining, web mining, statistical analysis, multimedia analytics, network analytics, mobile analytics, social network analytics, community detection

TABLE 11.2:   Big Data system architecture from the layered view

Layer	Function	
Application Layer	Query	
	Clustering	
	Classification	
	Recommendation	
Computing Layer	Integration	
	Management	File system
		SQL
		NoSQL
	Programming model	Dremel
		Pregel
		MapReduce
		Dryad
Infrastructure Layer	Computation	
	Network	
	Storage	

This layered view helps to provide a conceptually hierarchical structure to emphasize the complexity of a Big Data system and main functions for each layer can be delineated as follows.

- *Infrastructure layer.* This consists of a series of ICT resources, which usually can be organized and implemented by cloud computing infrastructures and virtualization technology. This layer will be furnished with an underlying framework to provide a standard interface for the upper-layer systems, and the interface can be composed of a suite of access protocols between layers. Within this framework, in order to meet the demand of Big Data and elevate the access efficiency of the resources, infrastructure resources should be optimized and allocated by maximizing system utilization, energy awareness and operational simplification, and so on.

- *Computing layer.* This layer lies between the infrastructure layer and top application layer, and seems like a middleware layer that integrates a variety of analytical tools running over the underlying ICT resources. The commonly used tools include the data integration, data management, and programming models. Data integration is used to collect data from distinct data sources and integrate the data set into a unified form after the necessary preprocessing operation. Data management is leveraged to provide some mechanisms and tools to implement persistent data storage and highly efficient scheduling, such as distributed file systems, and SQL or NoSQL data management. The programming model carries out the abstract application logic and performs the data analysis effectively. Some typical programming models include MapReduce [13], Dryad [14], Pregel [5], and Dremel [16].

- *Application layer.* The top layer of the Big Data architecture certainly completes various data analysis functions including query, statistical analysis, clustering, classification, and visualization of analytical results. Meanwhile, it also combines the basic analytical methods to develop various field-related applications. As pointed out by McKinsey [2], five potential Big Data application domains are health care, public sector administration, retail, global manufacturing, and personal location data.

11.2.3 Hadoop framework and its applications

Due to the huge success and popularity of Google's distributed file system and the MapReduce computing model in dealing with massive data handling, its recent implementation entity, Hadoop, has been widely adopted industrially to achieve diverse applications, such as spam filtering, web searching, click stream analysis, content mining, social networks, online recommendations, and so on. In addition, academic communities have paid much more attention to Hadoop and many research projects are also built on such a framework. Since the emergence of the Big Data movement Hadoop has been the mainstay of large-scale data processing technologies. In particular, Apache Hadoop provides an open-source computing software framework supporting massive data storage and processing, helping researchers and engineers to understand the internal core technology and mechanisms of distributed file system and MapReduce operations. Meanwhile, the Hadoop platform has provided many key techniques including hardware scalability, lower computation cost, high processing flexibility, and strong fault tolerance to facilitate Big Data analytics and management, and thus has become a critical foundation to construct domain-specific applications of Big Data.

11.2.3.1 Main modules of Hadoop framework

Undoubtedly the Apache Hadoop software library must fulfill the functions of data acquisition, storage and management, and data analytics according to the requirements of a Big Data processing system [9]. In the Hadoop framework, the key functioning modules include HDFS, MapReduce, HBase, and Chukewa, and these core library modules are hierarchically organized into a layered-protocol stack as shown in Figure 11.1. We will briefly introduce the functional modules according to the value-chain view of Big Data systems.

- *Data acquisition.* In Hadoop, this module covers two submodules: Flume and Sqoop. Flume is a distributed file system that can efficiently gather, assemble, and trans-

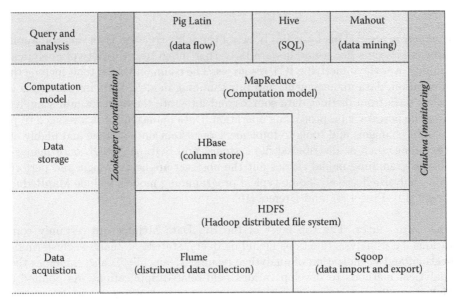

FIGURE 11.1: The layered structure of the Hadoop core software library nearly capturing all the functions of a Big Data system.

mit a large variety of data from various data sources into a centralized store. Sqoop accomplishes easy import and export of data among structured data scores and Hadoop.

- *Data storage.* This module includes two main functioning modules: HDFS and HBase, in order to effectively implement the data storage of a Big Data system. HDFS is a distributed file system running on commodity hardware, which refers to the design of the GFS system, and now has become the primary data storage substrate of Hadoop applications. In HDFS, the storage infrastructure contains two classes of nodes: NameNode and DataNode, in which a NameNode is used to manage the file system metadata and a DataNode stores the actual data. Moreover, any data file is divided into one or more blocks and these blocks are saved into a set of DataNodes, and any one block will be stored at several different DataNodes to avoid data loss. Apache HBase is a column-oriented database stores similar to Google's Bigtable; Hbase can serve as the input and output for MapReduce jobs performed in Hadoop and is accessed through Java API, REST, Avor, or Thrift APIs.

- *Computation Model.* This is implemented by Hadoop MapReduce which nearly imitates the Google's MapReduce and is made up of a single master *JobTracker* and one slave *TaskTracker* per cluster node. *JobTracker* is in charge of scheduling the slave jobs, monitoring, and re-executing the failed jobs. *TaskTracker* is responsible for the execution of tasks as directed by the master. Meanwhile, the MapReduce framework and HDFS are carried out on the same set of nodes.

- *Query and Analysis.* This layer is leveraged to perform the SQL-like structured data query and further data mining and analysis, including Pig Latin, Hive, and Mahout. Pig Latin is suitable for data flow tasks and creates the sequences of MapReduce programs, and Hive promotes easy data summarization and random queries, while Mahout performs data mining on top of the Hadoop framework and adopts the MapReduce paradigm. Mahout also provides many core algorithms used to cluster, classify, and spam filter.

- *Data Management.* This function is completed by Zookeeper and Chukwa, which can manage and monitor distributed applications on top of the Hadoop platform. Among them, Zookeeper provides a centralized service to maintain configuration, naming, distributed synchronization, and data grouping. Chukwa is utilized to monitor the system status and perform data analysis.

11.2.3.2   Improvements of the Hadoop framework

Although Hadoop has been a great success and has many advantages, it is in its infancy and still lacks some mature features found in traditional database management systems. As an example, Hadoop must parse each item when it imports and exports the data objects since it has no schema or index, and this results in the degradation of performance. Furthermore, Hadoop adopts the batch-processing paradigm and only provides a single fixed dataflow, and thus many complex algorithms are difficult to implement through the simple map and reduce operations in a job. At present, several typical approaches to improve the performance of the Hadoop framework are summarized in Table 11.3.

Regarding the improvements of Hadoop platforms, we recommend readers refer to [30,31] for complete knowledge of this topic. Again, as mentioned earlier, Hadoop is a batch-like platform to handle massive data, and thus it deserves to be greatly improved when we confront real-time applications. It is worth noticing that Storm [32] is a good candidate for

TABLE 11.3: Several approaches to improve the performance of the Hadoop framework

Items to Be Improved	Improvements	Cases
Flexible data flow	Provide loop-type algorithms	HaLoop [17], Twister [18]
Blocking operators	Avoid blocking operations	MapReduce Online [19]
I/O optimization	Reduce I/O cost	Hadoop++ [20], Hadoop DB [21]
Scheduling	Enhance resource utilization	Delay[22], dynamic scheduler [23]
Joins	Dispose of multiple inputs	Map-side [24], Reduce-side [25]
Performance tuning	Optimize applications	Automatic [26], Static [27]
Energy optimization	Provide energy-efficient algorithms	Covering-set [28], All-in [29]

processing unlimited streams of data, and can be applied widely in many applications, such as real-time analytics, online machine learning, continuous computation, distributed RPC, and so on.

11.3 Data Analyses inside Complex Real-World Systems

With the rapid development of computing hardware and data processing technologies after the 1990s, statistical analysis regarding many real-world systems can be leveraged to analyze in depth and/or mine the static or dynamic properties hidden behind these systems, such as small-world and scale-free properties. Meanwhile, Web2.0 and increasing information growth over the Internet present unprecedented challenges toward data processing techniques and analytics. At present, complex networks are inevitably correlated with Big Data technology and we will present some preliminary descriptions in brief as follows so as to lay down a solid foundation for the case studies.

11.3.1 Statistical analytics of real-world complex networks

Since the seminal works with regard to small-world and scale-free networks, complex networks have become an active topic within complex systems modeling and systems engineering over the past two decades. In particular, the analysis of real-world systems from the viewpoint of complex networks has attracted a lot of research efforts in the field of Big Data analytics and applications among academic communities [33–37].

11.3.1.1 Quantities characterizing complex networks

A network with complex topology is usually represented by a graph $G = (V, E)$ where V denotes a node set and E an edge set. $N = |V|$ and $|E|$ denote the number of nodes and of edges, respectively. Structural properties of a network can be characterized by a variety of parameters. Note that in this study, only undirected networks without redundant links (which means that there exists more than one link between two nodes) and self-connected nodes are considered.

Node degree. For node i, node degree k_i represents the number of connections it owns. Thus $<k>$, k_{max}, and k_{min} are defined as the average, maximal, and minimal node degree over the whole network, respectively.

The properties of a network can be expressed via its adjacency matrix $A = (a_{ij})_{N \times N}$. $a_{ij} = a_{ji} = 1$ if there exists an edge connecting node i and j, and 0 otherwise. In the case of the weighted network, for a link between vertex i and j, a weight or value w_{ij} is assigned to characterize the intensity of interaction of two nodes.

Degree distribution. The degree distribution $p(k)$, which indicates the probability that a randomly selected node has k connections, is the most important and essential statistical character of a large-scale, complex network. The cumulative degree distribution $P(k)$ is defined as

$$P(k) = \sum_{\kappa=k}^{k_{max}} p(\kappa) \tag{11.1}$$

Many real networks, such as the Internet, the World Wide Web, the movie actor collaboration network, and so on, appear to have a $p(k)$ that follows a power-law form, $p(k) \sim k^{-\gamma}$ and $2 \le \gamma \le 3$ [33]. Noticeably, for a network with power-law degree distribution $p(k) \sim k^{-\gamma}$, the cumulative degree distribution $P(k)$ also follows a power-law form and the scaling exponents have the following relation: $P(k) \sim k^{-(\gamma-1)}$ [34].

Characteristic path length, communication efficiency. The quantities used to measure how efficiently the information is exchanged over the whole network include the characteristic path length L and the global communication efficiency ε. For node pair (i, j), let l_{ij} denote the number of links in the shortest path connecting the two vertices; then L is defined to be the average value of the length of the shortest paths between any vertex pairs in the network:

$$L = \frac{1}{N(N-1)} \sum_{i,j} l_{ij} \tag{11.2}$$

However, this definition is valid only if the network G is totally connected, which means that there must exist at least a path connecting any pair of nodes with a finite number of steps. Otherwise, when from vertex i we cannot reach vertex j, $l_{ij} = +\infty$ and consequently L becomes invalid.

To characterize the ability of any two nodes to communicate with each other in nonconnected networks, global communication efficiency ε has been introduced [38]. The efficiency between vertex i and j is assumed to be inversely proportional to the shortest distance: $\varepsilon_{ij} = 1/l_{ij}$. With this definition, when there is no path between i and j, $l_{ij} = +\infty$; thus $\varepsilon_{ij} = 0$. Similarly, ε is defined as the average of ε_{ij} over all pairs of nodes:

$$\varepsilon = \frac{1}{N(N-1)} \sum_{(i,j)} \frac{1}{l_{ij}} \tag{11.3}$$

Clustering coefficient. Clustering of a node gives the probability that two nearest neighbors are connected with each other [39]. The clustering of i in a nonweighted network is defined as

$$c_i = \frac{1}{ki(ki-1)} \sum_{j,h} a_{ij} a_{ih} a_{jh} \tag{11.4}$$

If node i has less than two neighbors, c_i is set to 0. Clustering c_i measures the local cohesiveness of node i while the average clustering coefficient $cc = \sum_i ci/N$ measures the global density of interconnected triples in the network. In addition, the organizational structure of a network can be further studied via the degree-dependent average clustering coefficient $c(k)$ [40]:

$$c(k) = \frac{1}{Np(k)} \sum_{i:ki=k} ci \qquad (11.5)$$

which is the mean clustering for nodes with degree k.

Assortative mixing. The degree correlation is another important piece of information about the network. Recall that the network is said to be assortative if the nodes tend to connect to other nodes which have similar (dissimilar) properties. Previous studies found that social networks tend to be assortative while technological and biological networks are generally disassortative [41].

To characterize the degree–degree correlations between nodes, Newman [41] proposed the concept of assortativity coefficient (ac) based on the computation of the Pearson coefficient of the endpoint degrees of each link. ac is defined as

$$\mathrm{ac} = \frac{M^{-1}\sum_l k_i k_j - [M^{-1}\sum_l \frac{1}{2}(k_i + k_j)]^2}{M^{-1}\sum_l \frac{1}{2}(k_i^2 + k_j^2) - [M^{-1}\sum_l \frac{1}{2}(k_i + k_j)]^2} \qquad (11.6)$$

where i and j are two ending nodes of link l with degree k_i and k_j, respectively, and M is the total number of links. $-1 <= \mathrm{ac} <= 1$. If $\mathrm{ac} > 0$, the high-degree nodes are preferentially connected with other high-degree nodes, and the network is called an assortative network; if $\mathrm{ac} < 0$ it is called a disassortative network while high-degree nodes attach to low-degree nodes.

To probe the degree correlation of the network, one may also study the average degree of the nearest neighbors of one node, which is defined as [42]

$$knn, i = \frac{1}{ki} \sum_{j \in \tau(i)} kj \qquad (11.7)$$

The degree-dependent average nearest-neighbors degree $k_{nn}(k)$ is the mean of $k_{nn,i}$ restricted to the class of nodes with degree k. It is noticed that if $k_{nn}(k)$ is an increasing (decreasing) function of k, nodes with high degree tend to connect to other nodes with high (low) degree, and thus the network becomes assortative (disassortative).

11.3.1.2 Case studies

Here, we will analyze 10 real-world systems from the viewpoints of complex networks. (1) *EUair*: Network of European airlines [43]; (2) *USair*: Network of US airlines[44]; (3) *EUPowGrid*: Network of European power grid [44]; (4) *Celegans*: Network of the nematode worm *Caenorhabditis elegans* [45]; (5) *Kohonen*: Network composed of articles with topic "self-organizing maps" or references to "Kohonen T" [45]; (6) *SciMet*: Network composed of articles from citing Scientometrics [45]; (7) *SmaGri*: Network composed of citations to Small & Griffth and Descendants [45]; (8) *Yeast*: Network of protein interaction [45]; (9) *Facebook*: Social circles from Facebook [46]; (10) *PolBlogs*: Network of hyperlinks between weblogs on US politics [45]. Table 11.4 lists several important topological properties of these 10 real-world networks. In this table, N and M are the numbers of nodes and links, respectively. $<k>$ and $k_{\max}$ denote the average and maximal value of node degree, respectively. cc is the average clustering coefficient. ac is the assortative coefficient. $<l>$ and d are the average and maximal shortest distance between node pairs, respectively.

TABLE 11.4:   The topological properties of 10 real-world networks

Network	N	M	$<k>$	$k_{\mathbf{max}}$	cc	$<l>$	d	ε	ac
EUair	417	5906	14.16	112	0.42	2.76	7	0.40	-0.15
USair	332	4252	12.81	139	0.63	2.74	6	0.41	-0.21
EUPowGrid	1494	4312	2.89	13	0.11	18.88	48	0.07	-0.12
Celegans	297	4296	14.46	134	0.29	2.45	5	0.44	-0.16
Kohonen	3704	25346	6.84	740	0.25	3.67	12	0.30	-0.12
SciMet	2678	20736	7.74	164	0.17	4.18	12	0.26	-0.04
SmaGri	1024	9832	9.60	232	0.31	2.98	6	0.35	-0.19
Yeast	2375	23386	9.85	118	0.31	5.10	15	0.22	0.45
Facebook	4039	176468	43.69	1045	0.61	3.69	8	0.31	0.06
PolBlogs	1222	33428	27.36	351	0.32	2.74	8	0.40	-0.22

11.3.2   Attack robustness and optimization of real-world networked systems

Recently many theoretical and experimental works have been performed to analyze the attack robustness of a vast variety of different networked systems [33–35]. A network or system is said to be robust if as many as possible elements remain globally connected even after a fraction of nodes or edges have been removed. Albert et al. [47] found the *"robust yet fragile"* generic property of these networks: these networks display an unexpected degree of robustness to random failure; however, they are extremely vulnerable to intentional attacks. This work has triggered successively enormous interest on the effect of different topological properties on the attack robustness of networks, including the degree distribution, centrality, assortativity, interaction strength of the edges, and so on [48–57]. Meanwhile, many researchers also have devoted a great deal of effort to the study of how to construct optimal networks and how to make existing networks more robust against random and/or targeted attacks [58–72].

In a recent work [58], Schneider et al. propose a new measure for network robustness under malicious attacks on highly connected nodes. For an existing network with a prescribed degree distribution, based on a simple greedy algorithm, they also present a link-rewiring method which can significantly improve the robustness. It has been discovered that independent of the degree distributions of the networks, the most robust structures exhibit an "onion-like" topology in which high-degree nodes form a core surrounded by rings of nodes with decreasing degree.

11.3.2.1   Robustness measure—R

When nodes are gradually damaged due to random failures or targeted attacks, a network may be split into several unconnected parts. In percolation theory [58–60], the robustness of networks is usually measured by the percolation threshold f_c, the critical fraction of nodes attacked at which the whole network collapses completely. However, in realistic cases, this measure overlooks situations in which the networks suffer big damage but are not completely collapsing. Recently, Schneider et al. [58] proposed a novel measure, node robustness R, to evaluate the robustness of networks under attacks considering the size of the largest connected component during all possible malicious attacks, namely

$$R = \frac{1}{N} \sum_{q=1}^{N-1} S(q) \tag{11.8}$$

where N is the total number of nodes in the initial network and $S(q)$ is the relative size of the largest connected component after removing q nodes with the highest degrees. The normalization factor $1/N$ ensures $1/N \leq R \leq 0.5$. Two special cases exist: $R = 1/N$ corresponds to star-like networks while $R = 0.5$ represents the case of a fully connected network. Generally, the larger the value of R, the more robust is the network at resisting intentional attacks on high-degree nodes.

11.3.2.2 Optimization method

Based on the measure R, a link-rewiring method [58] is also developed to improve the robustness by exchanging only a small number of edges while keeping the degree of every node unchanged and the whole network connected. The optimization method is described as follows. Starting from a given network G_0, two randomly selected edges are swapped, that is, the edges e_{ij} and e_{mn}, which connect node i with node j, and node m with node n, respectively, become e_{in} and e_{mj}, if and only if the value of R is increased, that is, calculating the values of R before and after the swap, denoted to be R_o and R_n, respectively, only if $R_n > R_o$ is the swap accepted. This procedure is repeated with another randomly chosen pair of edges until no further substantial improvement is achieved.

A simple example is shown in Figure 11.2. The initial network G_0 has $N = 6$ nodes and $L = 6$ undirected edges (see Figure 11.2a), which gives us $R_0 \approx 0.22$. After performing the swap on edges $(1, 2)$ and $(5, 6)$, the resultant network G_1 (Figure 11.2b), with new edges $(1, 6)$ and $(2, 5)$, has an improved R value, $R_1 = 0.25$. Then a further increase of R can be obtained in network G_2 (Figure 11.2c) with $R_2 \approx 0.33$. Obviously, G_0, G_1, and G_2 have the same degree sequences.

11.3.2.3 Onion-like topology

To verify the effectiveness of the network optimization algorithm proposed earlier, we apply the algorithm to a scale-free network with $N = 50$ nodes, average node degree $<k> = 4$ and degree distribution $p(k) \sim k^{-3}$. Figure 11.3 shows the topologies of the network before and after optimization. As shown in Figure 11.3b, the optimized network exhibits a type of "onion-like" topology, that is, a core composed of high-degree hub nodes exists which is hierarchically surrounded by rings of nodes with decreasing degree. Also, the number of edges between nodes with equal degree is greatly increased, which are highlighted with bold black lines in Figure 11.3.

11.3.2.4 Numerical simulations and analysis

In this section, the proposed optimization algorithm is applied to two real-world networked systems, which are both air transport networks whose nodes represent airports and where a link exists between two nodes if there is a direct flight between corresponding airports. The first network is the air transport network in Europe [43] (labeled as *EUair*) with 417 nodes, 2953 links, and average node degree $<k> \approx 14.16$; the other one is the air

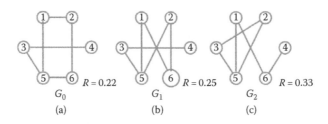

FIGURE 11.2: The R values of a network (a) before and (b,c) after the edge swaps.

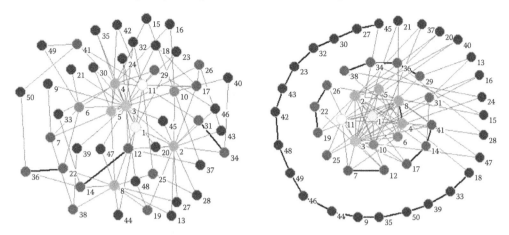

FIGURE 11.3: Visualization of network topologies of a scale-free network before and after optimization. The networks are size $N = 50$, average node degree $<k> = 4$, and degree distribution $p(k) \sim k^{-3}$. Nodes with similar degree have the same shading. Edges between nodes with equal degree are highlighted with bold black lines. (a) Initial network with $R = 0.1404$ and without any optimization; and (b) onion-like topology of optimized network with enhanced $R = 0.3828$.

transport network in the United States [44] (labeled as *USair*) with 332 nodes and 2126 links; thus the average node degree is $<k< \approx 9.45$.

Robustness analysis before and after optimization. An attack can be simulated by the removal of nodes and links from the network. An intentional attack is targeted to disrupt the network by removing the most "important" nodes; here, we choose high-degree nodes as important ones. After node and link removal, since only nodes belonging to the largest connected component are still functional, the relative size S of the largest component is used to probe the functional integrity and quantify the robustness of networks after an attack. $S = N'/N$ where N' and N denote the number of nodes in the largest connected component and that in the initial network, respectively. Obviously, the larger S is, the more nodes remain in the largest component, which indicates the network is more robust under attacks. The critical threshold f_c, the minimum fraction of nodes to be removed from the network which causes the whole network to collapse, that is, $S \approx 0$, is another important quantity measuring the performance of the system. The larger f_c is, the more robust is the network.

To compare the performance of optimized and initial networks, extensive numerical experiments have been performed. Figure 11.4 shows the performance of initial (curves with empty symbols) and optimized networks (curves with solid symbols) under both intentional and random attacks. RA represents random attack while IA stands for intentional attack on high-degree nodes.

The lines with square symbols in Figure 11.4 represent the relative size S of the largest connected component after a fraction f of high-degree nodes are attacked. Note that, before and after optimization, under intentional attacks, for both networks, the critical percolation threshold f_c, at which the network becomes fully fragmented and the largest connected component disappears, does not change much. Nevertheless, when $f < f_c$, it can be observed that suffering the same level of attack (i.e., the same values of f), the optimized networks, compared to the initial ones, have significantly higher values of $S(f)$. For example, for the *EUair* network (Figure 11.4a), when $f = 0.2$, $S(f)$ can be increased from 0.4 to 0.6, that is, after removing 20% of the high-degree nodes from the network, before optimization, only

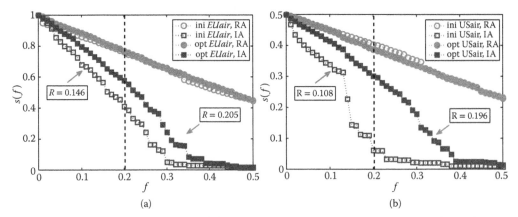

FIGURE 11.4: The relative size $S(f)$ of the largest connected component after a fraction f of nodes are attacked from the (a) *EUair* and (b) *USair* networks. *RA* represents random attack while *IA* stands for an intentional attack on high-degree nodes. The lines with empty symbols correspond to initial networks, while the lines with solid symbols represent optimized networks. Each point of the random attack case is averaged over 10 independent runs.

40% of the nodes in the original network remain connected while others (40%) break off from the main "body." However, after applying the optimization algorithm on the *EUair* network, the giant connected component consists of 60% of the nodes with only 20% of the nodes losing connection from it. In addition, when $f = 0.2$, the values of $S(f)$ increase from 0.062 to 0.315 for *USair* (Figure 11.4b). Moreover, when f is closer to f_c, the improvement of $S(f)$ becomes more pronounced. All these results demonstrate that the optimized networks become significantly more robust under intentional attacks, implying that intentional attacks can cause less damage to the robust network or many more hub nodes have to be attacked to significantly damage the system.

For comparison, the optimized networks' resilience to random attacks is also investigated. The lines with symbols of empty circles and solid circles in Figure 11.4 correspond to $S(f)$ after attacking a fraction f of randomly chosen nodes. Interestingly, although the robustness of optimized networks against intentional attacks is clearly improved, we observe that $S(f)$ remains nearly the same for both the initial and optimized networks, verifying that with small changes in the network structure (at low cost) the robustness of diverse networks can be improved dramatically whereas their ability to resist random attacks or failures remains unchanged.

Changes of topological properties during optimization. Will the optimization also affect other topological properties of the network? This question motivates us to investigate the changes of several important topological properties during the structure optimization aiming at maximizing R, including the natural connectivity (nc) [25], the average clustering coefficient (cc) [7], and the assortativity coefficient (ac) [9].

Considering that the network's invulnerability is rooted in the alternative path redundancy, Wu et al. [25] proposed a measure named natural connectivity to characterize the invulnerability of the network, which is defined to be

$$nc = ln(\frac{1}{N} \sum_{i=1}^{N} e\lambda i) \qquad (11.9)$$

where $\lambda_1 > \ = \lambda_2 > \ = \ldots, \lambda_N$ are the eigenvalues of the adjacent matrix of network G. A larger value of nc indicates the increase of the extent of alternative path redundancy between nodes, thus improving the invulnerability of the whole network.

As shown in Figure 11.5a, the values of R increase monotonously during the optimization process (i.e., for *USair*, R increases from 0.108 to 0.196; for *EUair*, R increases from 0.146 to 0.205). Meanwhile, it can be observed from Figure 11.5b that nc increases with R (43.428–44.260) for *EUair* while as R is increased, nc decreases for *USair*. However, the change is very small (from 35.428 to 34.906). This result demonstrates that in some extents these two measures R and nc almost coincide with each other in characterizing network robustness.

Next we study the average clustering coefficients of networks before and after the optimization. From Figure 11.5c it is clear that all the values of cc of these two networks decrease during the optimization procedure, which indicates the decrease of the intensity of nodes in corresponding networks (*USair*: from 0.625 to 0.386; *EUair*: from 0.775 to 0.691). Since initially *USair* and *EUair* both have "small-world" properties with higher values of clustering coefficient, the optimization introduces substantial changes of cc.

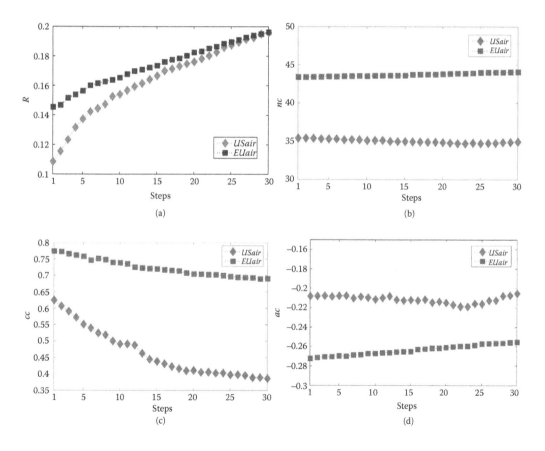

FIGURE 11.5: Changes of robustness measure R and three topological properties during optimization for *USair* (curves with ◆) and *EUair* (curves with ■) networks. (a) R: robustness measure; (b) nc: natural connectivity; (c) cc: clustering coefficient; (d) ac: assortativity coefficient.

To unveil the microcosmic connectivity pattern between nodes, the changes of assortativity coefficients of these networks are also explored. Previous empirical study showed that almost all the social networks show an assortative mixing pattern while many technological and biological networks are disassortative [41]. *USair* and *EUair* are both typical technological networks; thus, for the initial networks $ac < 0$. In addition, for *USair* and *EUair*, after optimization the values of ac increase but the properties of disassortativity remain unchanged.

Discussion. In this section, we focus on the problem of improving the robustness of an existing network against intentional attacks. An optimization model is established with the objective of maximizing the robustness measure R. An efficient method which can improve the robustness of a given network by rewiring links while keeping the degree of every node fixed is applied to two airport transport networks. The numerical simulation results demonstrate that, compared with the initial networks, the robustness of the optimized networks in resisting intentional attacks on high-degree nodes is greatly improved, while as for random attacks, there is no evident difference between the initial networks and the optimized ones.

Furthermore, we investigate the effect of the structure modification on other topology properties, including the natural connectivity, clustering coefficient, and assortativity coefficient. The changes of these properties during the optimization procedure are analyzed. Two robustness indexes R and nc are found to be consistent in characterizing the network's robustness under attacks. Moreover, the extent of clustering of these networks all decrease, thus in network design, the balance between improving a network's resilience to attacks and keeping a higher value of the clustering coefficient should receive careful consideration. Lastly, it is found that the structural modification does not significantly affect the two real-world technical networks.

11.4   Trust-Enhanced Cloud Service Selection Method

With the development of the Internet and IT, cloud computing has become the core of the new generation computing paradigm. Information services based on cloud computing are increasingly vital and popular. Effective selection and recommendation models, which offer optimal cloud services to users, are in high demand for the increasing number of cloud services on the cloud platform [73–79].

11.4.1   Trust modeling of cloud service selection

In recent years, a growing number of enterprises offer a large number of different kinds of cloud services depending on the open cloud computing environment. It is a challenging task to select the optimal cloud services from a set of cloud service candidates. In e-commerce, trust is considered a significant factor that affects customers' decision-making. Research on the social network finds that people tend to communicate with those who have similar characteristics, based on the homophily feature of the social network. The more interactions there are, the higher degree of trust there is. According to previous work, interaction frequency can reflect the trust degree to some extent on chat platforms such as QQ and MSN. Therefore, we can use interaction frequency to measure trust.

11.4.1.1   Direct trust degree

In the social network, users can be represented by nodes and the directed edge from one node to another stands for trust. Here, we classify the trust into three categories: direct, indirect, and hybrid trust. Direct trust means that there exists direct interaction between users. Considering all trust paths between two users, it is defined as the hybrid trust.

It is worth noting that in social psychology, it is not a simple linear functional relationship between the interaction frequency and trust degree. In this chapter, we propose a segment trust utility function to explain this relationship. The function we propose complies with the law of diminishing marginal utility between $f_{\min}$ and $f_{\max}$. In the study of economy, marginal utility measures the additional satisfaction obtained from consuming one additional unit of a good. Diminishing marginal utility is the principle that as more units of a good are consumed, the consumption of additional amounts will yield smaller additional utility. Under the same rule, when the interaction frequency is between the interval of $[f_{\min}, f_{\max}]$, the increasing rate of trust degree decreases gradually. The trust utility function $t_{p,q}$ used to calculate the trust degree between u_p and u_q is defined as follows:

$$t_{p,q} = \phi(x) = \begin{cases} k_1 x, 0 \leq x \leq f_{\min} \\ k_2 (x+\beta)^{k_3}, \quad f_{\min} \leq x \leq f_{\max} \\ k_4 x + b, f_{\max} \leq x \leq f_M \end{cases} \tag{11.10}$$

k_1, k_2, k_3, k_4, and b are expressed as follows:

$$\begin{aligned} k_1 &= f_{\min} \\ k_2 &= \alpha(f_{\min}+\beta)^{\frac{\ln\alpha-\ln(1-\alpha)}{\ln(\beta+f_{\max})-\ln(\beta+f_{\min})}} \\ k_3 &= \frac{\ln(1-\alpha)-\ln\alpha}{\ln(\beta+f_{\max})-\ln(\beta+f_{\min})} \\ k_4 &= \frac{\alpha}{f_M-f_{\max}} \\ b &= \frac{(1-\alpha)f_M-f_{\max}}{f_M-f_{\max}} \end{aligned} \tag{11.11}$$

11.4.1.2   Hybrid trust degree

In the social network, apart from direct trust, existing research interests focus on indirect trust, which usually involves more than one indirect trust path between two users. Hybrid trust is composed of the direct and indirect trust paths. We put the trust paths from u_p to u_q into a set $Path_{p,q}$. The paths in this set are all indirect trust paths between two users, and the trust transitivity distances are no longer than the maximum trust transitivity distance $L_{\max}$ we set. It can be defined as

$$Path_{p,q} = \{r_1, \ldots, r_N\} \; 0{<}L_{r_e} \leq L_{\max}, e \in \{1, .., N\} \tag{11.12}$$

According to Equation 11.13, we suppose there are M users between u_p and u_q in path r_e. The indirect trust degree $t_{ind}(p,q,r_e)$ between u_p and u_q in path r_e can be calculated as follows:

$$t_{ind}(p,q,r_e) = t_{p,1} t_{1,2} \ldots t_{M-1,M} t_{M,q} \tag{11.13}$$

To estimate the hybrid trust degree, we integrate the direct trust degree with the indirect trust degree. We use the parameter $\lambda(\lambda \in [0,1])$ as the weight of direct trust. The hybrid trust degree $T_{p,q}$ can be computed as

$$T_{p,q} = f(t_{p,q}, t_{ind}) = \begin{cases} \sum\limits_{i=1}^{N} \frac{t_{ind}(p,q,r_e)}{N}, L \neq 0 \\ \lambda t_{p,q} + (1-\lambda)\frac{t_{ind}(p,q,r_e)}{N}, L = 0 \end{cases} \tag{11.14}$$

11.4.2   Cloud service selection process

In this section, we estimate the basic similarities based on the Pearson correlation coefficient (PPC), and then we utilize the trust degrees to enhance the basic similarities.

11.4.2.1   Multi-QoS properties combination

QoS is widely adopted to describe nonfunctional characteristics of services. It is an objective description of service performance, and is also the expression of the service perceived by the users. As various QoS properties have different dimensions and ranges of values, we normalize their values to the interval [0, 1] and classify them into two categories: "cost" and "benefit." If one QoS property is a "benefit" attribute, for example throughput, the higher its value, the more probability for users to choose it.

Suppose $CS = \{cs_i | i \in \{1, \ldots, n\}\}$ is the set of n cloud services. $U = \{u_p | p \in \{1, \ldots, m\}\}$ is the set of m cloud service users. $QoS = \{Q^k | k \in \{1, \ldots, l\}\}$ is the set of l QoS matrices, with different QoS properties. $Q^k_{p,i}$ represents the observed QoS value of the cloud service cs_i invoked by u_p in the matrix Q^k, where $i \in \{1, \ldots, n\}$, $p \in \{1, \ldots, m\}$, and $k \in \{1, \ldots, l\}$. According to different attributes of QoS ("benefit" or "cost"), when $Q^k_{p,i} \neq \emptyset$, the normalized QoS value $Q^k_{p,i}$ can be calculated as follows:

$$Q^k_{p,i} = \begin{cases} \frac{Q^k_{p,i} - Q^k_{\min}}{Q^k_{\max} - Q^k_{\min}}, & Q^k_{p,i} \text{ is "benefit"} \\ \frac{Q^k_{\max} - Q^k_{p,i}}{Q^k_{\max} - Q^k_{\min}}, & Q^k_{p,i} \text{ is "cost"} \end{cases} \qquad (11.15)$$

Since QoS has the multiproperty feature each of which affects the QoS with varying weight, to support user selection of cloud services, we compose multiple QoS properties on the basis of their weights. This is defined as

$$Q_{p,i} = \sum_{k=1}^{l} w_k Q^k_{p,i} \qquad (11.16)$$

11.4.2.2   Similarities estimation

In traditional e-commerce recommendation systems, PPC has been widely applied to estimate the similarities between different users.

Given the QoS value $Q_{p,i}$, we can obtain the numerical distance between different users by employing PPC based on user-item regularly, which is widely employed in recommendation systems for similarity calculation. $Sim_{p,q}$ between u_p and u_q can be calculated as follows:

$$Sim_{p,q} = \frac{\sum\limits_{cs_i \in CS_{p,q}} \left(Q_{p,i} - \overline{Q}_p \right) \left(Q_{q,i} - \overline{Q}_q \right)}{\sqrt{\sum\limits_{cs_i \in CS_{p,q}} \left(Q_{p,i} - \overline{Q}_p \right)^2} \sqrt{\sum\limits_{cs_i \in CS_{p,q}} \left(Q_{q,i} - \overline{Q}_q \right)^2}} \qquad (11.17)$$

11.4.2.3   Trust-enhanced similarities estimation model

In the social network environment, trust between entities has a huge impact on the decision making of users. Especially in cloud service selection, trust degree will greatly improve the reliability of the selection based on similarity. However, in traditional selection methods, scholars only consider the scores in the user-item matrix and find the similarities among users. In the meantime, trusts within these users are ignored. We use the trust degree to enhance the PPC similarity $Sim_{p,q}$ and obtain trust-enhanced similarity $Sim_{p,q}^T$ as follows:

$$Sim_{p,q}^T = [(\delta - 1)T_{p,q} + 1]Sim_{p,q} \tag{11.18}$$

We can change the value of δ to adjust the enhancement strength that trust degree has on the similarities. When $\delta = 1$, $T_{p,q}$ has no influence on $Sim_{p,q}$, that is, we do not take trust into account. When $\delta > 1$, if $T_{p,q} = 0$, there is no trust between u_p and u_q, $Sim_{p,q}^T = Sim_{p,q}$. If $T_{p,q} = 1$, u_p and u_q have the maximum trust degree, $Sim_{p,q}^T = \delta \times Sim_{p,q}$. We explain in the experiment that the upper bound of δ is not set here since the effect of trust enhancement on the basic similarity will not increase unlimitedly with the value of δ growing.

11.4.2.4 Cloud service ranking based on predicted QoS values

After calculating the trust-enhanced similarities among cloud service users, we can obtain similar neighbors of the target user by ranking the Sim^T values. Sim^T values from the neighbor set and employ the *Top-K Sim^T* of users for predicting missing QoS values of the target user as follows:

$$Q_{p,i}^{pre} = \bar{Q}_p + \frac{\sum\limits_{q \in \varpi_p} Sim_{p,q}^T(Q_{q,i} - \bar{Q}_q)}{\sum\limits_{q \in \varpi_p} Sim_{p,q}^T} \tag{11.19}$$
$$\scriptstyle p \in m, i \in n$$

11.4.3 Experiment and evaluation

In this section, we conduct several experiments to adjust parameters in order to get the optimal solutions and prove the effectiveness of our method. By exploring three critical parameters, we show that our model is able to produce high-quality cloud service ranking results.

11.4.3.1 Impact of β

β is the parameter that adjusts the curvature of the function. With the value of β adjusting, the extent of the impact of additional interaction frequency on the trust degree changes. Obviously, while the three curves in Figure 11.6 present similar trends, the results are better with larger user numbers. We show that our trust utility function is better than the simple linear function, and confirm that interaction frequency and trust degree are usually not a linear relationship when the interaction frequency is in the interval of specific values.

11.4.3.2 Impact of $L_{\max}$

$L_{\max}$ is the maximum trust distance, which determines the building of the trust network. It determines how many intermediate users are allowed to propagate a trust relationship between two strangers in the network. According to the six degrees of separation, every two strangers in the world can be connected in less than six persons. Therefore, we set $L_{\max}$ to 0, 1, 2, 3, 4 and 5. Specifically, $L_{\max} = 0$ means this experiment does not consider indirect trust. These observations indicate the results are better as the number of users increases, and using indirect trust relationships has better results than only using direct trust. As Figure 11.7 shows, when the maximum trust distance is equal to 5, the accuracy drops

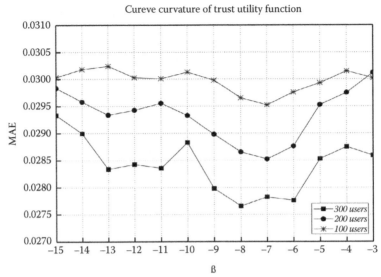

FIGURE 11.6: Impact of β on accuracy.

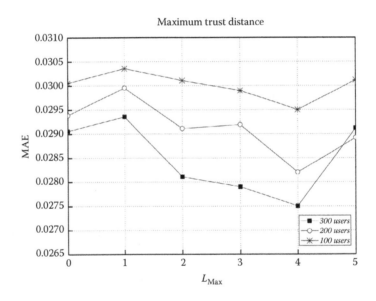

FIGURE 11.7: Impact of $L_{\max}$ of accuracy.

significantly. Therefore, while modeling trust relationships within a network, the maximum trust distance should be set with care in order to obtain the highest accuracy.

11.4.3.3 Impact of δ

δ is the maximum trust-enhanced factor which determines the strength that trust degree has on the similarities. It is the most important parameter in our method, as it controls the influence of the trust relationships. Figure 11.8 proves the effectiveness of utilizing a trust-enhanced similarity, as MAE has a better value at $\delta = 4$ than that when $\delta = 1$. Nevertheless,

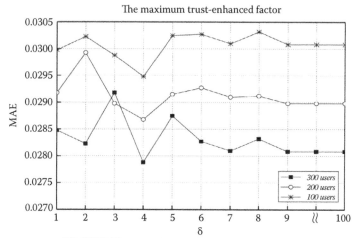

FIGURE 11.8: Impact of δ of accuracy.

the benefit of the trust relationship is not unlimited, since MAE remains steady when δ is larger than 9.

11.4.3.4 Summary

In this section, we propose a trust-enhanced cloud services selection approach which takes trust into the QoS value prediction. We combine various types of QoS properties together and use the trust intensity to enhance the similarities. The similarities integrate the numerical distance and experience usability. Then, we predict the missing QoS values based on the trust-enhanced similarity. Then, we rank the cloud services based on predicted QoS values. We prove our cloud services selection method has better QoS properties than other methods and obtain the optimum values of parameters by experiments.

With the advent of the Big Data era, traditional recommendations or selection methods are weak for massive data analysis and processing. In future work, we will focus on the statistical forecasting methods to predict missing QoS values to meet the challenges of the Big Data environment.

11.5 Conclusions

First, we reviewed recent progress on data analytics and their applications in the area of Big Data. Then, we introduced some basic concepts, definitions, and processing frameworks of Big Data. Subsequently, combined with network science theory and some cases, Big Data analytics have been applied to finding network topologies, attack resilience, and optimization on some real-world networked systems. As demonstrated in this chapter, information processing techniques greatly help to promote the rapid development of complex networks and complexity science. Finally, based on a Big Data background, we talk about the information services provided by the cloud computing platform, especially for effective selection and recommendation models offering the optimal cloud services to users under personalized environments and heterogeneous demands.

Acknowledgments

We greatly thank the National Natural Science Foundation of China for financial support under grants 61573199, 71201042, 61203138, 61374169. CYX acknowledges the fund from the Scientific Research Foundation for the Returned Overseas Chinese Scholars (Ministry of Education); SWS and CYX are also funded by the "131" Innovative Talents Program of Tianjin.

References

1. Gantz J and Reinse D. 2012. The digital universe in 2020: Big Data, bigger digital shadows, and biggest growth in the Far East, in *Proc. IDC iView, IDC Anal.* Future.

2. Manyika J, Chui M, Brown B et al. 2011. *Big Data: The Next Frontier for Innovation, Competition, and Productivity.* San Francisco, CA: McKinsey Global Institute pp. 1–137.

3. Cukier K. 2010. Data, data everywhere. *Economist 394:* 3–16.

4. Lohr S. 2012. The age of Big Data. *New York Times* [Online]. 11. Available: http://www.nytimes.com/2012/02/12/sunday-review/big-datasimpact-in-the-world. html?pagewanted=all&r=0.

5. Noguchi Y. 2011. Following digital breadcrumbs to Big Data gold, *National Public Radio, Washington, DC, USA* [Online]. Available: http://www.npr.org/2011/11/29/142521910/.

6. Noguchi Y. 2011. The search for analysts to make sense of Big Data, *National Public Radio, Washington, DC, USA* [Online]. Available: http://www.npr.org/2011/11/30/142893065/the search for analysts to -make-%sense-of-big-data.

7. White House 2012. Fact sheet: *Big Data across the Federal Government* [Online]. Available: http://www.whitehouse.gov/sites/default/files/microsites/ostp/big_data%_fact _sheet_3_29_2012.pdf.

8. Hinton GE, Osindero S, and Teh Y. 2006. A fast learning algorithm for deep belief nets. *Neural Computation* 18: 1527–1554.

9. Hu H, Wen Y, Chua T et al. 2014. Toward scalable systems for Big Data analytics: A technology tutorial. *IEEE Access 2:* 632–687.

10. Gantz Y, and Reinsel D. 2011. Extracting value from chaos, in *Proc. IDCiView*, pp. 1–12.

11. Zikopoulos P and Eaton C. 2011. *Understanding Big Data: Analytics for Enterprise Class Hadoop and Streaming Data.* New York, NY: McGraw-Hill.

12. Cooper M and Mell P. 2012. *Tackling Big Data* [Online]. Available: http://csrc.nist. gov/groups/SMA/forum/documents/june2012presentations/f%csm_june2012_cooper_ mell.pdf.

13. Dean J and Ghemawat S. 2008. MapReduce: Simplified data processing on large clusters, *Commun. ACM 51:* 107–113.

14. Isard M, Budiu M, Yu Y et al. 2007. Dryad: Distributed data-parallel programs from sequential building blocks, in *Proc. 2nd ACM SIGOPS/EuroSys Eur. Conf. Comput. Syst.*, Jun. pp. 59–72.

15. Malewicz G, Austern MH, Aart JC et al. 2010. Pregel: A system for large-scale graph processing, in *Proc. ACM SIGMOD Int. Conf. Manag. Data*, Jun. pp. 135–146.

16. Melnik S, Gubarev A, Long JJ et al. 2010 Dremel: Interactive analysis of web-scale datasets. *Proc. VLDB Endowment 3:* 330–339.

17. Bu Y, Howe B, Balazinska M et al. 2010. Haloop: Efficient iterative data processing on large clusters. *Proc. VLDB Endowment 3:* 285–296.

18. Ekanayake J, Li H, Zhang BJ et al. 2010 Twister: A runtime for iterative MapReduce, in *Proc. 19th Assoc. Comput. Mach. (ACM) Int. Symp. High Perform. Distrib. Comput.* pp. 810–818.

19. Condie T, Conway N, Alvaro P et al. 2010. MapReduce online, in *Proc. 7th USENIX Conf. Netw. Syst. Des. Implement.* p. 21.

20. Dittrich J, Quiane-Rui JA, Jindal A et al. 2010. Hadoop++: Making a yellow elephant run like a cheetah (without it even noticing). *Proc. VLDB Endowment 3:* 515–529.

21. Abouzied A, Bajda-Pawlikowski K, Huang J et al. 2010. HadoopDB in action: Building real world applications, in *Proc. Assoc. Comput. Mach. (ACM) SIGMOD Int. Conf. Manag. Data.* pp. 1111–1114.

22. Zaharia M, Borthakur D, Sarma JS et al. 2010. Delay scheduling: A simple technique for achieving locality and fairness in cluster scheduling, in *Proc. 5th Eur. Conf. Comput. Syst.* pp. 265–278.

23. Sandholm T and Lai K. 2010. Dynamic proportional share scheduling in Hadoop, in *Job Scheduling Strategies for Parallel Processing.* Berlin, Germany: Springer-Verlag 110–131.

24. Blanas S, Patel JM, Ercegovac V et al. 2010. A comparison of join algorithms for log processing in MapReduce, in *Proc. Assoc. Comput. Mach. (ACM) (SIGMOD) Int. Conf. Manag. Data.* pp. 975–986.

25. Lin J and Dyer C. 2010. Data-intensive text processing with MapReduce, *Synthesis Lect. Human Lang. Technol. 3:* 1–177.

26. Babu S. 2010. Towards automatic optimization of MapReduce programs, in *Proc. 1st Assoc. Comput. Mach. (ACM) Symp. Cloud Comput.* pp. 137–142.

27. Jahani E, Cafarella MJ, and Re C. 2011. Automatic optimization for MapReduce programs. *Proc. VLDB Endowmen 4:* 385–396.

28. Leverich J and Kozyrakis C. 2010. On the energy (in) efficiency of Hadoop clusters, *ssoc. Comput. Mach. (ACM) SIGOPS Operat. Syst. Rev. 44:* 61–65.

29. Lang W and Patel JM. 2010. Energy management for MapReduce clusters. *Proc. VLDB Endowment 3:* 129–139.

30. Lee KH, Lee YJ, Choi H et al. 2012. Parallel data processing with MapReduce: A survey, *Assoc. Comput. Mach. (ACM)SIGMOD Rec. 40*: 11–20.

31. Rao BT and Reddy L. 2012. *Survey on improved scheduling in Hadoop MapReduce in cloud environments*. arXiv preprint arXiv:1207.0780 [Online]. Available: http://arxiv.org/pdf/1207.0780.pdf

32. Apache Software Foundation. 2013. *Apache Storm* [Online]. Available: http://storm-project.net/

33. Albert R and Barabási AL. 2002. Statistical mechanics of complex networks. *Rev. Mod. Phys. 74*: 47–797.

34. Newman MEJ. 2003. The structure and function of complex networks. *Siam Rev. 45*: 167–256.

35. Boccalettia S, Latorab V, and Morenod Y. 2006. Complex networks: Structure and dynamics. *Phys. Rep. 424*: 175–308.

36. Kivela M, Arenas A, Barthelemy M et al. 2014. Multilayer networks. *J. Complex Networks 2*: 203–271.

37. Boccaletti S, Bianconi G, Criado R et al. 2014. The structure and dynamics of multilayer networks. *Phys. Rep. 544*: 1–122

38. Latora V and Marchiori M. 2001. Efficient behavior of small-world networks, *Phys. Rev. Lett. 87*: 198701.

39. Watts DJ and Strogztz SH. 1998. Collective dynamics of small world networks. *Nature 393*: 440–442.

40. Ravasz E and Barabási AL. 2003. Hierarchical organization in complex networks, *Phys. Rev. E 67*: 026112.

41. Newman MEJ. 2002. Assortative mixing in networks. *Phys. Rev. Lett. 89*: 208701.

42. Pastor-Satorras R, Vázquez A, and Vespignani A. 2001. Dynamical and correlation properties of the Internet. *Phys. Rev. Lett. 87*: 258701.

43. Cardillo A G, Gómez-Gardeñes J, Zanin M et al. 2013. Boccaletti, emergence of network features from multiplexity *Sci. Rep. 3*: 1344.

44. Batagelj V and Mrvar A, 2006. Pajek data sets, http://vlado. fmf. uni-ljsi/pub/networks/data/.

45. Lv LY. 2014. Link prediction data sets. http://www.linkprediction.org/index.php/link/resource/data.

46. Leskovec J and Krevl A. 2014. SNAP datasets: Stanford large network dataset collection, http://snap.stanford.edu/data.

47. Albert R, Jeong H, and Barabási AL. 2000. The Internet's Achilles heel: Error and attack tolerance of complex networks *Nature 406:* 378–382.

48. Cohen R, Erez K, Ben-Avraham D et al. 2000. Resilience of the Internet to random breakdowns. *Phys. Rev. E. 85:* 4626–4628.

49. Cohen R, Erez K, Ben-Avraham D et al. 2001. Breakdown of the Internet under intentional attack. *Phys. Rev. E. 86*: 3682–3685.

50. Callaway DS, Newmann MEJ, Strogatz SH et al. 2000. Network robustness and fragility: percolation on random graphs. *Phys. Rev. E. 85*: 5468–5471.

51. Holme P, Kim BJ, and Yoon CN. 2002. Attack vulnerability of complex networks. *Phys. Rev. E. 65*: 056109.

52. Iyer S, Killingback T, Sundaram B et al. 2013. Attack robustness and centrality of complex networks. *Plos One 8:* e59613.

53. Motter AE, Nishikawa T, and Lai YC. 2002. Range-based attack on links in scale-free networks: Are long-range links responsible for the small-world phenomenon? *Phys. Rev. E. 66:* 065103.

54. Albert R, Albert I, and Nakarado GL. 2004. Structural vulnerability of the North American power grid. *Phys. Rev. E. 69*: 025103.

55. Solé RV, Casals MR, and Murtra BC. 2007. Robustness of the European power grids under intentional attack. *Phys. Rev. E. 77*: 026102.

56. Berche B, Ferber CV, Holovatch T. 2009. Resilience of public transport networks against attacks. *Eur. Phys. J. B 71*: 125–137.

57. Wu J, Barahona M, Tan YJ et al. 2012. Robustness of random graphs based on graph spectra. *Chaos 22*: 043101.

58. Schneider CM, Moreira AA, Andrade JS et al. 2011. Mitigation of malicious attacks on networks. *P. Natl. Acad. Sci. USA 108*: 3838C3841.

59. Wu ZX and Holme P. 2011. Onion structure and network robustness. *Phys. Rev. E. 84*: 026106.

60. Tanizawa T, Havlin S, and Stanley HE. 2012. Robustness of onionlike correlated networks against targeted attacks. *Phys. Rev. E. 85*: 046109.

61. Tanizawa T, Paul G, Havlin S et al. 2006. Optimization of the robustness of multimodal networks. *Phys. Rev. E. 74*: 016125.

62. Tanizawa T, Paul G, Cohen R et al. 2005. Optimization of network robustness to waves of targeted and random attacks. *Phys. Rev. E. 71*: 047101.

63. Paul G, Tanizawa T, Havlin S et al. 2004. Optimization of robustness of complex networks. *PhysCondensed Matter*; DOI:10.1140/epjb/e2004-00112-3.

64. Paul G, Tanizawa T, and Havlin S. 2004. Optimization of robustness of complex networks. *Eur. Phys. J. B 38*: 187–191.

65. Paul G, Sreenivasan S, and Shlomo H. 2006. Optimization of network robustness to random breakdowns. *Physica A 370*: 854–862.

66. Valente A, Sarkar A, and Stone HA. 2004. Two-peak and three-peak optimal complex networks. *Phys. Rev. Lett. 92*: 118702.

67. Moreira AA, Andrade JS, and Herrmann HJ. 2009. How to make a fragile network robust and vice versa. *Phys. Rev. Lett. 102*: 018701.

68. Wang B, Tang HW, and Guo C. 2006. Entropy optimization of scale-free networks robustness to random failures. *Physica A 363*: 591–596.

69. Wang B, Tang HW, Guo C et al. 2006. Optimization of network structure to random failures. *Physica A 368*: 607–614.

70. Shi C, Peng Y, and Zhuo Y. 2012. A new way to improve the robustness of complex communication networks by allocating redundancy links. *Phys. Scripta 85*: 035803.

71. Motter AE. 2004. Cascade control and defense in complex networks. *Phys. Rev. Lett. 93*: 098701.

72. Chi LP, Yang CB, and Cai X. 2006. Stability of random networks under the evolution of attack and repair. *Chinese Phys. Lett. 12*: 263.

73. Adomavicius G, Huang GZ, and Tuzhilin A. 2014. Personalization and recommender systems. In *INFORMS Tutorials in Operations Research* Catonsville, Maryland: Institute for Operations Research and the Management Sciences (INFORMS). pp. 55–107.

74. Yao L, Sheng QZ, Segev A et al. 2013 Recommending web services via combining collaborative filtering with content-based features. *2013 IEEE 20th International Conference on Web Services (ICWS)*. Santa Clara, CA.

75. Huang AF, Lan MC, and Yang SJH. 2009. An optimal QoS-based web service selection scheme. *Inform. Sciences 179:* 3309–3322.

76. Ur Rehman Z, Hussain FK, and Hussain OK. 2001. Towards multi-criteria cloud service selection. In *International Conference on Innovative Mobile and Internet Services in Ubiquitous Computing*, Seoul, Korea, June 20-July 2, 44–48.

77. Zeng L, Benatallah B, Ngu AHH et al. 2004 QoS-aware middleware for web services composition. *IEEE Trans. Softw. Eng. 30*: 311–327.

78. Menasce D. 2002. QoS issues in web services. *IEEE Trans. Internet Compu. 6*: 72–75.

79. Xiao X and Ni LM. 1999. Internet QoS: A big picture. *IEEE Trans. Netw. 13*: 8–18.

Chapter 12

Visualizing Life in a Graph Stream

James Abello, David DeSimone, Steffen Hadlak, Hans-Jörg Schulz, and Mika Sumida

12.1 Introduction

When exploring a data stream it is natural to ask how to relate current stream snapshots to past snapshots. Depending on the data semantics and the task at hand different interpretations are possible. For example, in the case of microblog data (like Twitter) making sense of conversations and discussions related to a particular topic may entice users to join the discussion. For data analysts, a usual task is to discern how tweets–information–patterns spread with the possible goal of intuitively explaining their findings. In monitoring traffic scenarios, teasing out those communication patterns that deviate from a considered normal behavior can be used as proxies for intrusion detection. In general social networks, identifying influential nodes in a "volatile" graph stream is of considerable interest. We report here a useful approach to identify trends and exceptional nodes in a graph stream. The fundamental idea is to view a graph stream as a collection of "elementary" time-stamped events whose aggregation through time generates "salient" patterns whose activity rate is incrementally maintained. We are able to isolate group "herding" and "straying" as peculiar behaviors that can be subject to both human and computer verification. We quantitatively estimate the overall behavior of the detected salient edges as a convex combination of their "herding" or "straying" tendencies and their firing rates and recency "profiles." All our computations are accompanied by a visualization platform that integrates dynamic node link views of "recent" subgraphs with a tape view of their Top-K edge statistics (Figure 12.1). The approach discussed here has been coupled with a degree-of-interest (DoI)-based exploration system [1] to provide the user with the functionality to take a closer look at particular keywords of interest identified with our novel approach. Currently, such DoI-based systems are not equipped to operate on the graph streaming setting proposed here.

Our main contributions are:

- An adaptive and simple approach to graph stream processing that is based on the "firing rates" of edge cooccurrences.

- Use of the notion of "recency" as a mechanism to measure the decay of an edge or vertex in a graph stream.

- Isolation of "herding" and "straying" patterns in a graph stream as proxies of group behavior.

- Quantification of the notion of "persistent" and "statistically salient" behaviors in a graph stream which can be automatically verified by a human or a computational agent. This is possible by the incorporation of an incremental maximal matching where the matching edges are weighted by the sum of the overall frequency of their endpoints.

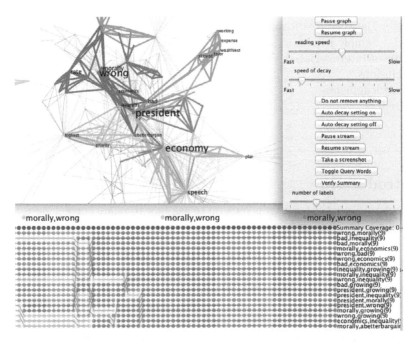

FIGURE 12.1: Complete view of the graph stream visualization interface featuring tweets from President Obama's speech on the U.S. economy on July 24, 2013.

- Incorporation of visual cues that correspond directly to the notion of "recency" and "firing rates." This is facilitated by coupling a force directed node link layout with an intuitive Top-K Tape representing the most "salient" graph stream elements in a dynamically adjusted time window.

- A first application of DoI functions to graph streams.

The chapter layout is as follows: Section 12.2 describes related work. Section 12.3 introduces the general data model and the fundamental graph stream statistics on which we base our cooccurrence graph stream processing. They are: recency, firing rate, and persistence. Section 12.4 describes the decay mechanisms used to maintain the most "recent" cooccurrence subgraph and presents the statistical mechanisms to extract the most "salient" edges in the stream. Section 12.5 describes the visualization of the recent subgraph as an animated graph visualization and its most salient edges as a collection of time plots that we call the **Top-K Tape**. Section 12.6 details the life cycle of a vertex in a graph stream. Section 12.7 describes the different states of the "life" of a graph stream edge and how they determine the "life persistence" of those vertices, which have cooccurred prominently in the graph stream. Prominent edge states are: trending, untrending, herding, and straying. Section 12.8 illustrates the application of our approach to the processing of Twitter data, however this work is by no means confined to microblog data. Section 12.9 illustrates the coupling with the DoI-based system. Finally, Section 12.10 concludes this chapter by outlining possible avenues for further research.

12.2 Related Work

During the last decade, the visualization of dynamic graphs has become a quite active and diverse interdisciplinary research field. Recent surveys of the area [2–4] discuss in a very comprehensive manner the existing variety of approaches and insights. We refer the reader to these publications for a more in-depth treatment of this area. When analyzing time-varying graphs, common approaches are to choose a single point in time in the sense of an animation (e.g., [5]) or to aggregate longer time spans into a super or union graph [6], a static structure that can be more easily visualized. A common concept regarding the use of animation to visualize dynamic graphs is the user's "mental map." Issues concerning the preservation of the "mental map" are discussed in [7–9]. Layout algorithms taking these issues into account are discussed in [10, Section 12.3]. However, these approaches have several limitations. The user has either to inspect each time point individually or lose the temporal context altogether. Alternatives show the structure over multiple time points in a single image to overcome these limitations. They either use small multiples in which the structure is shown individually for each time point [11–13] or embed a representation of time points into the node or edge representation [14–16]. As they try to convey every piece of information (all nodes, edges, and time points) they cannot scale to large dynamic graphs.

The research described here can be placed in the context of statistically driven extraction of salient subgraphs with bounded resources. We concentrate on the formulation of a principled approach based on the novel notions of firing rate and recency distribution. They are the result of viewing graph streams as time cooccurrence graphs from which only recent and Top-K prominent subgraphs are extracted and subsequently visualized. To our knowledge such an approach has not been pursued before.

12.3   General Definitions: Data Model and Statistics

Definition 1 *Graph streams as cooccurrence graphs. A graph stream, on a set of vertices (or nodes) V, is a collection of time-stamped pairs $\langle (x,y), t_i \rangle$ where x and y are elements of V and t_i indicates a time point when the pair of vertices x and y cooccur. For each edge $e = (x,y)$, we let $T_{e,t}$ denote the set of time points ($t_i \leq t$) in which the pair of vertices x and y cooccur. The cardinality $|T_{e,t}|$ is the frequency of cooccurrence of the edge $e = (x,y)$. We denote the collection of cooccurring node pairs up to time t by $E_t = \{e = (x,y) : T_{e,t} \text{ is nonempty}\}$, and the corresponding set of vertices V_t.*

Definition 2 *Firing rates. Following [17], each edge in E_t has a firing rate: $fr(e,t) = \frac{|T_{e,t}|}{t}$ and its corresponding firing sequence is $fr(e) = \langle fr(e,t) \rangle$. The firing rate of any subset E' of E_t is just the sum of the firing rates of the edges in E' up to time t; and its corresponding firing sequence will be denoted by $fr(E') = \langle fr(E',t) \rangle$. The firing rate of a vertex x (up to a particular time t) is the sum of the firing rates of its incident edges up to that time t. With these conventions, a graph stream is a graph sequence $\{G_t = (V_t, E_t)\}$ with a corresponding firing sequence $\langle fr(E_t) \rangle$. We will refer to $fr(E_t)$ as the firing rate of G_t. Note that firing rates can be thought of being analogous to velocities.*

Definition 3 *Instantaneous events. An instantaneous snapshot of a graph stream at time point t is the collection of edges that cooccur at time t. An instantaneous event is a maximal connected subgraph of an instantaneous snapshot. In other words, an instantaneous event at time point t is a maximal connected cooccurrence subgraph.*

Definition 4 *Average vertex firing rate. The vertex average firing rate at time point t is the average firing rate of all vertices in the cumulative graph stream at time t. If there are N_t different vertices in the graph up to time point t,*

$$AFR(t) = \frac{\sum_{i=1}^{N_t} FR(w_i, t)}{N_t}$$

We let $\sigma_v(t)$ denote the standard deviation of the set of firing rates of all vertices in the cooccurrence graph up to time t.

Since the arrival distribution of different edges is in general quite different, we keep track of what we call edge (or vertex) **recency**, which is defined as follows:

Definition 5 *Edge and vertex recency. If e is an edge in the graph stream,*

$$Recency(e,t) = \begin{cases} 1 & \text{if } e \text{ appears in the graph stream at time } t \\ \dfrac{1}{t - t_{last}} & \text{otherwise} \end{cases}$$

where t_{last} = the immediate previous time e appeared in the graph stream. The recency of a vertex is defined in an analogous way.

The aforementioned mathematical framework allows us to formulate questions related to the "behavior" of either vertices or edges in a graph stream $\{G_t\}$ in terms of their associated firing rates. The overall approach consists of comparing the firing sequence of an edge or a vertex with the firing sequence of the graph stream in which they reside. We **selectively label** from the graph stream, incrementally, those edges (or vertices) whose firing rate is substantially above the average firing rate (AFR) of the subgraph edges (or vertices).

Definition 6 *Vertex label select.* *The* label select *value of a vertex is a nonnegative integer when its firing rate is above* $AFR(t)$. *When the value is positive, it is equal to the number of standard deviations by which its firing rate exceeds* $AFR(t)$.

$$LabelSelect(w,t) = \begin{cases} -1 & \text{if } FR(w,t) \leq AFR(t) \\ \left\lfloor \dfrac{FR(w,t) - AFR(t)}{\sigma(t)} \right\rfloor & \text{if } FR(w;t) > AFR(t) \end{cases}$$

12.4 Recency and Top-K Filters

12.4.1 Recency filtering

In order to focus on relatively recent subgraphs, which we call "recency graphs," we systematically remove edges and vertices from the graph stream when they have been inactive for some data-driven time interval. It is based on another time-dependent parameter of the graph stream that we call $ScreenTime_{CG}(t)$, which is a function of the processor intake rate and the screen display capacity. The intention is to have an edge removal control mechanism. Namely, if $Recency(e,t) < \dfrac{1}{ScreenTime_{CG}(t)}$, we remove the edge. If all the edges connected to a vertex are removed, we remove the vertex from the recency graph.

12.4.2 Top-K Tape filtering

Although the recency graph gives a comprehensive snapshot of what the graph stream looks like at a particular time, it does not keep a record of past events. Once a vertex disappears from the recency graph, no record of it is left in the recency graph regardless of how prominent it may have been. There is no way to tell how long the vertices have been in the recency graph or whether and how often they have been labeled. We would like to know not only about the present state of the data stream but also have a description of important past events. In order to do this, we introduce the **Top-K Tape**, which encodes a window into the past activity of the recency graph.

Let $\mu_e(t)$ be the average edge firing rate for all edges in the cooccurrence graph up to time point t. Let $\sigma_e(t)$ be the standard deviation of the firing rate of all edges up to time t. If an edge's firing rate drops below $\mu_e(t) + 2\sigma_e(t)$, the edge is removed from the Top-K Tape.

We define the **Top-K Tape** (bottom of Figure 12.1) to be the set of all edges in the recency graph, up to time point t, with firing rate greater than $\mu_e(t) + 2\sigma_e(t)$. The Top-K Tape is represented by a series of horizontal rows of dots. Each of these rows represents an edge. Each column represents a separate time point. The tape has a width that represents a number of time units into the past from the current time. The current time is on the right of the tape and the most distant time point is on the left. The rows are sorted, in nonincreasing order, according to their firing rate on the overall graph stream. The Top-K Tape also has two special *virtual* edges, one representing the overall **Graph Stream Firing Rate**, and the other representing the **Top-K Firing Rate**. Both of these edges compete for a slot in the Top-K. These virtual edges are shown in Figure 12.2 in purple at the top row and in black at the start of the bottom row.

Definition 7 *Edge persistence.* *Persistent edges are those edges that have a high appearance rate for an extended period of time, whereas nonpersistent edges are those edges*

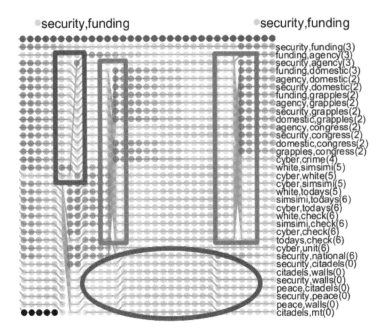

FIGURE 12.2: Tweets captured during a maritime security conference, March 2, 2015 through March 4, 2015.

that may appear in the graph stream in concentrated bursts, but not enough to be considered persistent. We define edge persistence to be the average length of contiguous edge lifetime segments. Edges with a higher persistence value have either longer total lifespans, or a lower number of contiguous lifetime segments. See Section 12.6.2 for how these quantities are displayed in the Top-K Tape.

12.5   Graph Stream Visualization

Based on these definitions, we have designed a visualization platform for graph streams as shown in Figure 12.1 that integrates two views. First, a dynamic node link view of "recent" subgraphs that combines an animation with a special supergraph of the past time points. And second, a tape view of the Top-K edge statistics serving as an overview of only the K most important edges within these recent subgraphs for multiple time points.

12.5.1   Recency graph visualization

The visualization of the recency graph as depicted in Figure 12.3 has the following visual attributes, which are determined by the previously defined parameters.

Edge recency color: Edges in the graph are colored using the diverging spectral color scheme from Colorbrewer2.org [18]. The color of an edge reflects its recency value. An edge is colored red when $Recency(e, t) = 1$. If its recency value decreases, its color begins to fade towards blue.

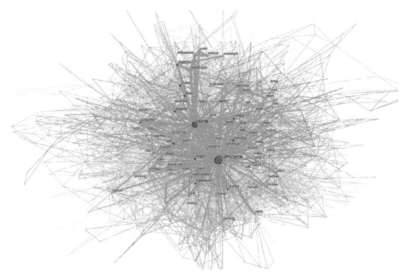

FIGURE 12.3: The recency graph of a stream of tweets on Hurricane Irene.

Edge thickness as edge firing rate: The thickness of an edge encodes that edge's present firing rate.

Vertex recency color: For vertices, we average the recency values of all its incident edges and use that value to determine its color.

Vertex size as vertex firing rate: The size of a vertex reflects its firing rate. If a vertex has a high firing rate, it will be larger on the screen than other vertices with lower firing rates.

Vertex label size: A vertex is unlabeled if its label select value is -1. Vertices with higher label select values are assigned larger fonts.

Visually, vertices with large textual labels are those whose average firing rate is at least one standard deviation above the overall firing rate of the graph stream.

12.5.2 Top-K Tape visualization

An edge appearing in the Top-K Tape has the following visual attributes as shown in Figure 12.4.

Dot color: As in the recency graph, dot color indicates recency. There is a direct correspondence between the color of a dot on the tape, and the edge occurrence it represents in the recency graph.

Dot size: The size of a dot is a variable controlled by the user to increase visibility.

Dot label: The dots at the right-most column are labeled with the two vertices of the edge they represent. The order of these vertices in the label is determined by the vertices' total frequency in the cooccurrence graph. Each dot label also contains an additional number in parenthesis. This number is the edge's persistence value in the cooccurrence graph.

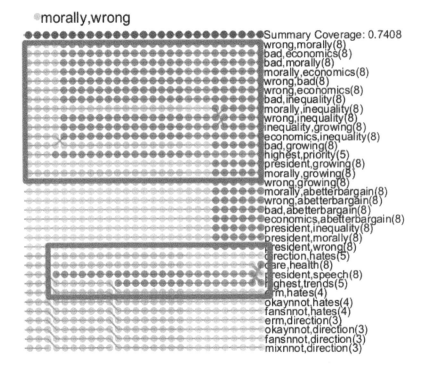

FIGURE 12.4: Speech on the economy by President Obama, July 24, 2013.

12.6   Life Cycle in a Graph Stream

The life cycle of a vertex can be described as follows. When a vertex first appears in the graph stream, it is added to the recency graph. If a vertex is inactive in the graph stream for too long, then it is removed from the recency graph. If that vertex appears again in the graph stream, it is treated like the first occurrence. Otherwise, the vertex becomes labeled when its firing rate exceeds $AFR(t)$. Note that a vertex can cycle between the two states "labeled" and "unlabeled" if its firing rate fluctuates between above and below $AFR(t)$. The life of a vertex in the recency graph is succinctly shown in the transition diagram in Figure 12.5. Orthogonally, an edge has its own life cycle in a specially selected Top-K substream, from which we can identify those vertices which cover prominently a substantial portion of the overall graph stream. The main mechanism for this identification is based on edge group behavioral patterns. They include *trending, untrending, herding, and straying*. Their details are discussed in the next section.

12.7   Top-K Edge Group Patterns

12.7.1   Trending and untrending

Definition 8 *Trending and untrending.* *The Top-K edges are ranked in nonincreasing order by their firing rate. If an edge increases in rank between times t and $t + 1$, we say that edge is trending. Likewise, if an edge decreases in rank between times t and $t + 1$, we*

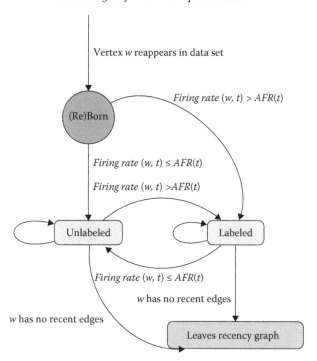

Vertex w reappears in data set

Firing rate (w, t) > AFR(t)

(Re)Born

Firing rate (w, t) ≤ AFR(t)

Firing rate (w, t) >AFR(t)

Unlabeled

Labeled

Firing rate (w, t) ≤ AFR(t)

w has no recent edges

w has no recent edges

Leaves recency graph

FIGURE 12.5: Vertex life in the recency graph.

*say that edge is untrending. Since it may be the case that an edge is trending/untrending, even though its velocity remains unchanged, we introduce next the notion of **significance** in trending/untrending.*

Definition 9 *Significant trending.* *An edge e exhibits significant trending if $rank(e,t) < rank(e,t+1)$ and $fr(e,t) < fr(e,t+1)$.*

Trending/untrending can be detected visually by examining the lifetime path of an edge in the Top-K Tape. If an edge is trending, it will be shown by a visual upward line between an edge's representative dots. If there is an increase in color while an edge increases in rank, this signifies significant trending. If there is a decrease in color while an edge decreases in rank, this signifies significant untrending (see dashed gray rectangle in Figure 12.2).

12.7.2 Herding and straying

Edges that commonly cooccur share similar trending/untrending behavior relative to one another. We introduce the concepts of **herding** and **straying** (see light gray square in Figure 12.6).

Definition 10 *Herding.* *Two edges, e and f, are defined to be herding if $diff(rank(e,t), rank(f,t)) = diff(rank(e,t+1),rank(f,t+1))$, where diff signifies the difference between two quantities.*

Certain edges that are herding across multiple subgraphs may break away from their herd. An edge, e, is said to be **straying at time point** t (refer to light gray rectangle in Figure 12.4) if e was herding with **at least** h other edges at time point $t-1$ and e is not herding with any edge at time point t.

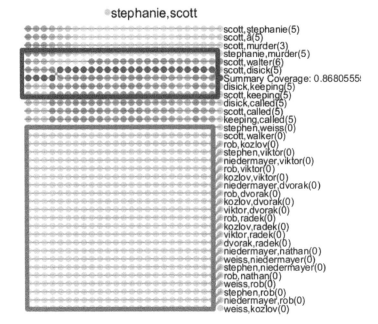

FIGURE 12.6: Tweets from the day following the Walter Scott shooting, April 9, 2015 through April 12, 2015.

12.7.3 Pattern identification

At any time, an edge participates in a combination of trending/untrending and straying or herding patterns, as defined in the previous subsections. The possible transitions between these patterns can be formalized by a set of finite state transitions (Figure 12.7). These patterns can be used to generate summaries for a graph stream.

Definition 11 *Pattern score.* *If an edge is trending/untrending over a time interval $t_{pattern}$, we assign it pattern points equal to its net difference in rank during $t_{pattern}$. If an edge is herding, we subtract points equal to the number of edges it is herding with. If an edge is straying, we add to its pattern score points equal to the size of the herd it strayed from.*

If an edge's score is strictly greater than 1, we place the edge into our data summary. At this point, we verify if the summary is meaningful, in the sense that it provides a large edge cover of the graph stream.

12.7.4 Verification and evaluation

One useful concept involved in analyzing a graph stream is persistence. The Top-K Tape records the history of recency and firing rate over time. However, for longer data sets, the Top-K Tape may become too long with respect to screen capacity, and not viable to view in its entirety. Thus, we would like to summarize the graph stream via patterns that we observe over the Top-K Tape's lifetime. These patterns (trending, untrending, herding, straying) can be used to generate a collection of edges that represent a summary of the Top-K Tape.

In early implementations, we had a human recording these events by hand. Later, we automated the process to have a computer agent record these pattern occurrences. To

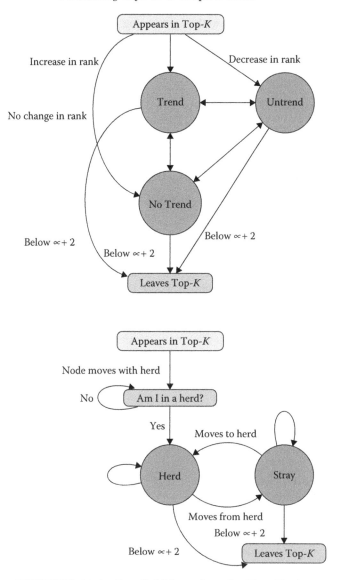

FIGURE 12.7: Parallel life cycles of a Top-K edge.

compare the two approaches, we needed to develop a verification metric for a summary of a graph stream. We defined a summary to be a collection of vertices. A "good" summary would be a collection of vertices that covers a large portion of the graph stream. A "great" summary would be a minimum vertex cover for the entire graph stream. With this in mind, we defined a normalized metric to compare pattern summaries. As finding a minimum vertex cover is *NP*-Complete, we will instead use a maximal matching, which has a 2-approximation ratio to the minimum vertex cover. However, in extremely large graph streams (in terms of number of vertices and edges), even taking the approximation may be too expensive. We take a different approach by incrementally taking the union of several smaller maximal matchings.

We incrementally construct a cover by selecting a maximal matching for the subgraph induced by the members of the Top-K Tape at the current time point t. We unite the maximal matching of this subgraph with our running vertex cover. We record the ratio of coverage size over the total number of vertices (see Algorithm 1).

Complexity. Ultimately, our proposed approach to graph stream processing depends on:

- The graph stream edge arrival rate

- The processing speed of the available computing platform

- The amount of RAM buffer space available, for incrementally maintaining:

    - A fixed number of cumulative vertex graph stream statistics that include vertex firing rate and their most recent active timestamps

    - The statistically selected Top-K edges and the corresponding new set of vertices used to extend our greedy maximal matching cover ratio

Assuming that the processing speed ps is at least twice the graph stream edge arrival rate, and that the available RAM is $O(|V_t|)$, we can process any graph stream $G_t = (V_t, E_t)$ in the worst case in time $O(|V_t|)$. This conforms to the semiexternal model of computation. However, in the arguably realistic scenario where, at each time point t, the size of each instantaneous subgraph is bounded, we can incrementally update the Top-K edge buffer in time proportional to its size.

12.7.5 Holes

Another useful pattern is what we call a **hole**. A hole at time point t occurs when the number of edges with velocity greater than $\mu_e(t) + 2\sigma_e(t)$ is less than the number of edges with velocity greater than $\mu_e(t+1) + 2\sigma(t+1)$. As the number of edges included in the Top-K is dependent on the velocity distribution of the graph stream, a hole signifies a change in the mean and standard deviation of this distribution.

12.8 Twitter Data Sample Results

In order to test our graph stream abstraction, we created an implementation to process Twitter data. We provide to the Twitter API a set of *query words* and obtain a collection of time-stamped tweets containing the input keywords. We first remove nonalphanumeric words, stop words, convert all letters to lowercase, and use a stemming algorithm [19] to represent words with similar stem by the same representative. Each word in a tweet is

---

Algorithm 1: Incremental summary ratio calculation algorithm

Data: $graphStream_t$, the graph stream up to time t
Result: Normalized Summary Ratio
  VertexCover = {}
  for each time point t **do**
    Let TK be the set of edges in the Top-K at time t
    ACOVER = approxCoverViaMaximalMatching(TK)
    VertexCover = VertexCover ∪ ACOVER
    record |VerticesCoveredBy(VertexCover)| / |Vertices($graphStream_t$)|
  end for

---

mapped to a vertex in the graph stream. If the set of words in a tweet is considered ordered according to their order of appearance in the tweet, it makes sense to consider two words w-connected by a tweet if they are separated by no more than w tweet words. The subgraph corresponding to a tweet is the graph with vertices consisting of all words in a tweet and edges drawn between all pairs of w-connected words. These cliques of w-connected words become our instantaneous events (see Definition 3).

The TwitterMap interface has several customization options for exploring a given data set. The user has the ability to pause/resume the incoming graph stream for more detailed observation. They can also change the speed at which tweets are read into the system, and the rate at which tweets will decay after they have been processed in the system. In addition to this, the user can control the number of labels present in the recency graph, and the visible size of both the vertices and the edges within the Top-K Tape. We remove query words from the analysis via a toggle button.

12.8.1 Sample results

Figure 12.2 shows tweets recorded during a major international conference on "Maritime and Cyber Security" under the query [maritime, security], on March 2, 2015 through March 3, 2015. Figure 12.4 shows a tape segment recorded during a speech by President Obama on the subject of economic reform, under the query [president, obama] on July 24, 2013. Figure 12.6 shows tweets recorded the days following the police shooting of an unarmed African American man under the query [walter, scott] on September 4, 2015 through November 4, 2015. Each of the shown tape segments differ from one another in terms of the resulting visible color scheme and the types of visible Top-K patterns.

These visible patterns give us an intuitive way to visually describe the graph stream over this time window. For example, the Obama tape segment has

- Few straying word pairs (light gray rectangle in Figure 12.4)

- Heavy herding (light gray rectangle in Figure 12.4)

- Little color variation (both rectangles in Figure 12.4)

The fact that most of the word pairs are not changing in rank shows that the conversation is dominated by a set of common words (see dashed dark gray rectangle in Figure 12.4). There are few straying words (see light gray rectangle in Figure 12.4,) most likely caused by a group of select tweets being heavily retweeted during this time window. This may be a characteristic of large live public events where many commentators are sharing their views over social media. We can also notice that the highest ranked words are also negative in connotation. We see word pairs such as "morally wrong" and "bad economics" ranked at the top of our view.

We can contrast the Obama set with the Walter Scott set. One may notice that the Walter Scott set features

- A varied color palette (dashed gray rectangle in Figure 12.6)

- Large size herding (light gray rectangle in Figure 12.6)

- A group of word pairs ranked above the Top-K dark virtual edge line (see top of Figure 12.6)

The varied color palette shows that this conversation exhibited a more diverse collection of activity patterns than the Obama speech set. The word pairs have more time between their occurrences, and thus we see more of the color scale (see light gray rectangle in Figure 12.6). The heavy herding tells us that this conversation is also dominated by several groups of words with similar behavior, most likely due to several tweets being heavily retweeted in this time frame. The group of words above the Top-K dark virtual line indicates that this particular set of words recently had a spike in activity (see top of Figure 12.6). We see that these word pairs are focused on the victim, Walter Scott, and his wife Stephanie.

The security tape segments exhibit the following characteristics:

- Varied color palette (see entirety of Figure 12.2).

- A mix of both straying and herding (notice the contrast between straying in the dashed rectangle and herding in the dashed oval in Figure 12.2).

- Black virtual edge is visible (see bottom left of Figure 12.2).

- A relative lack of rank stability (light gray rectangles in Figure 12.2).

The varied color palette still signifies that this conversation has a more diverse activity set of patterns than the Obama speech set (see dashed oval in Figure 12.2). The visibility of the black virtual edge is indicative of a rise in system acceleration (i.e., increasing velocity in a small period of time). Observe that the word pair "security funding" strayed away from its herd (see dashed rectangle in Figure 12.2), signifying a change in the focus of the conversation.

12.9 Degree-of-Interest-Based Visual Exploration

The previous sections have looked at the big picture of graph streams. Once the viewer is familiar with that big picture, the question remains, what else is there? Are there further interesting characteristics that are drowned out by the most prevalent features? This can be seen in the recency view in Figure 12.3, where the two most common keywords "hurricane" and "irene" are dominating the visualization, making it very hard to take a closer look at the other identified keywords. In this section, we take a closer look at these "undercurrents" of the graph stream.

The method we apply for doing so is based on DoI functions. We can use these functions to capture those aspects that interest us in the graph stream and increase the DoI of the involved vertices and edges, while at the same time decreasing the DoI of vertices and edges involved in aspects that are not of interest to us. Using the modular interface for specifying DoI functions that we have previously presented [1], we can zero in on features of interest in a step-by-step fashion depending on what we find in the process.

As a starting point, we use Figure 12.3 that shows the recency graph of the tweets from August 26, 2011 with the two dominant keywords "hurricane" and "irene." We then define a DoI function that assigns a low DoI to these two keywords (see DoI-module A in Figure 12.8), as well as to keywords with very low firing rates (see DoI-module B in Figure 12.8) to cut down on the clutter. The result of these removals can be seen on the right side of Figure 12.8. Smaller patterns emerge now and we can already see from this figure that people were scared, President Obama was on the news, and that there were many tweets about staying safe. One can also see a number of swear words that appear frequently in the context of the hurricane tweets. A secondary analysis using the linguistic inquiry and

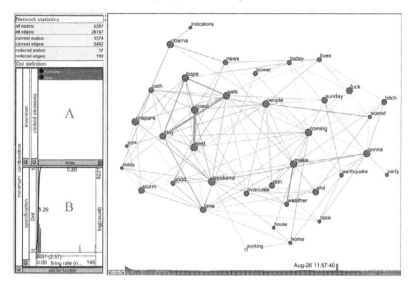

FIGURE 12.8: The first step of the DoI definition: building a DoI function that removes the dominant keywords by manual selection (DoI module A) and edges of low firing rate (DoI module B) by assigning them low DoI values. The result can be seen on the right side. $doi_1(x_i) = min(\{inv(select(x_i)), inter(firing\_rate(x_i)))\})$.

word count (LIWC)[*] confirms that these tweets are overwhelmingly emotionally negative with a 37.9 score on emotional tone. Scores below 50 denote emotional negativity, scores above 50 denote emotional positivity [20]. To put this score in context, according to the LIWC website the average score on emotional tone in social media lies at 63.35.

This first step allowed us already to form a more differentiated picture than the original image. The drawback of the defined DoI function is that it only focuses on keywords that occur often, but neglects those words that appear in the context of these frequent keywords— that is, words that are not as frequent by themselves but regularly tweeted alongside the frequent words. These contextual keywords can help to further derive meaning from the observed structures. Hence in a second step, we add them back in by applying a structural propagation of the DoI values derived from the first step to their neighboring vertices, which is depicted as DoI module C in Figure 12.9. Since this propagation also adds the keywords "hurricane" and "irene" back in, we have to subtract them again as shown by DoI module D in Figure 12.9. The result of applying this revised DoI function is depicted on the right side of the figure. While at this point, it does not generate additional insights, one can clearly see that certain keywords are now shown with more context – e.g., "obama" is now connected with "president" and "americans," and the hurricane apparently "approaches" the "east" "coast."

Prepared in such a way, we can now take a look at later time points using the very same DoI function to examine the patterns that lie beneath the overall Twitter hype around "hurricane" "irene," which we filtered out. Figure 12.10 shows two time points: 08/27/2011 (top) and 08/29/2011 (bottom). In the snapshot from 08/27/2011, one can see a pattern surrounding the "category" of the "storm," as its "strength" "weakens." This captures nicely the time point at which it was announced that hurricane Irene was downgraded to

---

[*] See http://www.liwc.net.

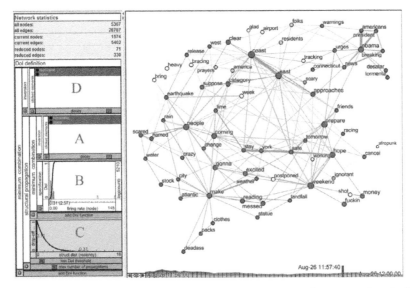

FIGURE 12.9: The second step of the DoI definition: enhancing the DoI function by adding a structural propagation (DoI module C) that distributes high DoI values to neighboring vertices. The result can be seen on the right side. $doi_2(x_i) = min(\{inv(select(x_i)), prop_s(doi_1(y_i))\})$.

a Category 1 hurricane. This is also reflected in the emotional tone of these tweets, which score with 59.5 a much higher and even slightly positive emotional tone than the day before. Yet this score plunges again right after the landfall of Irene on the East Coast that brought flooding and power outages in its wake. The emotional tone reaches a minimum of 34.92 on 08/29/2011, when then presidential candidate Michele Bachmann commented on the disastrous situation on the East Coast that hurricanes and earthquakes are God's warning to Washington. Mentions of this quote were frequently retweeted, so that a corresponding pattern shows in Figure 12.10 (bottom). The overall negative reception of this quote on Twitter together with the first mentions of the next "tropical" "storm" of the 2011 hurricane season "katia" seems to have brought the emotional tone down to this minimum.

From these examples, it becomes apparent that the DoI-based inspection of the streaming data is a valuable tool to define just the parts of the data that are of interest while cutting down on the noise (i.e., keywords of low interest cluttering the view) and the already known facts (i.e., the most dominant keywords). Since the DoI function can be adjusted at any time throughout the exploration, found facts—i.e., prominent keywords—can be added to the exclusion list to bring out even more subtle details.

12.10   Conclusions

We have introduced a simple and useful view of graph streams as cooccurrence graphs. The fundamental driving statistics are edge firing rate, recency, and edge persistence. They allow us to isolate edge group patterns like "herding" and "straying" that can be used as proxies of interesting graph stream behaviors. Salient vertices in the graph stream pop up as those vertices with persistent firing rates substantially above the average firing rate of the

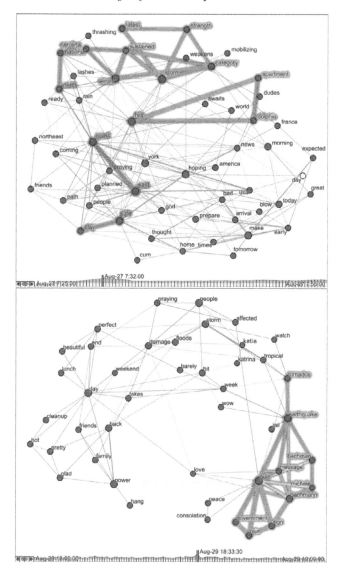

FIGURE 12.10: Additional results of the DoI function previously defined in Figure 12.9 to show frequent subgraphs (manually highlighted by thick shading) for two time points: 08/27/2011 (top) and 08/29/2011 (bottom).

entire graph stream. Their selection is certified by their coverage ratio of the entire stream. This ratio can be incrementally verified by either a human or a computer. Drowned-out "undercurrents" of the graph stream can be brought to light by using a modular DoI function that filters out the known patterns in the stream and enhances the unknown ones.

The approach suggested here presents several directions of future research. They include

- Summarization of graph streams by transition diagrams.

- Collaborative exploration of graph streams.

- Detection of verifiable graph stream properties. Candidates include: stream entropy, entropy norms, and discrepancy.

Acknowledgments

We acknowledge the partial support of the US National Science Foundation (DIMACS REU NSF Grant CCF-1263082, 2015) and of the federal state of Mecklenburg-Vorpommern within the EFRE project "Basic and Applied Research in Interactive Document Engineering and Maritime Graphics" (Grant ESF/IV-BM-B35-0006/12). The version of TwitterMap presented here was created with the Java visualization library Graph Stream (http://graphstream-project.org).

References

1. James Abello, Steffen Hadlak, Heidrun Schumann, and Hans-Jörg Schulz. A modular degree-of-interest specification for the visual analysis of large dynamic networks. *IEEE Transactions on Visualization and Computer Graphics*, 20(3):337–350, 2014.

2. Fabian Beck, Michael Burch, Stephan Diehl, and Daniel Weiskopf. The state of the art in visualizing dynamic graphs. In Rita Borgo, Ross Maciejewski, and Ivan Viola, editors, *EuroVis'14: State-of-the-Art Reports of the Eurographics/IEEE Symposium on Visualization*, pages 83–103. Geneva, Switzerland: Eurographics Association, 2014.

3. Steffen Hadlak, Heidrun Schumann, and Hans-Jörg Schulz. A survey of multi-faceted graph visualization. In Rita Borgo, Fabio Ganovelli, and Ivan Viola, editors, *EuroVis'15: State-of-the-Art Reports of the Eurographics/IEEE Symposium on Visualization*, pages 1–20. Geneva, Switzerland: Eurographics Association, 2015.

4. Natalie Kerracher, Jessie Kennedy, and Kevin Chalmers. The design space of temporal graph visualisation. In Niklas Elmqvist, Mario Hlawitschka, and Jessie Kennedy, editors, *EuroVis'14: Short Paper Proceedings of the Eurographics/IEEE Symposium on Visualization*, pages 7–11. Geneva, Switzerland: Eurographics Association, 2014.

5. Benjamin Bach, Emmanuel Pietriga, and Jean-Daniel Fekete. GraphDiaries: Animated transitions and temporal navigation for dynamic networks. *IEEE Transactions on Visualization and Computer Graphics*, 20(5):740–754, 2014.

6. Stephan Diehl, Carsten Görg, and Andreas Kerren. Foresighted graphlayout. Technical Report A/02/2000, Saarbrücken, Germany: Universität des Saarlandes, 2000.

7. Daniel Archambault, Helen Purchase, and Bruno Pinaud. Animation, small multiples, and the effect of mental map preservation in dynamic graphs. *IEEE Transactions on Visualization and Computer Graphics*, 17(4):539–552, 2011.

8. Kazuo Misue, Peter Eades, Wei Lai, and Kozo Sugiyam. Layout adjustment and the mental map. *Journal of Visual Languages and Computing*, 6(2):183–210, 1995.

9. Helen Purchase, Eve Hoggan, and Carsten Görg. How important is the "mental map"?—An empirical investigation of a dynamic graph layout algorithm. In Michael Kaufmann and Dorothea Wagner, editors, *GD'06: Proceedings of the International Symposium on Graph Drawing*, volume 4372 of *Lecture Notes in Computer Science*, pages 184–195. Berlin, Heidelberg, Germany Springer, 2007.

10. Ulrik Brandes, Natalie Indlekofer, and Martin Mader. Visualization methods for longitudinal social networks and stochastic actor-oriented modeling. *Social Networks*, 34(3):291–308, 2012.

11. Adam Perer and Jimeng Sun. MatrixFlow: Temporal network visual analytics to track symptom evolution during disease progression. In *AIMA'12: Proceedings of the American Medical Informatics Association Annual Symposium*, Chicago, IL, volume 2012, pages 716–725, 2012.

12. Bruno Pinaud, Guy Melançon, and Jonathan Dubois. PORGY: A visual graph rewriting environment for complex systems. *Computer Graphics Forum*, 31(3pt4):1265–1274, 2012.

13. Sébastien Rufiange and Michael J. McGuffin. DiffAni: Visualizing dynamic graphs with a hybrid of difference maps and animation. *IEEE Transactions on Visualization and Computer Graphics*, 19(12):2556–2565, 2013.

14. Purvi Saraiya, Peter Lee, and Chris North. Visualization of graphs with associated timeseries data. In John Stasko and Matthew O. Ward, editors, *InfoVis'05: Proceedings of the IEEE Symposium on Information Visualization*, pages 225–232. Chicago, IL: IEEE Computer Society, 2005.

15. Klaus Stein, Rene Wegener, and Christoph Schlieder. Pixel-oriented visualization of change in social networks. In Nasrullah Memon and Reda Alhajj, editors, *ASONAM'10: Proceedings of the International Conference on Advances in Social Networks Analysis and Mining*, pages 233–240. Odense, Denmark: IEEE Press, 2010.

16. Andrea Unger and Heidrun Schumann. Visual support for the understanding of simulation processes. In Peter Eades, Thomas Ertl, and Han-Wei Shen, editors, *PacificVis'09: Proceedings of the IEEE Pacific Visualization Symposium*, pages 57–64. Bejing, China: IEEE Computer Society, 2009.

17. James Abello, Tina Eliassi-Rad, and Nishchal Devanur. Detecting novel discrepancies in communication networks. In Geoffrey I. Webb, Bing Liu, Chengqi Zhang, Dimitrios Gunopulos, and Xindong Wu, editors, *ICDM'10: Proceedings of the International Conference on Data Mining*, pages 8–17. IEEE Computer Society, 2010.

18. Mark Harrower and Cynthia A. Brewer. Colorbrewer.org: An online tool for selecting colour schemes for maps. *The Cartographic Journal*, 40(1):27–37, 2003.

19. Martin F. Porter. An algorithm for suffix stripping. In Karen Sparck Jones and Peter Willett, editors, *Readings in Information Retrieval*, pages 313–316. San Francisco, CA: Morgan Kaufmann Publishers Inc., 1997.

20. Michael A. Cohn, Matthias R. Mehl, and James W. Pennebaker. Linguistic markers of psychological change surrounding September 11, 2001. *Psychological Science*, 15(10):687–693, 2004.

Index

Printed and bound by CPI Group (UK) Ltd, Croydon, CR0 4YY

23/10/2024

01777691-0007